Ordered Sets
and Lattices II

Recent Titles in This Series

152 H. Draškovičová, et al., Ordered Sets and Lattices II
151 I. A. Aleksandrov, L. A. Bokut′, and Yu. G. Reshetnyak, Editors, Second Siberian Winter School "Algebra and Analysis"
150 S. G. Gindikin, Editor, Spectral Theory of Operators
149 V. S. Afraĭmovich, et al., Thirteen Papers in Algebra, Functional Analysis, Topology, and Probability, Translated from the Russian
148 A. D. Aleksandrov, O. V. Belegradek, L. A. Bokut′, and Yu. L. Ershov, Editors, First Siberian Winter School in Algebra and Analysis
147 I. G. Bashmakova, et al., Nine Papers from the International Congress of Mathematicians 1986
146 L. A. Aĭzenberg, et al., Fifteen Papers in Complex Analysis
145 S. G. Dalalyan, et al., Eight Papers Translated from the Russian
144 S. D. Berman, et al., Thirteen Papers Translated from the Russian
143 V. A. Belonogov, et al., Eight Papers Translated from the Russian
142 M. B. Abalovich, et al., Ten Papers Translated from the Russian
141 Kh. Drashkovicheva, et al., Ordered Sets and Lattices
140 V. I. Bernik, et al., Eleven Papers Translated from the Russian
139 A. Ya. Aĭzenshtat, et al., Nineteen Papers on Algebraic Semigroups
138 I. V. Kovalishina and V. P. Potapov, Seven Papers Translated from the Russian
137 V. I. Arnol′d, et al., Fourteen Papers Translated from the Russian
136 L. A. Aksent′ev, et al., Fourteen Papers Translated from the Russian
135 S. N. Artemov, et al., Six Papers in Logic
134 A. Ya. Aĭzenshtat, et al., Fourteen Papers Translated from the Russian
133 R. R. Suncheleev, et al., Thirteen Papers in Analysis
132 I. G. Dmitriev, et al., Thirteen Papers in Algebra
131 V. A. Zmorovich, et al., Ten Papers in Analysis
130 M. M. Lavrent′ev, et al., One-dimensional Inverse Problems of Mathematical Physics
129 S. Ya. Khavinson; translated by D. Khavinson, Two Papers on Extremal Problems in Complex Analysis
128 I. K. Zhuk, et al., Thirteen Papers in Algebra and Number Theory
127 P. L. Shabalin, et al., Eleven Papers in Analysis
126 S. A. Akhmedov, et al., Eleven Papers on Differential Equations
125 D. V. Anosov, et al., Seven Papers in Applied Mathematics
124 B. P. Allakhverdiev, et al., Fifteen Papers on Functional Analysis
123 V. G. Maz′ya, et al., Elliptic Boundary Value Problems
122 N. U. Arakelyan, et al., Ten Papers on Complex Analysis
121 D. L. Johnson, The Kourovka Notebook: Unsolved Problems in Group Theory
120 M. G. Kreĭn and V. A. Jakubovič, Four Papers on Ordinary Differential Equations
119 V. A. Dem′janenko, et al., Twelve Papers in Algebra
118 Ju. V. Egorov, et al., Sixteen Papers on Differential Equations
117 S. V. Bočkarev, et al., Eight Lectures Delivered at the International Congress of Mathematicians in Helsinki, 1978
116 A. G. Kušnirenko, A. B. Katok, and V. M. Alekseev, Three Papers on Dynamical Systems
115 I. S. Belov, et al., Twelve Papers in Analysis
114 M. Š. Birman and M. Z. Solomjak, Quantitative Analysis in Sobolev Imbedding Theorems and Applications to Spectral Theory

(Continued in the back of this publication)

American Mathematical Society

TRANSLATIONS

Series 2 • Volume 152

Ordered Sets and Lattices II

by

H. Draškovičová
T. S. Fofanova
V. I. Igoshin
T. Katriňák
M. Kolibiar

N. Ya. Komarnitskiĭ
A. V. Mikhalev
V. N. Saliĭ
L. A. Skornyakov

American Mathematical Society
Providence, Rhode Island

Translation edited by SIMEON IVANOV

1991 *Mathematics Subject Classification.* Primary 06Axx, 06Bxx, 06Dxx, 06Exx; Secondary 03Gxx, 06Cxx, 08Bxx.

Library of Congress Cataloging-in-Publication Data (Revised for volume 2)
Ordered sets and lattices.
(American Mathematical Society translations, ISSN 0065-9290; ser. 2, v. 141)
"The present collection appears under the joint editorship of M. Kolibiar Bratislava and L. A. Skornyakov (University of Moscow) ... It is based on articles reviewed in the Referativnyĭ zhurnal: Matematika, from January 1978 to June 1982" – Pref.
Translated from the Russian.
Includes bibliographies.
1. Ordered sets. I. Lattice theory. II. Draškovičová, Hilda. III. Kolibiar, Milan. IV. Skorniakov, L. A. (Lev Anatol'evich)

Library of Congress Cataloging-in-Publication Data
QA3.A572 ser. 2, vol. 141 [QA171.48]   510 s 88-38112
ISBN 0-8218-3121-6 (alk. paper)   511.3'3
ISBN 0-8218-7501-9 (part II)

COPYING AND REPRINTING. Individual readers of this publication, and nonprofit libraries acting for them, are permitted to make fair use of the material, such as to copy a chapter for use in teaching or research. Permission is granted to quote brief passages from this publication in reviews, provided the customary acknowledgment of the source is given.

Republication, systematic copying, or multiple reproduction of any material in this publication (including abstracts) is permitted only under license from the American Mathematical Society. Requests for such permission should be addressed to the Manager of Editorial Services, American Mathematical Society, P.O. Box 6248, Providence, Rhode Island 02940-6248.

The appearance of the code on the first page of a chapter in this book indicates the copyright owner's consent for copying beyond that permitted by Sections 107 or 108 of the U.S. Copyright Law, provided that the fee of $1.00 plus $.25 per page for each copy be paid directly to the Copyright Clearance Center, Inc., 27 Congress Street, Salem, Massachusetts 01970. This consent does not extend to other kinds of copying, such as copying for general distribution, for advertising or promotional purposes, for creating new collective works, or for resale.

Copyright ©1992 by the American Mathematical Society. All rights reserved.
Printed in the United States of America.
The American Mathematical Society retains all rights
except those granted to the United States Government.
The paper used in this book is acid-free and falls within the guidelines
established to ensure permanence and durability. ⊚
This publication was typeset using $\mathcal{A}_{\mathcal{M}}\mathcal{S}$-TEX,
the American Mathematical Society's TEX macro system.

Contents

Preface	ix
H. Draškovičová, T. Katriňák, and M. Kolibiar, Distributive lattices, Boolean algebras, and related lattices	1
N. Ya. Komarnitskiĭ, Algebras of logic	63
T. S. Fofanova, General theory of lattices	95
V. I. Igoshin, Classes of lattices and related algebras	139
V. I. Igoshin, A. V. Mikhalev, V. N. Saliĭ, and L. A. Skornyakov, Concrete lattices	155
V. N. Saliĭ, Partially ordered sets. Semilattices. Generalizations of lattices	209

Russian Contents*

Х. Драшковичова [H. Draškovičová], Т. Катриняк [T. Katriňák] и М. Колибиар [M. Kolibiar], Дистрибутивные решетки, булевы алгебры и близкие к ним решетки, Univerzita Komenského, Bratislava, 1988, стр. 9–100

Н. Я. Комарницкий, Алгебры логики, Univerzita Komenského, Bratislava, 1988, стр. 101–148

Т. С. Фофанова, Общая теория решеток, Univerzita Komenského, Bratislava, 1988, стр. 149–214

В. И. Игошин, Классы решеток и близких к ним алгебр, Univerzita Komenského, Bratislava, 1988, стр. 241–321

В. И. Игошин, А. В. Михалев, В. Н. Салий и Л. А. Скорняков, Конкретные решетки, Univerzita Komenského, Bratislava, 1988, стр. 323–381

В. Н. Салий, Частично упорядоченные множества. Полурешетки. Обобщения решеток, Univerzita Komenského, Bratislava, 1988, стр. 215–239

---

*The American Mathematical Society scheme for transliteration of Cyrillic may be found at the end of index issues of *Mathematical Reviews*.

Preface

This volume contains a collection of articles offering a survey of current results in lattice theory. The accounts are based on the review articles which appeared in the Russian *Referativnyĭ Žurnal Matematika* from July 1982 to December 1985. The present book can be considered as a continuation of a similar one which appeared under the title *Partially Ordered Sets and Lattices* (Russian), Univerzita Komenského Bratislava, 1985, English translation, *Ordered Sets and Lattices*, Amer. Math. Soc. Transl., Ser. 2, vol. 141, 1989. References to the Bibliography of that book are given in the form [OSL, II, 342] which refers to a paper (or book) [342] from the Bibliography at the end of Chapter II. As in the previous book, an attempt is made to include all papers published in the given period as well as all accessible current papers concerning lattice theory. As far as possible, every item listed in the Bibliography contains numbers referring to the corresponding review article in *Referativnyĭ Žurnal Matematika* (RŽ Mat) and / or in *Mathematical Reviews* (MR), where more detailed information is available. Work on assembling the volume and the direction of the project has been carried out by Milan Kolibiar of Comenius University at Bratislava and by Lev A. Skornyakov of Moscow University.

It should be mentioned that this review book does not cover all topics concerning lattice theory. For instance, topics such as algebraic approach to logic, Boolean functions, vector lattices, ordered algebraic systems (in particular, ordered topological spaces), semigroup properties of semilattices, as well as all papers dealing with non-lattice-theoretic aspects of congruence lattices, subalgebras and so on, have found no entry into this volume.

The following notation is frequently used:

Symbol	Explanation
$\|A\|$	cardinality of a set A
$P(A)$	set of all subsets of a set A
Eq A, Eq (A)	lattice of all equivalence relations on a set A
Sub A, Sub (A)	lattice of all subalgebras of an algebra A
Aut A, Aut (A)	group of all automorphisms of an algebra A
End A, End (A)	semigroup of all endomorphisms of an algebra A
Con A, Con (A)	lattice of all congruence relations of an algebra A
$I(A)$	lattice of all ideals of a (semi-)lattice A
$F(A)$	lattice of all filters of a (semi-)lattice A
ZF	Zermelo-Fraenkel axiomatic set theory
CH	continuum hypothesis
GCH	generalized continuum hypothesis

Chapter I
Distributive Lattices, Boolean Algebras, and Related Lattices

KH. DRASHKOVICHOVA, T. KATRINYAK, AND M. KOLIBIAR
[H. DRAŠKOVIČOVÁ, T. KATRIŇÁK, AND M. KOLIBIAR]

Information on certain papers related to distributive lattices and Boolean lattices can be found in other chapters. Namely, §3.2 of Chapter III-3 contains information on distributive lattices of congruences.

The authors are grateful to V. N. Salii who wrote §2.10 of Chapter I-2.

Reference [III, 270] means that the coordinates of the corresponding paper can be found in the Bibliography of Chapter III, reference number 270.

§I.1. Boolean algebras

1.1. Monographs. Several monographs, textbooks, and survey articles devoted to the theory of Boolean algebras have appeared recently. The survey [44] is based on materials published in Referativnyĭ Zhurnal Matematika between July 1973 and December 1977 and contains these sections: on the definition of Boolean algebras, topics connected with mathematical logic, topological and homological topics, additional structures on Boolean algebras and investigation of certain special classes of Boolean algebras, Stone duality and related topics.

The book [220] by Gumm and Poguntke is elementary in character. It contains sections on switching algebra, sentential logic, Boolean terms, disjunctive normal form, applications to logic, Boolean rings, Boolean lattices, complete atomic Boolean algebras and their representations, Stone representation theorem, and free Boolean algebras. The book of Błaszcyk [111], devoted to topological aspects of Boolean algebras, is also introductory.

1991 *Mathematics Subject Classification.* Primary 06E05, 06E10, 06E15, 06E20, 06E30, 06D05, 06D10, 06D15, 06D20, 06D25, 06D30; Secondary 03G05.

The work [233] of Heindorf on the model theory for Boolean algebras considers such topics as axiomatizability, categoricity, decidability, etc. for the theory of Boolean algebras in nonelementary languages. The expressive strength of various languages for the theory of Boolean algebras is compared. The monograph [458] considers decidability and undecidability of theories of various signatures (types) connected with Boolean algebras. In particular, it contains a proof of the decidability of the theory of Boolean algebras with a distinguished finite group of automorphisms.

Nerode and Remmel study the lattice of recursively enumerable subalgebras of a recursive Boolean algebra, in the survey [344] of lattices of recursively enumerable structures.

The book [43] of Pavlovskiĭ is devoted to applied aspects of Boolean algebra. Its contents are Boolean functions and their classes, functional decompositions, applications to logic, construction of Zhegalkin polynomials, synthesis of Boolean graphs, and analogous topics.

Lidl and Pilz [310] consider applications of Boolean algebras.

Salii's book [5] is connected with Huntington's problem: is every complex uniquely complemented lattice distributive? A series of results is established, which show that, under rather weak additional conditions, distributivity follows from the existence of unique complements. A few open problems related to Huntington's problem are raised. The book contains interesting information of historical nature.

1.2. Characterization of Boolean algebras. Systems of axioms. Frink gave an elegant characterization of Boolean algebras in 1941: if $(A; \cdot, ', 0)$ is an algebra of the type $(2, 1, 0)$ such that $(A; \cdot)$ is a semilattice, 0 is a fixed element, and $x \cdot y = x$ if and only if $x \cdot y' = 0$, then, defining $x+y = (x' \cdot y')'$ and $1 = 0'$, we obtain a Boolean algebra. Frink's proof uses Zorn's Lemma. Padmanabhan [350] gave a simple proof of this statement in the first-order language. Beran [104] gave a system of three axioms for Boolean algebras and an analogous system for orthomodular lattices. An independent and selfdual equational basis for Boolean algebras as well as for distributive and modular lattices was given by Padmanabhan [351]. Necessary and sufficient conditions for a bounded semilattice to be a finite Boolean algebra are given in [250]. Carrega [128] showed that the class of Boolean algebras coincides with the class of orthomodular lattices that do not contain sublattices isomorphic to two finite lattices he mentioned.

Pühringer [368] gave an axiomatic definition of partial Boolean algebras obtained from a Boolean algebra by removing the zero, and he proved the completeness of this system of axioms. The result is applied to the proof of completeness of the so-called atomless algebras of individuals.

1.3. Products, Boolean degrees. Płonka [358] introduced the concept of a mixed product of three Boolean algebras. Sobolska [421] generalized this concept for the case of k Boolean algebras $B_1, \ldots, B_k$ ($k \geq 3$) and gave a finite basis for the variety K_k generated by all such products. The concept of cofinality for an infinite Boolean algebra A is considered in [300]. It is the least infinite cardinal number κ for which there exists a strictly increasing sequence $(B_\alpha : \alpha < \kappa)$ of subalgebras of A whose union is A. It is proved that if all A_i, $i \in I$, have cofinality $\geq \omega_1$, then the same holds for $\prod A_i$ assuming that the cardinality of I is less than the first uncountable measurable cardinal number. Furthermore, if B is a homomorphic image of the product $\prod(A_i : i \in I)$ of Boolean algebras, $|A_i| < |B|$ for all $i \in I$, and B is infinite, then $|B|^\omega = |B|$ if the cardinality of I is less than the first uncountable cardinal number, and this last restriction is essential. Dwinger [179] gave, in algebraic terms, a necessary and sufficient condition for the completeness of a Boolean degree of a complete Boolean algebra. Contessa [142] showed that the ultraproduct of complete Boolean algebras need not be complete. Earlier Pierce [356] proved that isomorphism types of special Boolean algebras form a commutative semiring S, in which the addition is induced by the operation of taking a direct sum and the multiplication by the operation of taking a tensor product. For every element $x \in S$ either (i) all elements x^n ($n \in N$) are different, or (ii) $x, x^2, \ldots, x^n$ are distinct and $x^n = x^{n+1}$. In [357] he established that, in S, case (ii) is possible for every number n. This solves an important problem.

Ershov [15] introduced the concept of a D-degree of an algebra, which generalizes that of a Boolean degree (instead of a Boolean algebra a distributive lattice is used). Omarov [42] studied elementary properties of D-degrees and elementary properties of free products in the class of Boolean lattices. He gives a criterion for Boolean lattices to be k-saturated.

1.4. Special Boolean algebras. Abian [71] gave two methods for construction of free Boolean algebras (these methods coincide in the case of finitely many free generators). One of them gives the Stone representation of the free Boolean algebra constructed. It is proved in [25] that an ergodic Boolean algebra (cf. [8]) is either the algebra 2^A for a set A or a direct union of τ copies of a homogeneous Boolean algebra for some cardinal number τ.

Weese [457] considered various versions of extensions of the theory of Boolean algebras by introducing a symbol for an automorphism or distinguishing a group of automorphisms. A survey of results on the decidability of such theories is given in that paper and two new theorems proved. Burris [125] investigated decidability problems for Boolean algebras with a distinguished group G of automorphisms. He proved that the variety $BA(G)$ of such algebras for a finite G is finitely generated and a discriminator variety, and hence has the decidable first-order theory. He proved that, for a special group G, $BA(G)$ is decidable with distributive congruence lattices but it is not a discriminator variety. This is the first example of this kind. A further result of the paper is in contrast with these results: if G is not locally finite, then the first-order theory of $BA(G)$ is hereditarily undecidable. Mart'yanov [31] showed that the elementary theory of Boolean algebras of the type $(+, \cdot, ', \alpha)$, where $+, \cdot, '$ denote the ordinary Boolean operations and α is a symbol for an automorphism, is unsolvable. Dulatova [12] proved the undecidability of the class of atomic Boolean algebras with an automorphism as well as the undecidability of the class of atomic Boolean algebras with an additional predicate that distinguishes an atomic subalgebra such that, for every element a of the algebra, the distinguished subalgebra has the least element in the set of all elements $\geq a$. This answer a problem raised by Rubin.

Shelah [413] constructed a rigid Boolean algebra of cardinality $\aleph_1$ that does not contain an infinite chain and satisfies certain additional conditions.

A subset A of a Boolean algebra is called j-chained if, for every subset $F \subset A$ that contains j elements, condition $\bigwedge(a: a \in F) \neq 0$ holds. A Boolean algebra B is called σ-j-chained if its nonzero elements can be partitioned into a countable number of j-chained sets. B is called a ccc-algebra if it does not contain an uncountable set of pairwise orthogonal elements. Bell [102] gave an example of a Boolean ccc-algebra that is not σ-2-chained, and an example of a σ-j-chained Boolean algebra that is not σ-$(j+1)$-chained. The algebras in these examples are ZF-definable subalgebras of $P(\omega)/F$, where F is the ideal of finite subsets. The latter example provides a positive answer to a question asked by van Douwen [166].

A Boolean algebra B is called homogeneous if, for all $u, v \in B - \{0, 1\}$, there exists an automorphism f such that $f(u) = v$. An algebra B is called rigid if it does not possess nontrivial automorphisms. A principal ideal $(b]$, $b \in B$, is called a factor of B. The existence of Boolean algebras that have neither homogeneous nor rigid factors was proved earlier in [87]. Štěpánek [423] suggested a simpler construction which shows that every Boolean algebra C admits a complete embedding in a complete Boolean algebra with neither homogeneous nor rigid factors and with certain additional conditions. Answering a problem from Štěpánek's paper, Koppelberg and Monk [301] showed that, for every Boolean algebra A, there exist extensions $C \supset B \supset A$ such that B and C are homogeneous Boolean algebras, every endomorphism (automorphism) of A can be extended to an endomorphism (automorphism) of B, and no nontrivial one-to-one endomorphism of B can be extended to an endomorphism of C. Furthermore, there exist complete Boolean algebras $B \subset C$ such that C is homogeneous and no nontrivial automorphism of B can be extended to an automorphism of C. A Boolean algebra B is called weakly homogeneous if for any distinct nonzero elements $a, b \in B$ there exist nonzero elements $a_1 \leq a$ and $b_1 \leq b$ for which the principal ideals $(a_1]$ and $(b_1]$ are isomorphic. Using trees, Brenner [117] gave a construction of Boolean algebras that gives simple examples of Boolean algebras of high cardinality that are rigid or weakly homogeneous and that have no homogeneous factors, thus answering a problem of Rubin [385].

It is shown in Chapter VIII of Shelah's monograph [411] how one can construct "many" nonisomorphic countable models of a nonsuperstable theory. In [414] he described the combinatorial part of this method in a more explicit form, strengthened results of the monograph, and constructed more diverse models with properties of pairwise nonembeddability, where the embeddability was understood in a certain wide sense. The most interesting are applications for the combinatorial methods developed for construction of various Boolean algebras with rigidity-type conditions. In particular, a few questions about Boolean algebras raised by van Douwen, Monk, and Rubin [167] were answered.

Rubin [386] described a construction of uncountable Boolean algebras based on certain additional set-theoretical hypotheses. Using $\diamond$, he constructed a Boolean algebra B of cardinality $\aleph_1$ with the following properties: (a) B has precisely $\aleph_1$ subalgebras; (b) every uncountable subset of B contains a chain isomorphic to the chain of rationals, an infinite independent set and three distinct elements a, b, and c, such that $a \wedge b = c$; (c) B is retractive and not embeddable in an interval algebra. Also, this paper showed that, assuming CH, the theory of Boolean algebras is undecidable in the second-order language.

A family $\{a_i : i \in I\}$ of elements of a Boolean algebra B is called free if, for every finite subset K of pairwise elements of I, we have $\bigwedge(\varepsilon_i a_i : i \in K) \neq 0$, where $\varepsilon_i a_i$ is either a_i or its complement. Argyros [82] studied Boolean algebras B in which every free family of elements of B had cardinality less than that of B. In particular, the following theorem was proved. Let λ be an uncountable regular cardinal number and B a Boolean algebra generated by a family $D = \{a_\xi : \xi < \lambda\}$ such that, for every filter F of B, $|F \cap D| < \lambda$. Then every free family in B has cardinality less than $|B|$. Using this theorem, he gave an example of a topological space that produced an affirmative answer to a question asked by Shelah [412].

Hansoul [224] studied special classes of Boolean algebras, primitive and quasiprimitive, to arrive at a synthesis of results of Pierce [356] and Hanf [223]. The set of all classes of isomorphic countable quasiprimitive Boolean algebras with the operations of product and free product forms a semiring, also studied in [224].

Lacava [306] studied properties of principal Boolean algebras. This concept was introduced by Mangani and Marcja [318] to characterize saturated models. A Boolean algebra is called a tree algebra if it is a field of subsets of a tree T generated by the set of subsets $\{s \in T : t \leq s\}$, $t \in T$. An interval algebra is a field of subsets of a linearly ordered set L generated by the set of intervals $[a, b)$, where $a \in L$ and $b \in L \cup \{\infty\}$. Brenner and Monk [118] studied embeddings of tree algebras in interval algebras and vice versa. For example, they proved that every tree algebra is embeddable in an interval algebra and there exist interval algebras not embeddable in tree algebras. Chains in tree algebras were studied. Bonnet [115] studied the class of superatomic interval Boolean algebras of cardinality $\kappa \geq \omega_1$. On this class he introduced and studied several quasi order relations setting $B \preceq B'$ if B is embeddable in B', or elementarily embeddable in B', or a homomorphic image of B', etc.

Miller and Prikry [330] proved the compatibility of the following statement with the negation of CH: every complete Boolean algebra of cardinality not exceeding c (continuum) is isomorphic to a quotient algebra of $P(\omega)$ modulo an ideal that is at most ω_1-generated. Boolean algebras of the form $P(\omega)/I$, where I is an ideal of $P(\omega)$, were studied by Just and Krawczyk [258]. They proved, assuming CH, that for ideals I_1 and I_2 of a certain special class there exists a mapping $P(\omega)/I_1 \to P(\omega)/I_2$ that is simultaneously an isomorphism of Boolean algebras and an isometry. This gave a positive answer to a problem of Erdös [188]: if I_1 is the ideal consisting of the sets of density 0 and I_2 the ideal of the sets of logarithmic density 0, are $P(\omega)/I_1$ and $P(\omega)/I_2$ isomorphic?

Venkataraman [448] considered properties of Boolean algebras $A(G)$ and $F(G)$ which consist of all almost periodic functions or all functions from a group G into a Boolean algebra B, respectively.

Koppelberg [299] studied, among other things, the following topics. Let B be a Boolean algebra, $g \in \operatorname{Aut}(B)$, $|B| = \kappa \geq \omega$, fix $g = \{x \in B : g(x) = x\}$. Does there exist a subgroup $G \subset \operatorname{Aut}(B)$, such that $|\operatorname{fix} g| < \kappa$ for every $g \in G - \{\operatorname{id}\}$? Does there exist a nontrivial automorphism g of B such that $|\operatorname{fix} g| < \kappa$? The author considers mainly the classes of complete and homogeneous Boolean algebras.

Todorčević [436] studied special subalgebras of the Boolean algebra of subsets of a set that possess a separation property and considered minimal algebras of this type.

Representations of Boolean algebras by stable tolerance relations on a group were considered in [V, 182] mentioned in Chapter V-3.

1.5. Extensions of Boolean algebras. The study of constructive extensions of computable algebraic systems was initiated by Ershov [16]. Madison [316] gave an example of a countable extension of a countable atomless Boolean algebra that is not constructive with respect to any of the uncountable number of its indicators. This example produced a totally nonconstructive extension.

Talamo [428] studied special extensions of Boolean algebras. He says that a subalgebra B' of a Boolean algebra B generates B if each element of B is a supremum of a set of elements of B'. An ideal of B is called σ-complete (principal) if it contains suprema of its countable (arbitrary) subsets. An algebra B is called σ-saturated if each of its σ-complete maximal ideals is principal. A σ-saturated Boolean algebra C is called a σ-saturation of its subalgebra B if the following holds: if $X \subset B$ and there exists $\Sigma^B X$, then there exists $\Sigma^C X$ that equals $\Sigma^B X$. A σ-saturation D of an algebra B is called minimal if for every σ-saturation C of B there exists an injective homomorphism $D \to C$ that preserves all suprema of sets existing in D and is an identity mapping on B. The following statements were proved. (1) Let S be an uncountable set and B a σ-algebra generated by all finite or countable subsets of S. Then B has no minimal σ-saturation. Moreover, there exist nonisomorphic σ-saturations of this algebra. (2) If B is a complete Boolean algebra of cardinality less than the first measurable cardinal, B is σ-saturated.

Czelakowski [151] studied a generalization of partial Boolean algebras, introduced homomorphisms, embeddings, and weak embeddings of such partial algebras, constructed a direct limit of an ordered system of Boolean algebras, and found conditions equivalent to weak embeddability and embeddability of a generalized partial Boolean algebra into a Boolean algebra.

1.6. Boolean algebras with additional operations. Tembrowski in [432] and [433] studied three varieties of Boolean algebras with an additional binary operation, called B_3-algebras, B_4-algebras, and B_5-algebras, respectively. These classes are polynomially equivalent to classes of quasitopological Boolean algebras, topological Boolean algebras, or monadic algebras, respectively. The study of B_3-algebras was motivated by algebraic semantics of a Boolean strengthening of propositional calculus with identity. The author investigated various types of filters and their connection with congruences on these algebras.

Lipparini [311] studied closure algebras, i.e., algebras $\mathscr{A} = (A; +, \cdot,', 0, 1, k)$, where $(A; +, \cdot,', 0, 1)$ is a Boolean algebra and $k: A \to A$ is a mapping that satisfies all formal properties of the closure operator in a topological space. The class of these algebras is axiomatizable in the first-order logic. An algebra $\mathscr{A}$ is called existentially complete if every system of equations and nonequalities with parameters from A, which has a solution in an extension of $\mathscr{A}$, has a solution in A. Definitions of F-, $\vee$-, $\vee_\alpha$-, and $\vee_D$-theories can be found in [198]. Lipparini characterized existentially complete closure algebras. He proved that the theory of closure algebras and the theories of $\vee$-, $\vee_\alpha$-, and $\vee_D$-algebras possess no model companions, while the theory of F-algebras does.

Boolean algebras with a single additional operation were studied by Montagna [334]. A mapping σ of Boolean algebras is called a semimorphism if $\sigma 0 = 0$ and $\sigma(x + y) = \sigma x + \sigma y$. A well-founded algebra is a Boolean algebra with a semimorphism σ satisfying the condition $\sigma x \geq x$ for $x \neq 0$. The class C of all well-founded algebras contains the variety of diagonalizable algebras introduced by Magari. An example of a well-founded algebra is provided by the subalgebra of the Boolean algebra of all subsets of an arbitrary set connected with the relation $\in$ (the $\in$-algebra). It is shown that the variety generated by the clan C is the class P of all Boolean algebras endowed with the semimorphism operation σ. Each algebra of clan P is a homomorphic image of a subalgebra of a suitable $\in$-algebra. There exist diagonalizable algebras not embeddable in a complete well-founded algebra.

A necessary and sufficient condition for a well-founded algebra to possess a well-founded completion was given.

See also [II, 39] and [II, 270] mentioned in Chapter II-2 and [II, 165] in Chapter II-4.

1.7. Generalizations of Boolean algebras. Rybakov [51] proved the hereditary undecidability of the theory of free algebras of countable rank in varieties of topo-Boolean and pseudo-Boolean algebras and certain subvarieties. (A topo-Boolean algebra is a Boolean algebra with a unary operation of "interior" that satisfies the same identities as in topology, while a pseudo-Boolean algebra is a distributive lattice with the operations of relative pseudocomplementation and dual pseudocomplementation.) Malaï and Ratsa [28] investigated the structure of the set of derived operations of a three-element pseudo-Boolean algebra.

Quasi-Boolean algebras are studied in [170]; these are algebras of the form $(A; \vee, \wedge, \sim, 1)$, where $(A; \vee, \wedge)$ is a distributive lattice with the greatest element 1 and $\sim$ is a unary operation on A that satisfies the following conditions: $\sim\sim a = a$ and $\sim (a \vee b) = \sim a \vee \sim b$. Let U be a set, ρ a symmetric relation on U, and $Q(U, \rho)$ the lattice of subsets of the set ρ closed under the operation $\sim R = \rho - \check{R}$, where $\check{R}$ is the relation inverse to $R \subset U \times U$. It is proved that every quasi-Boolean algebra is isomorphic to a quasi-Boolean algebra $(Q(U, \rho); \cap, \cup, \sim, \rho)$. An outline was given of the proof of a theorem on representations by means of relations on Boolean algebras with an additional unary operation, an involution compatible with the Boolean operations. Mardaev [30] considered finitely generated implicative lattices and pseudo-Boolean algebras.

Sendlewski [408] proved that there exist exactly three pretabular varieties of N-lattices (quasi-Boolean algebras), and that each of them is locally finite.

See [VI, 276] at the beginning of §3 of Chapter VI for a generalization of Boolean algebras.

1.8. Subalgebras and other special subsets. Problems of existence of complements or quasicomplements of elements of the lattice Sub B of subalgebras of a Boolean algebra B were considered in [178]. An element $C \in \operatorname{Sub} B$ is called a quasicomplement of $A \in \operatorname{Sub} B$ if it is maximal with respect to the property $A \cap C = \{0, 1\}$. It was proved that the quasicomplementability relation need not be symmetric. A characterization of complemented countable subalgebras of the Boolean algebra $P(\omega)$ was given. Assuming the Martin Axiom of (MA), a Boolean algebra B of cardinality 2^{ω} was constructed; it has countable subalgebras A and C that are mutual quasi-complements in Sub B. On the other hand, MA implies that no two subalgebras of $P(\omega)$ of cardinality less than 2^{ω} are mutual quasicomplements.

The invariant depth $B = \max\{\aleph_0, \sup\{|X| : X \text{ is a well-ordered subset of } B\}\}$ in the class of Boolean algebras was studied in [327]. The behavior of depth with respect to direct products and free products was considered. In particular, (1) if depth $B = k$ and $\operatorname{cf} k = \aleph_0$, then depth is accessible (see [20] for the definition of $\operatorname{cf} k$ for a cardinal k); (2) if $\operatorname{cf} k > \aleph_0$, a Boolean algebra A has no chains of type k and a Boolean algebra B has no chains of type $\operatorname{cf} k$, then the free product $A * B$ has no chains of type k.

Let $(A; +, \cdot, -, 0, 1)$ be a Boolean algebra, and I and J its ideals. We say that I and J are (elementarily) equivalent if $(A; +, \cdot, -, 0, 1, I)$ and $(A; +, \cdot, -, 0, 1, J)$ are elementarily equivalent models. It was proved in [257] that if A is a superatomic Boolean algebra (that is, each of its homomorphic images is an atomic Boolean algebra), then each set of its pairwise nonequivalent ideals is countable. If either the Boolean characteristic of A is $(\infty, 0, 0)$ or A is countable and superatomic, then A has $2^{\aleph_0}$ ideals.

1.9. Connections with topology. Let B_i ($i \in I$) be Boolean algebras, X_i their Stone spaces, $B = \prod(B_i : i \in I)$ the direct product of B_i, and X the Stone space of B. It is known that the Čech-Stone compactification of the disjunct sum $\sum X_i$ is the Stone space of B. Bernardi [108] showed that the Boolean algebra B' formed by those elements $b \in B$, for which the set $\{i \in I : b_i \neq 1\}$ is finite, has the single-point compactification of Alexandrov of the disjunct sum $\sum X_i$ as its Stone space. Moreover, Boolean algebras B'' that satisfy the condition $B' \subset B'' \subset B$ correspond to those compactifications of $\sum X_i$ that are Stone spaces.

Using the Martin Axiom, Błaszczyk [110] constructed a dense subspace X of the real line R such that its Čech-Stone compactification βX is strongly rigid, that is, every endomorphism of the corresponding Boolean algebra is the identity automorphism.

Dimov [161] described the Stone duality analogously to the axiomatic description of the Pontryagin duality given by I. Prodanov. It is contained in the following theorem. Let B be the category of Boolean algebras and Boolean homomorphisms, C the category of zero-dimensional compacta and continuous mappings, and $T: B \to C$ the Stone duality functor. If K is a complete subcategory of the category of zero-dimensional Hausdorff spaces and continuous mappings and there exists a duality $F: B \to K$, then K coincides with C and $F \approx T$. Diers [158] introduced a class of locally indecomposable categories that are equivalent to categories of bundles of sets, endowed with a certain algebraic structure, on Boolean topological spaces. This generalized Stone's theorem on representations of Boolean algebras by Boolean topological spaces. Numerous examples were given. Hansoul [225] extended and generalized results from the papers [255] and [356] on the Stone-type topological representations of Boolean algebras and distributive lattices with additive operators (DLO). He introduced the concept of a Stone multi-algebra and proved that DLO with endpoints (NDLO) are characterized up to isomorphism by their dual Stone multi-algebras. It was proved that the category of NDLO-lattices and homomorphisms is dually isomorphic to the category of Stone multi-algebras. Among the remaining results we note the following: if DO (CBO) is the category of DLO-lattices (complete Boolean algebras) and homomorphisms (complete homomorphisms), then DO is faithfully representable in CBO, i.e., there exists a faithful covariant functor ext: DO → CBO such that (1) for every DLO L there exists a canonical embedding ext: $L \to \text{ext} L$; (2) for every morphism $\rho\colon L \to L'$ in DO, ext(ρ) is the only morphism in CBO that satisfies the relation $\text{ext}_{L'} \circ \rho = \text{ext}(\rho) \circ \text{ext}_L$. Mori [336] gave a topological proof of the (known) theorem that the cardinality of an infinite superatomic (i.e., all subalgebras of which are atomic) Boolean algebra coincides with the cardinality of its Stone space. Simon and Weese [419] use Boolean spaces to construct special Boolean algebras.

Lutsenko [26] investigated certain properties of a topology on a Boolean algebra introduced in [283]. Peters [353] studied the interconnection between two topologies on Boolean algebras: the topology introduced in [1] and the topology generated by a (0)-continuous submeasure [57].

Recent results on the structure and characterization of Boolean algebras of idempotent elements in topological semifields (that is, commutative real algebras endowed with an order and topology that satisfy suitable compatibility conditions) were described in the survey [67]. These algebras occur among von Neumann operator algebras. Takeuti [427] studied the Boolean algebra B of projections of a Hilbert space and the Boolean topos of sheaves on B.

Pashenkov [45] studied a certain pair of adjoint functors from the category of models into the category of Boolean algebras, on which an inner topology was defined in a certain way and a subset was chosen.

Balcar and Simon [86] surveyed results on cardinal invariants of Boolean spaces obtained by Balcar and his collaborators in the preceding five years. The survey includes the following topics: properties of Boolean algebras depending on the property of disjoint extension, characters of points in extremally disconnected compact spaces, and independent systems in Boolean spaces. In [165] the bounds are studied for those cardinal numbers that are values of cardinal functions on infinite compact F-spaces or (by the Stone duality) on weakly countable complete Boolean algebras.

See also [139] and [II, 141] mentioned in Chapter II-2 and [V, 157] and [V, 246] in Chapter V-5.

1.10. Connections with universal algebra. Dobbertin ([162]–[164]) worked out an abstract version of Vaught's theorem on isomorphism of countable Boolean algebras considering isomorphism classes of Boolean algebras with the direct sum as multiplication. The set BA of all isomorphism classes of countable Boolean algebras forms a commutative monoid $(BA, \times, 0)$, where $\times$ is the direct product operation and 0 the class of single-element Boolean algebras. The author introduced concepts of a V-monoid and V-near-rings. If $+$ is the coproduct operation, then $(BA; \times, 0)$ and $(BA; +, \times)$ are examples of such algebras. Several theorems that generalize properties of countable Boolean algebras were proved. For example, (1) Trnková's result (see vol. I of this survey, p. 15): if $nB \cong mB$ for $m < n$, then $nB \cong (n+1)B$ for a countable Boolean algebra B, was generalized; (2) the Cantor-Bendixson derivative and the superatomicity of the product and coproduct of arbitrary families of Boolean algebras may be defined in the universal Vaught monoid V; (3) every element $x \in V$ whose rank with respect to the sum is less than $\aleph_1$ belongs to BA (the algebra BA can be identified with a certain hereditary submonoid of V).

Let A be a finite algebra, B a Boolean algebra, and $A[B]$ a Boolean degree of A. It was proved in [328] that for a nontrivial quasiprimal algebra A (that is, the ternary discriminator is a polynomial in A) with a lattice reduct (that is, A has polynomial lattice operations) the following conditions are equivalent: (1) $C \cong A[B]$ for some complete Boolean algebra B; (2) C is a diagonal subalgebra in A^I for some I (that is, C consists of the elements of the form δ_a, $a \in A$, such that $\pi_i(\delta_a) = a$ for all $i \in I$) and the algebra C is complete.

Annelis [81] introduced a group structure on the set of equivalence classes of a Boolean algebra associated with a sequence of integers.

1.11. Connections with model theory. A classification of countable complete ω-stable theories based on consideration of Boolean algebras $B(M)$ of definable subsets of their countable models or, equivalently, on consideration of their Cantor-Bendixson spectrum was given in [319] and [320]. In [321] the same authors investigated pseudo-$\aleph_0$-categorical theories, that is, countable complete theories of the first order such that any two countable models of them have isomorphic Boolean algebras $B(M)$ of definable subsets. They proved that $\aleph_1$-categorical theories are pseudo-$\aleph_0$-categorical and every countable atomic Boolean algebra is isomorphic to $B(M)$ for some superstable pseudo-$\aleph_0$-categorical theory.

Let M be a subset of a Boolean algebra A and $M^d = \{a \in A : a \wedge b = 0$ for all $b \in M\}$. An ideal $I \subset A$ is called regular if $I^{dd} = I$. Regular ideals of A form a complete Boolean algebra $\overline{A}$. Heindorf [232] considered model theories of Boolean algebras in monadic languages of the second order. Let L_I (L_R) denote such a language in which set variables are interpreted as ideals (regular ideals) of a corresponding Boolean algebra. The author's approach to the theory of models of the language L_R consists of its reduction to the theory of models of the first order of pairs $(\overline{A}, A)$. The author showed that, for countable Boolean algebras, L_R and L_I have the same strength, while they are different in general. The L_R-theory of the class of all Boolean algebras is axiomatizable. Using the continuum hypothesis he proved that the L_R-theory and L_I-theory of the class of all Boolean algebras are undecidable. For every Boolean algebra there exists an L_R-equivalent Boolean algebra of cardinality at most 2^ω.

Results about Boolean algebras are contained in the paper [193] devoted to topological model theory.

De Oliveira [349] considered the beginnings of elementary model theory for cyclic Boolean algebras, proving the infinite axiomatizability in the logic of the first order, a few preservation theorems, and other model-theoretic properties of classes of cyclic and similar Boolean algebras.

1.12. Connections with measure theory. Tulipani [440] studied decidability questions for theories of classes of structures of the form (A, F, μ), called generalized measure spaces, where A is a Boolean algebra, F an ordered real-closed field, and $\mu\colon A \to \Phi$ is finitely additive measure. Using the method of interpretations he established the decidability of such structures. The most interesting result is the undecidability of the case when A is an infinite atomic algebra and F the field or real numbers (it is shown that its theory is hereditarily undecidable). The proof is given by reduction to the theory of finite graphs. (This result is somewhat unexpected, as the theory of Boolean algebras and the theory of real-closed fields are decidable.) In [439] the same author studied general measure spaces (B, F, m), where B is a Boolean algebra, F an ordered field, and $m\colon B \to \Phi$ a finitely additive measure satisfying the condition $m(1_B) = 1_F$. The author used the first-order language with a predicate distinguishing elements of B from those of F. Let T be the theory of the above structures in this language, T' an extension of T asserting that F is real-closed and m satisfies certain additional conditions. It was proved that T' is a model completion of T and T' is complete, decidable, and admits elimination of quantifiers. Analogous results were obtained for another extension T_1 of T. As an example, a simplified proof of Nikodým's theorem on extension of measures was obtained.

Let B be a σ-complete Boolean algebra with a continuous outer measure. Poroshkin [48] gave necessary and sufficient conditions for B to be a measure algebra.

Let $d\mu$ be a semimetric induced by a finitely additive measure μ on a Boolean algebra B, defined by the relation $d\mu(a, b) = \mu(a\Delta b)$ (Fréchet-Nikodým). It was proved in [83] that the factor algebra $B\mu$ of B modulo the ideal of μ-zero sets is $d\mu$-complete if and only if μ is countably additive and $B\mu$ is complete. Other conditions for the completeness of $B\mu$ were given as well.

Savel'ev [54] studied special outer measures on relatively complemented distributive lattices with values in topological spaces and uniform spaces.

Structures of the form $(A; \circ, \lesssim)$ were investigated in [403], where A is a set, $\circ$ a binary operation on A, and $\lesssim$ a reflexive and transitive relation on A ("measurement structures"). Representability of these structures by probabilistic measures was studied, and a one-to-one correspondence between probabilistic measures and σ-structures of qualitative probability for σ-complete Boolean algebras established.

1.13. Problems of logic.

1.13.1. *Problems of decidability.* Kokorin and Pinus [22] suggested a program of finding borderline situations when the decidability of a certain formalization is obtained simultaneously with the undecidability of its "insignificant" extension. In this direction Dulatova [13] considered problems of decidability for classes of finite Boolean algebras with an additive measure, enriched by various natural predicates. It is known that the theory of Boolean algebras is decidable in the first-order logic. Weese [456] gave a self-contained description of mostly known results on the undecidability of the theory of Boolean algebras in various extensions of the first-order logic, proved the undecidability of this theory in the second-order logic and weak second-order logic, the logic with the Hertig quantifier, the logic with the Marlitz quantifier Q_1^2, in stationary logic, and in logic with quantifiers over ideals. First-order theories of Boolean algebras with a distinguished group G of automorphisms were considered as well. It was proved that if G is not locally finite, then such a theory is undecidable. The undecidability of the theory of Boolean algebras as well as the theories of certain other classes of structures in stationary logic was also proved in [406].

Let $L = \{+, \cdot, -, C, 1, U\}$ be the language of the theory of Boolean algebras with an additional unary predicate U. Rubin [384] proved that the first-order theory of Boolean algebras in which U is interpreted as "being a subalgebra" is undecidable in L. Koppelberg [297] proved that the first-order theory of the class K of L-structures $(B, +, \cdot, 0, 1, A)$ where B is a complete Boolean algebra, A a complete subalgebra, and the embedding of A in B is complete, is decidable. The same author [298] proved that the theory of orthogonally complete Boolean algebras with a distinguished orthogonally complete subalgebra is undecidable.

If the unary quantifier $Q^d x$ is added to the first-order language, the language $L(Q^d)$ is obtained. If κ is an infinite cardinal number, the κ-interpretation of the quantifier $Q^d x F$ in the theory of Boolean algebras means the following: there exists a set A of cardinality not less than κ consisting of pairwise disjoint elements such that $F(a)$ holds for all $a \in A$. The language $L(Q^d)$ may be considered as a sublanguage $L(Q^2)$, where Q^2 is the binary quantifier of Ramsey. Molzan [332] suggested a possible step to a solution of the decidability problem of Boolean algebras with an ω-interpretation of the quantifier Q^2 in $L(Q^2)$. Let T be the $L(Q^2)$-theory of all superatomic Boolean algebras and T' the $L(Q^d)$-theory of superatomic Boolean algebras which have an ω-interpreted infinite quantifier. Using the elimination of quantifiers the author proved that T and T' are decidable. The classes of models of T and T' are equal, and two models of T in $L(Q^2)$ have the same theory if and only if they have the same theory in $L(Q^d)$.

Rybakov [52] proved the decidability of the universal theory of free algebras in the variety of topo-Boolean algebras which correspond to modal logics λ for $\lambda = S4 + \sigma_k$, $k < \omega$. Thus, he established the algorithmic decidability of the admissibility problem for each of the logics of the form $S4 + \sigma_k$. He proved that free algebras of every variety $\mathrm{eq}(S4 + \sigma_k)$, $k \geq 2$, do not have a basis for quasi-identities with finitely many variables and have a hereditarily undecidable elementary theory.

In §1.4 decidability problems are also considered.

1.13.2. *Constructivity, recursivity.* Pinus [46] introduced a notion of F-constructivization of a model, where F is a certain logical formalism; it generalizes the notions of constructive and strongly constructive models. There are a number of propositions concerning the existence of constructivizations of Boolean algebras with certain conditions on the recursivity of ultrafilters. He proved that for certain Boolean algebras there exist F-constructivizations, where F is a formalism stronger than the first-order predicate calculus. Uzakov [66] proved that a constructible atomic Boolean algebra has a recursive representation whose set of atoms is not hyper-immune. Morozov [32] showed that every countable Boolean algebra is strongly constructible; he found criteria of saturation and homogeneity for countable Boolean algebras. In [33] the same author classified countable homogeneous Boolean algebras over their isomorphism types and obtained a criterion of strong constructibility for countable homogeneous Boolean algebras. An automorphism φ of a constructive model (B, ν) is called recursive if there exists a general recursive function f such that $\varphi\nu = \nu f$. Morozov [34] proved that for every atomic constructive Boolean algebra (B, ν) there exists a constructivization μ such that the group of all recursive automorphisms of the constructive Boolean algebra (B, μ) is isomorphic to the group of all permutations of ω that shift only finitely many elements. Odintsov [41] found examples of constructive Boolean algebras with a nonconstructible factor algebra modulo an atomless ideal. He gave a complete classification of atomic Boolean algebras from the point of view of existence of a nonconstructible Boolean algebra which possesses a given factor algebra over an atomless ideal.

S. S. Goncharov [10] proved that there are no universal recursively enumerable Boolean algebras for the classes of recursively enumerable atomic, recursively enumerable atomless, and recursive atomic Boolean algebras. V. A. Goncharov [9] established the existence of a noncomplemented recursively enumerable independent subset in a constructive Boolean algebra.

Herrmann [V, 358] studied special pairs of Boolean algebras and showed that in the lattice of recursively enumerable sets there are so many elementary definable pairs of this type that, in their theory, a certain class of graphs that contains all finite graphs is elementarily definable. It follows that the elementary theory of the lattice of recursively enumerable sets is undecidable. (See also V-5. "Creative" pairs of subalgebras of recursively enumerable Boolean algebras were studied in [415]. It was proved in [251] that the four-element Boolean lattice can be represented by recursively enumerable tt-degrees with preservation of 0 and 1. An analogous result holds for the pentagon N_5 and lattices M_n, where $n \geq 3$.

Morozov [35] proved that a strongly constructible atomic Boolean algebra is determined by its group of recursive automorphisms up to a recursive isomorphism. However, the property of constructibility of a Boolean algebra cannot always be recognized from the group of all of its automorphisms. Connections between constructivizations of Boolean algebras and their groups of recursive automorphisms were studied by this author in [36]. The lattice of recursively enumerable filters of an atomless Boolean algebra was studied in [169].

Let B be a free recursive Boolean algebra with a countable number of generators and $L(B)$ the lattice of recursively enumerable subalgebras of B. Guichard [V, 346] proved that every automorphism of $L(B)$ is induced by a recursive automorphism of B.

1.13.3. *Other topics.*
Elementary theory of Boolean algebras with a family of ideals was studied in [438]. The method explained by the author permits elimination of quantifiers in this theory and gives a criterion for the elementary equivalence of models of the theory. Kir′yakov [21] used the Boolean algebra of sequential types to analyze logical formulas. Zakhar′yashchev [18] applied constructions connected with pseudo-Boolean algebras to axiomatize extensions of the intuitionist propositional calculus.

Smoryński [420] and Montagna [335] studied fixed-point algebras, that is, pairs (B, A) of Boolean algebras such that (1) each element of B is a self-map of A; (2) Boolean operations on B act pointwise; (3) every constant self-map of A is an element of B; (4) every element of B possesses a fixed point. These algebras stem from the logic of provability. Hanf and Myers [V, 354] proved that, for the first-order theories, Boolean algebras of sentences of any two undecidable languages or of any two functional languages are recursively isomorphic. They produced a very simple criterion for the undecidability of a language and the functionality of a language.

A Boolean algebra is called definably complete if every set of its elements definable in the first-order language has a supremum. The first-order calculus of individuals (denoted by CI) is the first-order theory of definably complete Boolean algebras with the least element eliminated. Hendry [234] studied five types of consistent extensions of CI, proved that four of them are complete, and conjectured that the fifth is complete as well, and that these are all the complete extensions of CI. This conjecture was confirmed in *Mathematical Reviews* 83m:03042.

Rehder [377] established a one-to-one correspondence between homomorphisms of a Boolean algebra and its operators of "asymmetric conjunction". These operators are connected with the operator "and next" in logic.

Dziobiak [184], [183] proved that the lattice join of two finitely axiomatizable quasivarieties, each of which is generated by finite Heyting (finite topological Boolean) algebras, need not be finitely axiomatizable. Epstein [187] generalized results from the 2-valued logic to the n-valued logic connected with Boolean algebras.

Consider the following statements: (1) Every complete Boolean algebra is an injective object in the category of all Boolean algebras. (2) Every Boolean algebra has a prime ideal. In the ZF axiomatics (1) $\Rightarrow$ (2) holds. Bell [101] showed that the opposite implication does not hold. He conjectured that (1) is equivalent to the Axiom of Choice.

Boolean algebras are considered in the papers of Wolniewicz [459], [460] devoted to the notion of situation after Wittgenstein.

See also [V, 237] in Chapter V-5.

1.14. Applications of Boolean algebras. The theory of Boolean-valued models was used in [23] to solve the problem of inner characterization of the subdifferential of an operator acting in an arbitrary Kantorovich space. In [24] Kusraev showed that positive order-continuous operators preserving order segments in Kantorovich spaces admit embedding in a suitable Boolean-valued model of set theory, turning into order-continuous functionals. This circumstance makes it possible to deduce various facts about operators from the corresponding theorems for functionals. Gordon [II, 10] characterized a class of injective modules in the framework of Boolean-valued models of set theory. Nordahl [346] constructed an eight-element Boolean lattice which permits one to find the best way of choosing donor blood for transfusion in the cases of certain diseases of newborns.

Kiss [291] found a short and elementary proof of the following result of Burris and McKenzie (see [126], pp. 67–106): Every variety, Boolean-representable by a finite number of finite algebras, is a join of an abelian and a discriminator variety.

Boolean algebras appear also in the paper [422], which is devoted to quantum systems. See also Chapter II.

Chromik [138] studied pairs of graphs $G_1 = (V_1, R_1)$ and $G_2 = (V_2, R_2)$, where $V_1 \cap V_2 = \varnothing$, with pairs of strong homomorphisms $G_1 \to G_2$, $G_2 \to G_1$. For every element f of a free Boolean algebra B_2 with two generators he constructed a graph $G_f = (V_1 \cup V_2, R_f)$ and, on the set $G = \{G_f : f \in B_2\}$ he defined an algebra $D = (G; \cap, \cup, ')$ in a natural way, showing that the mapping $f \to G_f$ is a homomorphism of B_2 onto D.

1.15. Miscellany. The papers [194], [450], [449] are devoted to games on Boolean algebras. Systems of linear equations on Boolean algebras were investigated in [329], [2].

By definition, a Boolean function $f: B^n \to B$ preserves a constant $a \in B$ if $f(a, \ldots, a) = a$. Tošić [437] proved several interesting results for such functions. For example, f preserves at least one constant if and only if

$f(0, \ldots, 0) \le f(1, \ldots, 1)$. A primitive Boolean function either preserves all constants or at most one constant. A concept of a partial semiderivative of a Boolean function was introduced and studied in [342].

We say that a subset X of an algebra A is elastic if, for every mapping $p: X \to A$, the congruence on A generated by all pairs $(x, p(x))$, $x \in X$, does not coincide with A^2. Properties of elastic subsets of Boolean algebras and other algebraic systems were investigated in [149].

A concept of a J-fuzzy Boolean algebra was introduced in [147]; this is the set of mappings of a partially ordered set J into a Boolean algebra satisfying certain natural conditions. Connections between Lukasiewicz J-algebras and fuzzy Boolean algebras and also duality for fuzzy Boolean algebras were studied. In [381] a combinatorial property of a system B of subsets of a finite set is established, ensuring that, if B has sufficiently many elements, a certain subset of B forms a Boolean algebra.

Bunyatov and Kasimov [7] proved an Eilenberg-Zil'ber type theorem, establishing an isomorphism between singular complexes of a polygonal product of Boolean closure algebras and the tensor product of singular complexes of the given Boolean closure algebras. The role of distributive laws is clarified in this situation. Beznosikov [4] pointed out certain conditions for the existence of an extension of a σ-continuous at 0 external homomorphism with values in a regular Boolean algebra. In [132] it was proved that the class of Boolean lattices forms a Mal'tsev variety.

Balcar and Franěk [85] generalized the classical theorem of Fikhtengol'ts, Kantorovich, and Hausdorff on independent subsets of the family of all subsets of a set: in each infinite complete Boolean algebra B of cardinality k there exists an independent subset of cardinality k. It follows that there exist 2^k ultrafilters on B and every Boolean algebra C that satisfies the inequality $|C| \le |B|$ is a homomorphic image of B. The proof presupposes ZFC.

Let $B(R)$ be a Boolean σ-algebra of Borel subsets of the real line R. Bulatović [123] defined mappings of $B(R)$ into a pseudo-Boolean algebra A, called random functions on A. The following was proved. Let F be a Boolean σ-algebra and h a σ-homomorphism of F into a Boolean σ-algebra A. If X is a random function on A, then there exists a real F-measurable function y (that is, a random variable in the classical sense) such that $X(B) = h(y^{-1}(B))$ for all $B \in B(R)$.

It was proved in [196] that, for an infinite cardinal number k, the cardinality of the set of nonisomorphic Boolean subalgebras of the algebra $P(k)$ is 2^{2^k} and there exist 2^{2^c} nonisomorphic countable complete subalgebras of $P(c)$. The second statement answers a question raised by Ulam. Kung [V, 419] gave a characterization of an abstract Poisson algebra in terms of the factor algebra of a free Boolean algebra obtained by means of a binomial identity.

Poroshkin [47] proved that the possibility for a Boolean algebra B to possess a one-sided continuous function (numerical, vector, or with values in a special topological space) separating 0 is equivalent to the following Horn-Tarski condition: (1) B can be represented as the union of a sequence $S_0, S_1, \ldots$ of sets neither of which contains infinite disjoint subsets. He discussed the possibility of defining other types of functions on an algebra B with property (1) and the question of the separability of a certain topology on B.

Monk [333] studied two set-theoretical invariants of a Boolean algebra: $\operatorname{ind} A = \sup\{|X| : X$ is an independent subset of $A\}$ and the set of cardinal numbers free $\operatorname{cal} A$, where $\kappa \in$ free $\operatorname{cal} A$ if and only if every κ-element subset of A contains κ independent elements. He studied the behavior of these invariants under such operations as products, weak products, free products, unions of chains, subalgebras, or homomorphic images. Special attention was given to Boolean algebras which possess a countable separability property or satisfy chain conditions. A few open problems were mentioned in this paper.

Shelah [412] showed that a Boolean algebra B, whose cardinality is a singular limit cardinal number, contains an antichain of cardinality $|B|$. The same holds if $|B|$ is a singular cardinal number with countable cofinality. On the other hand, he proved that the statement "there exists a Boolean algebra of cardinality $2^{\aleph_0}$ which contains neither an uncountable antichain nor an uncountable chain" is compatible with ZF and $2^{\aleph_0} > \aleph_1$. He studied Boolean algebras satisfying chain κ-conditions. Lutsenko [27] considered the following problem: let two isomorphic Boolean algebras of cardinality τ be representable as unions of chains of their subalgebras of lesser cardinality. Do these chains have isomorphic cofinal subchains? He gave cases with a negative answer and cases with a positive answer.

Let C (respectively, R) be a Boolean algebra of Borel sets of real numbers factored modulo the ideal of meager (respectively, of measure 0) sets. Gavalec [200] proved that for every cardinal number $\alpha \geq 2^\omega$ the algebra $RO(\alpha^\omega)$, where α has a discrete and 2^α Tikhonov [$\equiv$ product] topology and $RO(\alpha^\omega)$ denotes the Boolean algebra of regular open subsets of the space α^ω, is completely generated by the union of two complete subalgebras which are independent and isomorphic to the algebra C (respectively, algebra R).

Oberst [348] considered applications of the Chinese Remainder Theorem to Boolean algebras. Boolean algebras also appeared in [296], devoted to the algebraic theory for formal languages.

§I.2. Distributive lattices

2.1. Congruences, endomorphisms, products. See remarks concerning Huhn [III, 209] in III-3, 3.2 concerning representation of distributive semilattices with zero by compact congruences of a lattice. Conditions for congruence extensions in a 0-dimensional compact distributive lattice were found in [II, 156]. See III-3, 3.3 for information on the papers [III, 264] and [III, 39] devoted to endomorphisms and automorphisms of distributive lattices.

It has been conjectured for a long time that every distributive algebraic lattice is representable as the congruence lattice of a suitable lattice. Huhn [240] expressed the opinion that this problem was connected with the theory of free distributive products of lattices and decided to use this idea. As a first step he proved that if D_1 and D_2 are finite distributive lattices and D_1 is isomorphic to a 0-∨-subsemilattice of D_2 (let a^+ denote the image of a under this isomorphism), then there exists a distributive lattice D which contains D_1 as a 0-sublattice and D_2 as a $(0, 1)$-sublattice such that $a \leq a^+$ in D for every $a \in D_1$. Moreover, D can be chosen in such a way that, for any elements $a \leq b \leq c$ from D_1, b is relatively complemented in the interval $[a, c]$ of D only when it is relatively complemented in the interval $[a, c]$ of D_1.

Lattices from certain classes of distributive lattices can be characterized as coproducts (= free products) $A * L$ of certain sublattices A and L. (Such classes are Post algebras and their various generalizations.) Yaqub [461] studied conditions under which a coproduct of two distributive lattices is α-complete, α-representable (that is, it is an α-homomorphic image of an α-ring of sets), or an α-ring of sets modulo an α-ideal. He proved that if L is finite, then $A * L$ has one of the above three properties exactly when A has it. If A is a Boolean algebra, then $A * L$ is α-complete if and only if at least one of the lattices A and L is finite and the other lattice is α-complete. He applied these results to Post algebras and their generalizations.

Let L be a free (in the category of abelian l-groups) product of abelian l-groups G_i, $i \in I$. It was proved in [360] that the sublattice generated by the set $\bigcup(G_i : i \in I)$ can be decomposed into a free product (in the category D_e of distributive lattices with a fixed element e) of lattices G_i, $i \in I$. The problem was raised of whether an analogous statement held for any variety of l-groups. A characterization of free products in D_e was given. The paper [361] is dedicated to free products; it is a survey of its author's results on free products in various varieties of l-groups, including certain new results. The new results are about the representability of a free product in the variety generated by all linearly ordered groups. It was proved that a free product of finitely generated groups is not decomposable into a direct product in any proper variety of l-groups.

Boolean decompositions of bounded distributive lattices were studied in [226]. Representations of a bounded distributive lattice L as a Boolean product are in one-to-one correspondence with special equivalences on the Priestley dual space for L, factorizations over which lead to the trivial ordering. The Priestley dual space for the Boolean product of bounded distributive lattices is isomorphic to the cardinal sum of the Priestley dual spaces for germs.

Fraser [195] introduced the semilattice tensor product $A \otimes B$ of distributive lattices A and B and showed that if $A \otimes B$ is projective in the category $\mathscr{D}$ of distributive lattices, then A and B are projective in $\mathscr{D}$ as well, and he gave examples of (infinite) projective distributive lattices A and B for which $A \otimes B$ is not projective. Lakser [308] showed that if A and B are nontrival distributive lattices, then $A \otimes B$ is projective in $\mathscr{D}$ if and only if A and B are projective in $\mathscr{D}$ and each of them has the greatest element.

2.2. Various degrees of distributivity. Generalizations of distributivity.

Important results on the structure of completely distributive complete lattices are contained in Dwinger's papers [180], [181]. Let D_c denote the class of these lattices. The main goal of the first of these papers is a characterization of complete homomorphic images of a lattice $L \in D_c$ as certain substructures of L. The results show that many properties of finite distributive lattices hold for D_c as well. Let $L \in D_c$. By means of a special binary relation σ special subsets of L are defined, called σ-sets. It is proved that if $L, L_1 \in D_c$, then L_1 is a complete homomorphic image of L if and only if L_1 is isomorphic to a σ-set of L. The set of all σ-sets of L forms a lattice with respect to set-theoretical inclusion, and this lattice is anti-isomorphic to the lattice of all complete congruence relations on L. Necessary and sufficient conditions are found for L to be represented as a direct or subdirect product of a family of its σ-sets. It is proved that L is a subdirect product of its maximal σ-sets that are chains. This theorem strengthens a result of Raney of 1953. In the second paper the author described the structure of a free dense-in-itself completely distributive complete lattice $F_{DS}(S)$ as a certain σ-subset of $F_{D_c}(S)$. It was proved that if S is infinite, then $F_{DS}(S)$ is a subdirect product of copies of a segment of the real line. The same author [182] studied unary operations $a^u = \bigwedge(x : x \not\leq a)$ and $a_v = \bigvee(y : y \not\geq a)$ on completely distributive complete lattices. Bandelt [90] characterized coproducts in the category $D(\alpha, \beta)$ of bounded (α, β)-distributive lattices with morphisms preserving 0, 1, all infima of subsets of cardinality $< \alpha$, and all suprema of subsets of cardinality $< \beta$. Here α and β are infinite cardinal numbers and β is strongly α-inaccessible (that is, if $\kappa < \alpha$ and $\lambda < \beta$, then $\lambda^\kappa < \beta$). In [III, 62] he studied the concept of an $\mathscr{M}$-distributive lattice (completely distributive lattices are obtained as a special case; see III-1, 1.3). In another paper [91] he and Erné investigated conditions for $\mathscr{M}$-embeddability of an $\mathscr{M}$-distributive lattice in powers of a unit interval of the real line, as well as the possibilities of a representation by systems of subsets closed under special closure operations.

For these subjects see also [III, 357] and [III, 250] in III-1, 1.3 and [III, 128] in III-1, 1.2 and III-3, 3.2.

Negru [40] says that a lattice is quasidistributive if it satisfies the identities $(x+y)(z+u) = x(z+u)+y(z+u)+z(x+y)+u(x+y)$, $xy+zu = (x+zu)+(y+zu)+(z+xy)(u+xy)$ and proves that a free quasidistributive lattice with three generators is infinite. He gave examples of nontrivial identities which follow from the identities of quasidistributivity. Every modular sublattice of the lattice of Post classes is quasidistributive. Conditionally distributive lattices, that is, lattices L in which $x \wedge (y \vee z) = (x \wedge y) \vee (x \wedge z)$ for all elements $x, y, z \in L$ satisfying the condition $y \wedge z \neq 1$, were studied in [231]. There was given a representation of such lattices by families of sets. A dual concept for the concept of conditional distributivity was studied by Fourneau [III, 160], who used the term "weakly distributive lattice". He gave examples of such lattices, characterized them, and proved that the lattice $I(L)$ of ideals is weakly distributive if and only if L is weakly distributive. Thus every weakly distributive lattice can be embedded in a complete weakly distributive lattice. Gerstmann [203], [204] introduced concepts of infinitely n-distributive lattices, and completely n-distributive lattices, whose relation to A. Huhn's concept of n-distributivity is analogous to the relation of infinite distributivity and complete distributivity to ordinary distributivity. These lattices were studied in connection with closure operators in sets. The results were applied to a characterization of n-distributivity of the subalgebra lattice of an algebra. It was proved that infinite n-distributivity and distributivity are equivalent for the subalgebra lattice of an idempotent algebra. A theorem on complete n-distributivity of the subgroup lattice of an abelian group was proved.

2.3. Extensions of distributive lattices. Let B be a subalgebra of an algebra A. Then A is called an essential extension of B if the identity congruence Δ_B has the only extension to a congruence of A (of course, to Δ_A). An algebra A is called a strong extension of B if each congruence of B is a restriction of an A-congruence. If such an A-congruence is unique, A is called a perfect extension of B and B a perfect subalgebra of A. Hansoul and Varlet [III, 191] studied these concepts for lattices, in particular, for distributive lattices. For example, they proved that minimal perfect sublattices of a distributive lattice are its maximal chains (if they exist), and if B is a sublattice of a distributive lattice A, then B is perfect in A if and only if A is the only sublattice of A that contains B and is closed under relative complementation (whenever it is defined).

A lattice is called essentially metrizable if it is an essential extension of a countable lattice. It was proved in [206] that the following conditions are equivalent for every completely distributive lattice L: (1) the interval topology of L is metrizable; (2) L is essentially metrizable; (3) L is an essential extension of a countable chain.

Call a lattice L reductive if each of its nontrivial (that is, nonsingleton) intervals contains a nontrivial chain interval. By definition, L is completely nonreductive if it does not contain nontrivial chain intervals. In [207] the reductivity of the following lattices was proved: (1) a modular lattice of finite width; (2) a complete modular lattice in which each element is a finite join of ∨-irreducible elements. A distributive lattice is reductive if and only if it has an essential extension that is a product of chains. Every distributive lattice is decomposable as a subdirect product of a reductive lattice and a completely nonreductive lattice.

It was proved in [III, 296] that every completely distributive complete lattice has a complete linear extension. There was given an example of a completely distributive lattice without this property (see III-1, 1.3). As proved in [312], every distributive lattice with the least element is embeddable in the positive cone of a conditionally complete vector lattice.

Hagendorf [222] studied special extensions of chains. Let $A \leq B$ mean that a chain A is isomorphically embeddable in a chain B. A chain C is called cut-free if, for every cut $C = A + B$, $C \leq A$ or $C \leq B$; C is called break-free if all of its sufficiently large intervals are cut-free. Certain properties of cut-free and break-free chains were studied, and a necessary and sufficient condition for a chain to be cut-free was found. Let $A < B$ mean that $A \leq B$ and simultaneously $B \nleq A$. C is called an immediate extension of A if $A < C$, and $A < B < C$ holds for no B. It was proved that every chain has an immediate extension and a simple method of its construction was given. Some of the methods considered were used to study arbitrary relations.

A Cauchy structure is defined as the set of all Cauchy filters of a certain uniform convergence space. Methods of Cauchy structures and convergence structures on lattices were applied in [88] to study completions of distributive lattices. Of the great number of results of this paper we mention the following. Let a and b be elements of a lattice G and $a \leq b$. Define $a \downarrow b = \{x \in G : x \wedge b \leq a\}$. An element $g \in G$ is called an exact supremum of a set $X \subset G$ if, for all $a < b$, $X \subset a \downarrow b$ implies $g \in a \downarrow b$. An exact infimum is defined dually. By definition, a lattice is exactly complete if each of its subsets has an exact supremum and an exact infimum. It was proved that a lattice is exactly complete if and only if it is complete and infinitely distributive. Every distributive lattice G has a uniquely determined exactly complete extension $C(G)$ that satisfies the same universal formulas (with constants from G) as G does.

2.4. Dimension. Combinatorial topics. Romanovich [49] suggested a notion of dimension $\mathrm{lodim}\, L$ of an arbitrary lattice L, which, in the distributive case, coincides with the definition of an analogous dimension by Gavalec [200], and gave a formula that expresses $\mathrm{lodim}\, L$, where L is a special (in particular, lexicographic) product of bounded lattices L_i, in terms of $\mathrm{lodim}\, L_i$. Español [189] generalized the Krull dimension for a distributive lattice as follows: if X is a partially ordered set, then $\dim X \leq n$ if and only if some two elements coincide in each $(n + 1)$-chain. He characterized the Krull dimension of a distributive lattice. Isbell [244] introduced a graded dimension gdim of a locale (see the beginning of §2.5) and investigated its properties. For example, he proved that the dimension gdim is monotone and $\operatorname{gdim} X \geq \operatorname{ind} X$ for every locale X. A regular locale with a σ-local finite basis is metrizable.

An analog of a known theorem of Erdős and Rado on canonical colorings was proved for finite distributive lattices in [366].

A set D of subsets of a finite set is called a lower set if $a \in D$ and $b \subset a$ imply $b \in D$. Berge [107] proved that every lower set is a disjoint union of pairs $\{a, b\}$ such that $a \cap b = \varnothing$ and a singleton $\{\varnothing\}$ when $|D|$ is odd. A corresponding claim for finite distributive lattices together with a few generalizations and analogous results concerning the lattice of subsets of a finite set was proved in [154]. This proof is simpler than Berge's proof. Maximum antichains in products of chains were studied by Griggs [218]. Gansner [199] considered partitions of finite distributive lattices in disjoint chains in terms of parenthesizations, applying the technique of Greene and Kleitman. Speed [VI, 360] gave a formula for the Möbius function on a finite distributive lattice.

2.5. Connections with topology. A generalization of Grothendieck's interpretation of classical Galois theory was given in [256]. Grothendieck's concept of a topos was compared with the classical concept of a topological space. This comparison takes the form of a double generalization of the idea of space. First, the set of open sets of a topological space is replaced by an arbitrary lattice that satisfies the distributive law $a \wedge \bigvee(a_i : i \in I) = \bigvee(a \wedge a_i : i \in I)$. Such a lattice is called a locale. A category of (generalized) spaces and continuous mappings is defined in such a way that it is dual to the category of locales. The paper lays a foundation for studying such spaces. Second, it is shown that the distinction between a general topos and sheaves over the new concept of space consists of the possibility of action of a space groupoid.

Beznea [109] proved that a bounded distributive lattice is complete if and only if its Priestley space satisfies a certain condition. Katz [281] extended Priestley's theory of representation of distributive lattices of generalized deductive systems that satisfy structural rules (for example, those in Scott [405]) common for all ordinary deductive logics.

It is known that the category D_{01} of bounded distributive lattices is dually equivalent to the category of spectral spaces (Stone spaces) and the category of compact totally disconnected with respect to order ordered topological spaces, while the latter two categories are isomorphic. Grätzer [216] found a duality for distributive upper semilattices with zero, which generalizes the duality obtained by means of spectral spaces for D_{01} (see also [III, 8]). An alternative representation of distributive upper semilattices by ordered topological spaces was given in [352]. The paper [416] of Simmons mentioned in §2.8 is also related to Stone duality.

A lattice L is called catalytic in a variety V of lattices if, for every lattice $M \in V$, the set $\mathrm{Hom}(L, M)$ is a lattice with respect to the pointwise order. Priestley [365] showed that the following three conditions are equivalent for a bounded distributive lattice L:

(1) L is catalytic in the variety of bounded distributive lattices;

(2) the dual Priestley space of L is a bialgebraic lattice with the Hausdorff interval topology;

(3) L is a retract of a bounded distributive lattice that is freely generated by a certain ordered set.

As Lawson and Gierz [205] observed earlier, the spaces in (2) are precisely the compact 0-dimensional topological lattices.

Tiller [435] studied connections between special Hausdorff spaces and continuous lattices. A good introduction to the theory of continuous lattices was given in [III, 267] (see III-1, 1.3).

See also [III, 28] in III-1, 1.2.

A complete lattice satisfying the infinite distributive law $x \wedge \bigvee x_i = \bigvee (x \wedge x_i)$ is called a frame, and a lattice homomorphism preserving infinite joins is called a frame mapping. A standard example of frames is provided by lattices of open subsets of topological spaces while frame mappings are the binary relations converse to continuous mappings of topological spaces. This determines a contravariant functor from the category of topological spaces to the category of frames. Thus a transfer of topological notions to the wider category of frames may be interesting.

The paper [168] is dedicated to problems connected with the separation axioms T_1 and T_2. Concepts of T_1- and T_2-frames were introduced and studied. Isbell [243] considered the following problem: does there exist a T_0-topology on a frame L such that operations of binary intersection and infinitary union are continuous? He showed that if L is isomorphic to a topology, the answer is in the affirmative, while in the case of the Boolean lattice of regular open subsets of the real line the answer is in the negative. A survey of the theory of stably continuous frames, taking into account their significance for the theory of special locally quasicompact spaces and their stable quasicompactifications and their role in the context of the spectral and quasispectral theory of continuous lattices and Lawson duality, was given by Hofmann [III, 203]. A concept of a biframe was introduced in [89]. The known dual connection between topological spaces and frames was extended to a connection between bispaces and biframes. In this direction the concepts of regularity, complete regularity, and compactness were investigated.

The category of locales is dual to the category of frames. Thus, if we are interested in objects only, the concepts of a "frame," "locale," and "complete Heyting algebra" coincide. They differ only when investigated together with their morphisms. A standard example of a locale is the lattice of open subsets of a topological space. In papers dedicated to locales, concepts generalizing certain constructions known in topology were introduced and studied. Pultr [373] considered metrizable locales. For example, he proved that all regular Lindelöf locales are normal, that regular locales with a σ-discrete basis are always normal, and that a product of at most a countable number of metrizable locales is metrizable. In addition, generalization of Bing's criterion and a generalization of the Smirnov-Nagata metrization theorem were obtained. In [374] and [375] the same author considered the uniformizability of locales; these papers are a significant contribution to the theory of uniform locales. The author defined a concept of uniformity for locales and established a classical result that uniformizability is equivalent to complete regularity. He proved a few remarkable metrization theorems.

In the survey [253] Johnstone demonstrated advantages of studying topological problems via locales and gave applications of such studies.

A constructive study of points in locales was pursued in [338]. For example, it was proved that if a topos satisfies the De Morgan law, then completely prime filters (that is, points) in each compact regular locale L are precisely the maximal regular filters in L. A constructive version of the Hahn-Banach theorem was given (the classical theorem is not constructive).

See information on papers by Greco [III, 177] and [III, 180] in III-1, 1.3.

2.6. Connections with general algebra and other algebraic systems. An algebra A is called equationally compact if every system of equations has a solution in A whenever every finite subsystem of it has a solution in A. If one considers only the equations with a single variable and constants from A, then A is called equationally compact with respect to a single variable. Earlier Kelly [282] and Beazer [94] showed that infinitely $\wedge$- and $\vee$-distributive complete lattices and only they are equationally compact with respect to a single variable. Fleischer [192] gave a new proof of this result. These and other criteria for equational compactness were also obtained by Nauryzbaev, Omarov, and Mamedov [III, 20], [38], [39], [29].

Certain properties of a tensor product $A \otimes B$ of distributive lattices A and B were studied in [100]. It was proved that if A and B satisfy the descending chain condition, then $A \otimes B$ has the same property and each element of $A \otimes B$ is a finite join of $\vee$-irreducible elements.

Cornish [144] proved that the class L_p of reducts of lattices with respect to the ternary polynomial $p = x \wedge (y \vee z)$ is a variety with a four-element basis. If D is the variety of distributive lattices, then D_p is a variety in whose description by identities one needs four variables.

Romanowska [383] studied the algebra $A = (L^{|P|}; +, \cdot)$, where L is a finite distributive lattice, P a finite partially ordered set, $L^{|P|}$ the set of all mappings of P to L, and A a distributive lattice. On $L^{|P|}$ one defines a binary operation $\circ$ as follows: $f \circ g(p) = \sum (f(p) \cdot g(q) + f(q) \cdot g(p) : q \leq p)$ for $p \in P$. Properties and interconnections between these operations were considered.

Cohn [141] studied rings R that possess the following property: for every nonzero $c \in R$ the set of principal left ideals containing c is a distributive sublattice of the lattice of all left ideals of R. Bandelt and Petrich [92] studied semirings S that possess a congruence κ such that the factor ring S/κ is a lattice (of course, distributive) and κ-classes are subrings of S.

Sets with special systems of subsets that satisfy certain properties (convex structures) were studied in [446] and [447]. An example of such a structure is provided by the system of convex sublattices of a distributive lattice. A number of results was obtained and applied to complete distributive lattices.

See also [69].

2.7. Connections with logic. The survey of Esakia [70] is devoted to algebraic systems appearing in logic. The work reviewed in RZhMat in 1973–1977 was reflected. The author considered in detail cylindric, polyadic, Boolean algebras with additional structure, quantum logics, Post algebras and their generalizations, distributive lattices with additional operations, and problems of functional completeness.

Schmerl [396] proved that every $\aleph_0$-categorical distributive lattice of finite $\wedge$-breadth has a finitely axiomatizable theory. The theory of ordered sets of breadth n can be interpreted in the theory of distributive lattices of $\wedge$-breadth n. The theory of distributive lattices of $\wedge$-breadth 2 is undecidable.

Kumar [305] studied the class of recursively enumerable linear orders. The lack of antisymmetricity implies that the corresponding strict inequality relation may not be recursively enumerable (the equality relation need not be recursively enumerable either). There are other negative results as well.

Normann [347] investigated in detail the lattice of recursively enumerable degrees of continuous functionals. He proved that this lattice is distributive and every finite distributive lattice can be embedded in it. He investigated the problem of possible realization of propositions concerning the upper semilattice by recursively enumerable degrees of continuous functionals. He proved that for such quantifier-free propositions it suffices to consider recursively enumerable degrees of continuous functionals of type at most 3.

It is known that, for every pair of recursively enumerable sets x_1 and x_2, there exist disjoint recursively enumerable sets y_1 and y_2 such that $y_1 \subseteq x_1$, $y_2 \subseteq x_2$ and $y_1 \cup y_2 = x_1 \cup x_2$. Lachlan [307] called distributive lattices separated if they possess this property. He proved that the elementary theory of finite separated distributive lattices is decidable. Gurevich [221] proved that the elementary theory of finite separated distributive lattices is decidable. Gurevich [221] proved that the elementary theory of all separated distributive lattices is undecidable.

Kanovich [19] showed that algorithmic mass (a.m.) problems form a distributive lattice with respect to an arbitrary natural reducibility type. For the a.m. problem of the lowest level (separation problems for recursively enumerable sets) he constructed an infinite antichain in the lattice of T-degrees of a.m. problems. Thus, every finite distributive lattice is embeddable in the lattice of r-degrees (r = m, wm, btt, wbtt, tt, wtt, T).

Baranskii [3] proved that there exist distributive algebraic lattices A such that for every group G there is an algebraic lattice B elementarily equivalent to A and satisfying the condition $\operatorname{Aut}(B) \cong G$.

Dzgoev [11] constructed an example of a nonconstructible abelian group whose Cartesian square is constructible, and an analogous example in the class of distributive lattices.

2.8. Categorial problems. Homology and homotopy. Diers [157] gave conditions for a functor $U: A \to B$ under which each object in a category B is isomorphic to a continuous section of a certain sheaf with values in B and fibers in a category A. These results were applied to representations of distributive lattices as sections of corresponding sheaves. Simmons [416] studied a pair of triples (or monads in Mac Lane's terminology) induced by contravariant functors T and O between the categories LAT of bounded distributive lattices and TOP of topological spaces (in the sense of Stone's duality).

A category C is called universal if the category of all nonoriented graphs and their compatible mappings is isomorphic to a full subcategory of C. In [76] the universality of the following categories was studied: D_n of distributive lattices with n nullary operations, BD_n of bounded distributive lattices with n nullary operations, distinct from 0 and 1, P_n of ordered sets with n fixed points. It was shown that the categories D_n, BD_n, and P_n are universal if and only if $n > 2$, $n > 1$, and $n > 1$, respectively. It follows, for example, that for every infinite cardinal number α there exists a set $\{D_i : i < 2^\alpha\}$ of pairwise nonisomorphic distributive lattices with three nullary operations such that $|D_i| = \alpha$ and D_i do not have nontrivial endomorphisms for all i.

It was proved in [80] that, for any countable bounded distributive lattices B and C, there exist $2^{\aleph_0}$ nonisomorphic countable bounded distributive lattices L such that B is a sublattice and C a homomorphic image of L and $L \cong L + L + L$, but $L \not\cong L + L$ (here + denotes the coproduct in the category of countable bounded distributive lattices with $\{0, 1\}$-homomorphisms). As proved earlier by Trnková and Koubek, there exists a Boolean algebra B such that $B \cong B + B + B$, and $B \not\cong B + B$, but such an algebra is necessarily uncountable.

Todua [62] introduced homology and cohomology groups for distributive lattices. It was proved that if the coefficient groups are dual, then the groups themselves are dual. The concept of connectedness for distributive lattices was introduced and necessary and sufficient conditions for the connectedness of distributive lattices given. He defined in [63] homologies and cohomologies of a pair of distributive lattices over a pair of coefficients and established a duality between homologies and cohomologies defined over conjugate pairs of groups as well as an exact sequence of cohomologies that corresponds to an exact triple of coefficients. In the next paper [64] he introduced and studied homology and cohomology groups for a bounded distributive lattice L, which are based on finite multiplicative coverings, and proved that these groups are isomorphic to the groups introduced in [62]. In the case when L is a finite lattice, he pointed out a connection between these groups and groups introduced by J. Folkmann and G. C. Rota. Using these groups, Todua characterized the dimension of distributive and modular lattices, finite flat distributive lattices, and finite Boolean lattices. He computed these groups for Boolean lattices and free Boolean lattices.

Bunyatov and Baĭramov [6] constructed a K-theory on the category of distributive lattices and their homomorphisms. Bunyatov [5] constructed a theory of homotopy groups of distributive lattices.

2.9. Fuzzy structures. Distributive lattices are often used in research of fuzzy structures. Sometimes complete and even completely distributive lattices are used. For example, this case occurs in the theory of fuzzy topological spaces. Thinking that the condition of complete distributivity was too restrictive for applications, in particular, in probabilities, Höhle [236] developed a theory of $\mathscr{G}$-fuzzy topological spaces, where $\mathscr{G}$ is an arbitrary Boolean algebra (treated as the set of truth values). He developed in detail a convergence theory for $\mathscr{G}$-fuzzy filters in $\mathscr{G}$-fuzzy spaces and, by means of special ultrafilters, defined a property of the fuzzy compactness type, which he called probabilistic compactness. He introduced and studied the notions of probabilistically complete and probabilistically precompact $\mathscr{G}$-fuzzy uniform spaces.

Šešelja and Vojvodić [410] introduced and described fuzzy generalizations of equivalences and studied a connection between fuzzy generalized equivalence relations and naturally defined fuzzy partitions. Here the interval $[0, 1]$ with an additions operation $a \to 1 - a$ was taken for the distributive lattice.

Weak predicate logic WL was applied to fuzzy logic in [462]. The role of the Boolean algebra used in classical logic was played by the interval $[0, 1]$. Logical connectives $\vee$, $\wedge$, and $\supset$ were interpreted as by Heyting, and the negation defined by the equality $\neg x = 1 - x$. It was established that the structure introduced is a model for a fragment of the Zermelo-Fraenkel system obtained by weakening of the Axiom of Substitution.

Liu [313] introduced certain fundamental concepts of fuzzy algebra, for example fuzzy invariant subgroups and fuzzy ideals, and established some of their properties. He [III, 252] studied operations on L-fuzzy ideals.

A number of papers on uniformity and metrization on fuzzy sets illustrated the importance of investigating the intersection operation on union-preserving mappings of completely distributive lattices. Huttlon [241] obtained a formula for such mappings: $(f_1 \wedge f_2)(a) = \bigwedge(f_1(a_1) \vee f_2(a_2) : a_1 \vee a_2 = a)$. Liu [III, 253] and [314] showed that this formula holds if and only if

$f_1(0) = f_2(0)$ and, in the general case,

$$(f_1 \wedge f_2)(a) = f_1(a) \wedge f_2(a) \wedge \bigwedge(f_1(a_1) \vee f_2(a_2) : a_1 \vee a_2 = a)$$

holds.

Wang [454] studied order homomorphisms of fuzzes. A fuzz is defined as a complete distributive lattice with an order-reversing involution $'$. A mapping $f: L_1 \to L_2$ of fuzzes is called an order homomorphism if (1) $f(0) = 0$; (2) $f(\bigvee A_i) = \bigvee f(A_i)$ for all $A_i \in L_1$; and (3) $f^{-1}(B') = (f^{-1}(B))'$ for all $B \in L_2$. Certain elementary properties of such homomorphisms were obtained. Sufficient conditions were found, under which an order homomorphism is a Zadeh type function. Continuous, closed, and open order homomorphisms of fuzzes were studied.

2.10. On matrices over distributive lattices. Now we touch upon matrices over lattices. This subject is connected with many topics. First, the semiring of matrices over a distributive lattice L is isomorphic to a semiring of endomorphisms of a free D-module (a polygon). Second, every matrix over a complete lattice L can be considered as an L-valued selfmap of a set (see below). Third, matrices over a Boolean algebra form a relative or, which is the same, a relation algebra (see Chapter II). Finally, for every relation on a set X there is a Boolean $(X \times X)$-matrix over a two-element chain 2. This last direction is beyond the scope of our survey. We note only the monograph [289] which touches upon various aspects of the theory of Boolean matrices and is accompanied by an extensive bibliography. Of odd results related to semigroups of matrices over a Boolean algebra we note the equivalence of the following properties of a matrix A: (1) A is a right zero divisor; (2) $IA \neq I$, where I is a matrix all of whose entries are 1; (3) $E \leq AA^*$, where A^* is the transpose of A [315]. Skornyakov [60] established the equivalence of the following properties of an $n \times n$ matrix A over a distributive lattice D with 0 and 1: (1) A is right invertible; (2) A is left invertible; (3) the rows of A are orthogonal; (4) the columns of A are orthogonal; (5) $A^k = E$ for some $k \geq 1$; (6) the selfmap φ_A of the set of n-dimensional rows over D defined by the equality $\varphi_A(x_1, \ldots, x_n) = (x_1, \ldots, x_n)A$ is one-to-one; (7) φ_A is an injection; (8) φ_A is a surjection. Some of these equivalences (sometimes only in the case when D is a Boolean algebra) have been known earlier ([124], [131], [209], [387], [390], [456]). The equivalence of (1) and (2) holds even if D is an arbitrary commutative semiring [378]. The equivalence of (3) and (4) for the case when D is a pseudo-Boolean algebra was announced by Saliĭ [56]. The equation $AX = X$ for matrices over a Boolean algebra was investigated by Gilezan [208]. Skornyakov and Egorova [61] suggested a description of all normal subgroups of the group of all invertible matrices of order 3 over a distributive lattice with 0 and 1. Egorova [14] announced the following result: a semiring of $n \times n$ matrices over a distributive lattice with 0 and 1 is determined up to isomorphism by its subsemiring of matrices with permanent equal to 1. The semigroup of matrices over various subsets of the set of real numbers considered as a chain was considered from various sides in [150], [227], [284], [285], [286], [287], [288], [290]. Observe that the case when D is the segment [0, 1] plays an important role in the theory of fuzzy sets. See [160] for the connection between fuzzy sets and Boolean matrices. Skornyakov [59] suggested a criterion for the existence of an eigenvector with a given eigenvalue for matrices over a distributive lattice with 0 and 1. Properties of eigenvectors for matrices over a Boolean algebra had been earlier considered by Rutherford [389]. He [388] proved the Hamilton-Cayley theorem for matrices over an arbitrary commutative semiring (in particular, over a distributive lattice).

2.11. Miscellany. Picu [355] found a few conditions for the distributivity of a lattice. Zharinov's monograph [17] is devoted to applications of the theory of distributive lattices to complex analysis problems. The main goal was to find extensions and stronger forms of the renowned theorem of Bogolyubov on the "edge of the wedge". This theorem was restated as exactness of a certain homological sequence for various classes of holomorphic functions. Fourier hyperfunctions and ultrahyperfunctions are studied as well. The study is based on a homological characterization of distributive lattices of submodules and the duality of the structures of locally convex subspaces.

Embeddings of BCK-algebras in distributive lattices were studied in [145]. Kaiser [260] studied bounded distributive lattices D in which, for every $x, y \in D$ such that $x \wedge y = 0$, there exists an element $u \in D$ with a complement u^* such that $x \leq u$ and $y \leq u^*$ ("K-normal lattices"). Let $B(D)$ be a Boolean sublattice of D that consists of complemented elements. A K-normal lattice is called regular if $x = \bigwedge(u \in B(D) : u \geq x)$ for every element x. The lattice D^{**} of closed subsets of the ultrafilter space of the Boolean lattice $B(D)$ is always regular. It is called the regularization of D. A connection of all these concepts with logic is studied in the case when D is considered as the lattice of sentences of a logic. Distributive correlation lattices (see III-1, 1.5) were considered in [404], especially their fixed points with respect to the operation σ.

Varieties of pseudocomplemented distributive lattices form a chain $B_{-1} \subset B_0 \subset B_1 \subset \cdots \subset B_n \subset \cdots \subset B_\omega$. It was proved in [74] that B_3 contains a proper class of pairwise nonisomorphic algebras with finite endomorphism monoids, while very infinite algebra in B_2 has infinitely many endomorphisms. B_4 contains a proper class of pairwise nonisomorphic algebras such that the endomorphism monoid of each of the algebras from this class consists of the identity and a finite number of right zeros. Each algebra from B_3 with the endomorphism monoid of this type is, however, infinite.

See also [II, 143] in II-3, [II, 88] in II-4, and [II, 76], [II, 79], [II, 80], [II, 218], [II, 279] in II-5.

A number of papers was devoted to lattices of special subsets of algebras (ideals, subalgebras, etc). The monograph [93] contains a structural analysis of κ-complete ideals over an uncountable cardinal number κ. In [425] lattices compactly packed by prime ideals are studied, that is, lattices whose every ideal I, which is contained in the join of a family of prime ideals, is contained in one of the prime ideals of this family. It was proved, for example, that a distributive lattice with 1 is compactly packed by prime ideals if and only if it is Noetherian (that is, each ideal is principal). The same holds for join-distributive semilattices, l-groups, semi-Brouwerian algebras, and Boolean algebras. Göktas [214] considered lattices with operations of multiplication by numbers that satisfy certain natural conditions ("translation" lattices). Various types of ideals in these lattices were introduced and their properties investigated. Let L be a distributive lattice and $P(L)$ the family of all prime filters of L. A set-theoretic characterization of families of the form $P(L)$ (as subsets of 2^L) was suggested in [133].

See also [III, 105], [III, 104], [III, 58], [III, 103], [III, 237], [III, 238], [III, 106], [III, 368] in III-3, 3.1.

Let D be a finite distributive lattice. A permutation in D (called a D-complementation) was constructed in [354], which for Boolean algebras produces the complementation operation. It was proved that the D-complementation can be used in the theory of matroids (axiomatization of the notion of matroid, construction of the Tutte polynomial, etc.).

Let E be a finite set and D a lattice of its subsets containing $\varnothing$ and E. In [197] submodular functions on D were studied, that is, real-valued functions f that satisfy the condition $f(X)+f(Y) \geq f(X \cup Y)+f(X \cap Y)$ for any $X, Y \in D$. Extremal points of the set $B(f) = \{x \in R^E : x(X) \leq f(X)$ $(\forall X \in D)$, $x(E) = f(E)\}$ were found and upper and lower estimates for f computable in polynomial (in $|E|$) time were given.

Let L be a distributive lattice. Consider the following multiplication operation on $L \times L$: $(s, t) \cdot (x, y) = (P(s, t, x, y), Q(s, t, x, y))$, where P and Q are lattice polynomials. The following problem was investigated in [140]: for which P and Q $(L \times L; \cdot)$ is a semigroup from a certain class? It was proved that if L has no least (respectively, greatest) element, then a semigroup with identity is obtained if and only if L contains the greatest (least) element. In this case $P = s \wedge x$ and $Q = t \wedge y$ ($P = s \vee x$ and $Q = t \vee y$). If $|L| = 2$, all multiplications of this type are found for which $(1, 0)$ is an identity element. All such multiplications are found also in the case when the semigroup is idempotent with identity $(1, 0)$.

Sengupta [409] studied homomorphisms of a complete distributive lattice into the nonnegative real axis with the natural order. The set H of all such homomorphisms can be endowed with a structure of a distributive lattice. If a metric and a suitable measure are introduced, H becomes a locally convex vector space. An analog of a known theorem of Choquet was proved for it. This mathematical scheme was interpreted as a basis for a construction of a qualitative description of the choice process. The exposition was illustrated by examples of a construction and analysis of group choice, taking into consideration forming of coalitions and a possibility of negotiations.

See also [III, 303] in III-1, 1.6.

The following papers were devoted to tolerances in distributive lattices: [III, 98] (see III-1, 1.2) and [III, 274], [III, 275], and [III, 276] (see III-1, 3.2).

§I.3. Pseudocomplemented semilattices, p-algebras, and double p-algebras

For Heyting and De Morgan algebras see Chapter II.

First we recall certain notations. By S we denote a pseudocomplemented semilattice (ps), that is, $S = (S; \wedge, {}^*, 0, 1)$, where $a \wedge x = 0$ if and only if $x \leq a^*$. If S is a lattice, it is called a pseudocomplemented lattice (pl) or a p-algebra. If S is a ps, then $B(S) = \{x \in S : x \in x^{**}\}$ forms a Boolean algebra of closed elements and $D(S) = \{x \in S : x^* = 0\}$ denotes the filter of dense elements. If S is a p-algebra and the dual lattice $\overline{S}$ is a pl with the operation $+$ of dual pseudocomplementation, then S is called a double p-algebra. In this case $\overline{D}(S) = \{x \in S : x^+ = 1\}$ denotes the ideal of dually dense elements.

3.1. Representation by triples. In [OSL, I] we have already discussed a description of the structure of ps and pl by triples. In particular, we mentioned a paper [277] by Katriňák and Mederly, where various ps are defined in the following way: each $x \in S$ can be represented in the form $x = x^{**} \wedge d$ for some $d \in D(S)$. For every $a \in B(S)$ there exists a congruence $a\overline{\varphi}(S) \in \text{Con}(D(S))$ defined as follows: $d \equiv e(a\overline{\varphi}(S)) \Leftrightarrow a^* \wedge d = a^* \wedge e$. A triple $(B(S), D(S), \overline{\varphi}(S))$ is called associated with S. The isomorphism of triples is defined in a natural way. It was proved in [277] that decomposable ps S and S_1 are isomorphic if and only if their associated triples are isomorphic. An abstract triple $(B, D, \overline{\varphi})$ is defined, where B is a Boolean algebra, D a semilattice with 1, and $\overline{\varphi}: B \to \text{Con}(D) - (0, 1)$ an isotone mapping. The principal theorem from [277] states that an arbitrary abstract triple is isomorphic to a triple associated with a certain decomposable ps (see also [OSL, I]).

It is known that the mapping $x \to x^{**}$ in ps is a modal operator, or a multiplicative closure operator m, that is, $m(x \wedge y) = m(x) \wedge m(y)$, $m(m(x)) = m(x)$, $m(x) \geq x$ and $x \leq y$ implies $m(x) \leq m(y)$. Murty and Raman [341] studied semilattices with 1 and a modal operator. For such semilattices S they introduced an associated triple $(B_m(S), D_m(S), \overline{\varphi}_m(S))$ with respect to m and a concept of a semilattice decomposable with respect to m. Furthermore, if $D_m(S)$ is a neutral element of the lattice of filters $F(S)$, then the authors showed that such semilattices S and S_1 are isomorphic if and only if their associated triples with respect to the operators are isomorphic as well. They described homomorphisms, subalgebras, and congruences in the language of the triples (see also Cornish [OSL, V, 121] and Mederly [OSL, I, 250]).

We observed in [OSL, I] that there exist three methods for construction of ps and p-algebras from abstract triples. Murty [339] showed that a distributive p-algebra can be constructed by means of generalized admissible mappings. On other constructions see Katriňák [273] and Katriňák and Mederly [277]. Hoo [237] generalized a construction of Nemitz [OSL, I, 397] by means of a-admissible mappings to a-implicative semilattices. In the subsequent paper [239] Hoo and Ramana Murty found an error in [237]. A correction is possible if "a-admissibility" is replaced by a "strong a-admissibility" and the modularity of the semilattice is required.

Katriňák and Mederly [277] made an attempt to characterize all admissible ps and p-algebras. In the latest paper [340] Murty and Engelbert observed that this description was erroneous. Katriňák [464] corrected that error. The paper [340] contains also a solution to two problems from [277]: (1) a construction of quasi modular p-algebras from [277] was simplified: (2) for decomposable p-algebras S, which satisfy the condition $a\overline{\varphi}(S) = \theta(a\varphi(S))$, where $a\overline{\varphi}(S) \in F(D(S))$ for every $a \in B(S)$, we have $(a \vee b)\varphi(S) = a\varphi(S) \vee b\varphi(S)$ if and only if $D(S)$ is a neutral element of the lattice $F(S)$.

Recently Blyth and Varlet (see Chapter II) defined the MS-algebras $(L; \vee, \wedge, °, 0, 1)$, which generalize De Morgan and Stone algebras: L is a bounded distributive lattice that satisfies the conditions $x \leq x^{\circ\circ}$, $(x \wedge y)^\circ = x^\circ \vee y^\circ$, $1^\circ = 0$. (L is a De Morgan algebra if and only if $x = x^{\circ\circ}$ for all $x \in L$, and L is a Stone algebra if and only if $x \wedge x^\circ = 0$ for all $x \in L$.) In [112] they claimed that the method of triples could be generalized to MS-algebras from a suitable subvariety K_2. However, their proof contained an error, which they corrected in [113]. To construct an algebra from K_2 one needs to introduce a quadruple (K, D, φ, γ), where K is a Kleene algebra, D a distributive lattice with 1, $\varphi \colon K \to F(D)$ a suitable homomorphism, and γ a suitable modal operation on D. Katriňák and Mikula [278] characterized those algebras from K_2 that can be constructed from a triple (K, D, φ) alone (see II.6 on MS-algebras).

3.2. Congruences and homomorphisms.
Some authors continued studying congruences on ps, p-algebras, and double p-algebras. Algebras whose congruence lattice is Boolean, Stone, or relatively Stone, drew attention too.

We have already discussed in [OSL, I] Beazer's paper [96]. Sankappanavar [392] characterized classes of ps with semimodular or distributive congruence lattice. Bosbach [116] proved that a $\vee$-distributive algebraic lattice is the congruence lattice of an implicative semilattice.

Varlet [444] investigated regularity properties for ps, p-algebras, and double p-algebras. The regularity property is defined for subsets, congruences, and algebras. Namely, a subset C of an algebra A is called regular if C is a class of exactly one congruence on A. A congruence on A is called regular if all of its classes are regular. An algebra A is called regular if all of its congruences are regular. The author studied problems of recognition of regular congruences, the role of such congruences in the congruence lattice of an algebra, and a description of completely irregular algebras, that is, algebras whose only regular congruence is the universal congruence.

In the latest papers [275] and [276] Katriňák and El-Assar studied congruence lattices of algebras with a distributive congruence lattice. Using subdirect decompositions (see also Tanaka [431] and Hashimoto [228]) they characterized algebras whose congruence lattice is (1) atomic; (2) Boolean; (3) Stone; and (4) completely Stone (that is, satisfies the infinite identities).

As a corollary, they obtained certain results of Beazer [95] concerning ps, p-algebras, and double p-algebras, of Sankappanavar ([391], [393]) for ps and almost double Heyting algebras, and of Janowitz [248] for orthomodular lattices. It is known (see [OSL, I, 189, 195]) that every congruence θ on a quasi modular p-algebra S can be represented as a pair (θ_1, θ_2) of congruences from $\text{Con}(B(S)) \times \text{Con}(D(S))$. This makes it possible to give a new description of p-algebras that satisfy one of the properties (1)–(4). Also, using weak projectivity, Haviar and Katriňák [229] described lattices with the completely Stone congruence lattice representable in the form 2^P for some partially ordered set P. (Recall that a description of lattices with properties (1)–(4) had been known: Tanaka [431], Grätzer and Schmidt [217], Crawley [148], Katriňák [270], and Iqbalunnisa [242]). Also, the authors proved that, for distributive lattices, if the congruence lattice is relatively Stone, then it is Boolean. This problem for Boolean lattices was considered also by Beazer [99]. In addition, he gave another proof of Katriňák's [270] result: the lattice $I(L)$ of the ideals of a lattice L with 1 is Stone if and only if L is Stone and $B(L)$ is a complete Boolean algebra.

We have noted above Sankappanavar's paper [393]. He studied Heyting algebras whose dual lattice is a p-algebra. He proved that the class of these algebras satisfies CEP (congruence extension property) and described congruence-free and subdirectly indecomposable algebras from this class.

Goldberg [215] found necessary conditions for the congruence lattice of a distributive double p-algebra to be a chain (see also Beazer [OSL, I, 68]).

Kiss, Márki, Pröhle, and Tholen published a survey [292] devoted to such questions as CEP, amalgamation, epimorphisms and injectivity in various classes of algebras, including p-algebras. The paper contains rich information and a vast bibliography.

Prata dos Santos ([362] and [363]) generalized results of Varlet [OSL, I, 403] for ps and p-algebras.

Let (A, F) and (B, G) be universal algebras, with F and G generating classes P and Q of operations on A and B. A mapping $\varphi: A \to B$ is called a weak homomorphism if, for every $n > 0$ and every $f \in P_n$, there exists $g \in Q_n$ such that (6) $g(\varphi(x_1), \ldots, \varphi(x_n)) = \varphi(f(x_1, \ldots, x_n))$ for all $x_1, \ldots, x_n \in A$ and, conversely, for any n, $g \in Q_n$, there exists $f \in P_n$ such that (6) holds. Głazek and Katriňák [210] showed that a mapping φ is a weak homomorphism in a variety of algebras if (roughly speaking) φ has this property for all subdirectly irreducible algebras A and B. Furthermore, they established that if $\varphi: L \to L'$ is a weak isomorphism of distributive p-algebras and L' is not a Boolean algebra, then φ is an isomorphism. Also, they studied weak homomorphisms of double Stone algebras. They proved that, besides homomorphisms and dual homomorphisms, there exist two new weak homomorphisms of these algebras. The latest results can be found in Kolibiar's paper [295].

An algebra A is called affine complete if every operation on A that preserves the congruences is algebraic. Beazer ([97] and [98]) investigated affine complete (double) Stone algebras. In particular, he proved that if L is a Stone algebra and $D(L)$ possesses the least element, then the following statements are equivalent: (a) L is affine complete; (b) $D(L)$ is affine complete; and (c) $D(L)$ does not contain nontrivial Boolean intervals. If $D(L)$ is replaced by $K(L)$ for a double Stone algebra L that satisfies the condition $\varnothing \neq K(L) = [a, b]$, we arrive at a description of affine complete double Stone algebras.

3.3. Endomorphisms. We consider a series of papers devoted to the study of the endomorphism monoids of algebras in a category. It is known (Schein [395], Magill [317], and Maxson [322]) that, for Boolean algebras A and B,

$$\text{End}(A) \cong \text{End}(B) \Rightarrow A \cong B.$$

Adams, Koubek, Priestley, and Sichler considered those varieties of algebras that satisfy this condition. It was proved in [75] that the category of Stone algebras has this property. In contradistinction to this result, the majority of varieties behave badly, that is, the monoid $\text{End}(A)$ does not contain any information on the structure of A. More precisely, a category C of algebras is called universal if every category of algebras and their homomorphisms is embeddable in C as a full subcategory. There exist weaker forms and modifications of this concept, such as almost universality. In particular, for an (almost) universal category C the following holds: for every infinite cardinal number k there exist 2^k pairwise nonisomorphic algebras of cardinality k from C with isomorphic endomorphism monoids. It was proved in [75] that all varieties $B_3, \ldots, B_\omega$ of distributive p-algebras are almost universal. Recall that varieties of distributive p-algebras form an infinite chain $B_{-1} \subseteq B_0 \subseteq B_1 \subseteq \cdots \subseteq B_\omega$, where B_{-1} is the trivial variety, and B_0 and B_1 are the varieties of Boolean and Stone algebras, respectively. If $\text{End}(L) \cong \text{End}(K)$ and $L \not\cong K$ hold in B_2, then $\text{End}(L) \cong \text{End}(M)$ implies either $M \cong L$ or $M \cong K$ for every algebra $M \in B_2$. In [74] and [77] subcategories of B_3 were studied. It was proved in [74] that there exists a class (which is not a set) of pairwise nonisomorphic algebras from B_3 with a finite endomorphism monoid.

It was shown in [II, 55] that the Heyting variety is 0-universal. It follows that Heyting algebras are not determined by their endomorphism monoids. There exists a finitely generated universal variety of double distributive p-algebras (see [302]). In the latest papers [78] and [79] it was proved that the variety of De Morgan algebras is universal and its subvariety of Kleene algebras is almost universal.

3.4. Varieties, subdirectly indecomposable algebras, projective and free objects. Balbes [84] and Jones [254] characterized a free ps. Schmid [399] obtained a new description of these algebras. Let P be the variety of all p-algebras and P_k the variety of all p-algebras satisfying the identity

$$(L_k)(x_1 \wedge \cdots \wedge x_k)^* \wedge (x_1^* \wedge \cdots \wedge x_k)^* \vee \cdots \vee (x_1 \wedge \cdots \wedge x_k^*)^* = 1.$$

Katriňák [271] constructed P- and P_k-free algebras with base X for every set X. He proved that, for $|X| \geq 2$, these algebras are infinite and the word problem is solvable in them. Tamura [429] found a decision procedure for the word problem in the variety P. He worked within the framework of the Gentzen formal system.

Düntsch ([171], [172], and [174]) investigated projective Stone algebras. He showed that such algebras are double Stone algebras (that is, $x^* \vee x^{**} = 1$ and $x^+ \wedge x^{++} = 0$) and double Heyting algebras. Conversely, a double Stone algebra is projective in the variety of Stone algebras if L is regular (that is, $x^* = y^*$ and $x^+ = y^+$ imply $x = y$) or $B(L)$ and $D(L)$ are a projective Boolean algebra and a projective distributive lattice, respectively.

Düntsch also studied projective double Stone algebras. He proved in [177] that a finite double Stone algebra L is projective if and only if L is a projective Stone algebra and $D(L) \cap \overline{D}(L) = \varnothing$. In [176] he found a sufficient condition for a regular double Stone algebra to be projective.

A subdirectly irreducible algebra is called splitting in a variety if there exists the greatest subvariety of this variety that does not contain this algebra. This concept was introduced by McKenzie in [OSL, III, 147]. Katriňák [274] characterized splitting p-algebras in the varieties P and P_0 as bounded homomorphic images of finitely generated free p-algebras. He gave examples of such p-algebras.

Adams and Katriňák [73] solved a problem concerning the characterization of subdirectly irreducible double distributive p-algebras. In particular, they produced an example of a finitely subdirectly irreducible double distributive p-algebra that is not subdirectly irreducible.

Urquhart [442] proved that the lattice of varieties of double distributive p-algebras is continual and is not generated by its finite elements. Dziobiak [186] strengthened this result proving that every proper filter of this lattice is continual. He also proved that in this lattice there exists an infinite number of coverings for every variety generated by a finite algebra A, where A is not a finite Boolean algebra.

Tropin [65] proved that a free lattice is embeddable in the lattice of quasi-varieties of distributive p-algebras. In the latest paper [185] Dziobiak showed that lattices of quasivarieties of distributive (double) p-algebras and (double) Heyting algebras do not satisfy nontrivial lattice identities.

It is known that if B is a Boolean algebra, the distributive p-algebra $\overline{B}$, obtained from B by adjoining a new identity element, is subdirectly irreducible. Algebras $\overline{B}$ are also subdirectly irreducible ps. Recently Schmid [401] has shown that the quasivariety of ps generated by the algebras $\overline{B}$, where B is any Boolean algebra, can be characterized by a single quasi-identity. Further, Schmid [402] found an algorithm for a subdirect decomposition of finite ps. The estimate of this algorithm is $O(n^2)$, where n is the number of elements of the ps. In the general case the estimate is $O(n^4)$; see Demel [155].

A bounded (distributive) lattice is called a (distributive) Ockham algebra if it possesses a unary operation $^\circ$ that is a dual endomorphism of the lattice (see also II, 5). Sankappanavar [394] investigated the Ockham algebras in which $^\circ$ is the pseudocomplementation operation. He proved that these algebras form an arithmetic variety. In particular, for distributive p-algebras that are De Morgan algebras he generalized a result of Romanowska [II, 245] and described nonregular subdirectly irreducible algebras from this class. Urquhart [441] described the lattice of varieties of Ockham algebras, which turned out to be uncountable. Each element of this lattice is generated by its finite elements.

Let $B_{-1}, \ldots, B_n, \ldots, B_\omega$ be varieties of distributive p-algebras. Let $L \in B_n$. An expression of the form

$$p(a_1, \ldots, a_m, x_1, \ldots, x_n) = q(a_1, \ldots, a_m, x_1, \ldots, x_n),$$

where p and q are p-algebraic polynomials and $a_1, \ldots, a_m \in L$, is called an equation with parameters from L. Furthermore, L is called algebraically (existentially) closed in B_n, if every finite system of equations with parameters (open formula) in L, which has a solution in some extension $L' \supseteq L$, $L' \in B_n$ of the p-algebra L, has a solution in L. Schmid ([398], [400]) gives a description of algebraically closed ps and p-algebras in each of the classes $B_0, \ldots, B_\omega$. Clark [465] characterized algebraically existentially closed Stone and double Stone algebras. Schmid [397] studied properties of the model companion T_n^* of the elementary theory T_n of the class B_n.

Zimmermann and Köhler [463] studied the Mal'tsev product of two finitely based varieties of implicative semilattices.

3.5. Categorial duality, topological questions. The known Stone representation of any Boolean algebra as the set of open-closed subsets from a suitable compact and totally disconnected topological space X is the basis for the Stone theory, presented in the book by Johnstone [252] (see Isbell [245]). The book contains the fundamentals of the theory of lattices and Heyting algebras, theory of locales, Stone-Čech compactification, and is dedicated especially to the theory of topological representations and duality for distributive lattices, rings, ordered rings, and continuous lattices.

Besides the Stone representation, for distributive lattices with 0 and 1 there exists a representation suggested by Priestley [OSL, I, 398]. This representation constructs for each distributive lattice A with 0 and 1 a partially ordered set P of all prime ideals of A endowed with a compact topology and totally disconnected with respect to order. (For Boolean algebras the Priestley representation coincides with the usual Stone representation.) Thus one can obtain a dual category (of ordered topological spaces with isotone continuous mappings as morphisms) for the category of distributive lattices with 0 and 1 and homomorphisms preserving 0 and 1. Besides Priestley, other authors have transferred this theory to distributive lattices with 0 and 1 and an additional structure. For example, they have considered representations for (double) p-algebras, (double) Heyting algebras, De Morgan algebras, Post algebras, etc. All these topics were discussed in the survey [364] by Priestley.

Davey [152] investigated dualities for (double) Stone algebras and relative Stone algebras.

Topological representations of lattices are also used as a method for solving various algebraic, categorial and logical questions. We illustrate this on a few examples. Adams, Koubek, Priestley, and Sichler ([74]–[78] and [79]) solved problems related to the universality of the categories of distributive (double) p-algebras, Heyting algebras, and De Morgan algebras, applying the Priestley duality. Using a concept dual to that of a congruence, Adams and Katriňák [73] obtained an example of a finitely subdirectly indecomposable distributive double p-algebra that is not subdirectly indecomposable. Adams [466] proved that the intersection of two principal congruences of a Kleene algebra is principal too (see [II, 5]). Adams and Beazer [467] described a class of De Morgan algebras possessing this property (see [II, 5] and the paper [II, 249]). Hansoul and Vrancken-Mawet [226] (see also [II, 281]) characterized the subalgebra lattice of Heyting algebras. Also, they obtained a description of Heyting algebras with a (modular) distributive subalgebra lattice. Adams and Clark [72] considered universal terms of distributive p-algebras and Heyting algebras. (Recall that $f(x_1, \ldots, x_n)$ is called A-universal, where A is an algebra, if, for every $b \in A$, there exists $a_1, \ldots, a_n \in A$ such that $b = f(a_1, \ldots, a_n)$.)

3.6. Miscellany. Tsirulis (Cirulis) [68] considered the mapping $x \mapsto (x \neg a) \to a$ in an implicative semilattice. The concept of ps was generalized by Jayaram [250], who defined quasicomplemented semilattices. He proved that these semilattices are 0-distributive (see Varlet [443]). Thakare and Pawar [434] generalized results on prime ideals of distributive semilattices to 0-distributive semilattices. Jayaram [249] characterized semi-atoms in 0-distributive semilattices.

A subalgebra T of an implicative semilattice S is called total if, for every $t \in T$, it contains $s \to t$ for all $s \in S$. Köhler [294] proved that the lattice of all total subalgebras is a maximal distributive sublattice of the subalgebra lattice of S. Let $l(A)$ be the greatest length of a chain in a finite lattice A. Chen, Koh, and Teo [135] proved that $l(S(L)) = l(B(L)) + l(D(L)) + 1$, where L is a finite Stone algebra and $S(L)$ its subalgebra lattice.

A Stone algebra is called hereditarily atomic (superatomic) if each of its subalgebras (homomorphic images) is atomic. (It is known that these concepts coincide for Boolean algebras.) Budde [122] showed that a hereditarily atomic Stone algebra is superatomic. Conversely, Düntsch [175] constructed an example of a regular double Stone algebra that is superatomic, but not hereditarily atomic.

A complete lattice L is called a torsion lattice if for every $x \neq 1$ the interval $[x, 1]$ contains atoms. Călugăreanu [127] showed that a complete modular pseudocomplemented torsion lattice is atomic. Ramalho [376] showed that $D_n(L) \subseteq G_n(L) \subseteq \mathrm{Rad}_n(L)$ in a p-algebra L, where $D_n(L)$ is the filter generated by all n-dense elements. Here $x \in L$ is called n-dense if $x = (x_1 \wedge \cdots \wedge x_n)^* \vee (x_1^* \wedge \cdots \wedge x_n)^* \vee \cdots \vee (x_1 \wedge \cdots \wedge x_n^*)^*$ for certain $x_1, \ldots, x_n \in L$. Also, $G_n(L)$ and $\mathrm{Rad}_n(L)$ denote the intersection of all n-normal and all prime filters (see [OSL, I, 96]).

Katriňák [272] simplified the proof of Lambrou's theorem [OSL, I, 227] on complemented p-algebras.

Düntsch [173] constructed an example of a distributive lattice that produced a negative answer to a question of Chen and Grätzer [134] on the partially ordered set of prime ideals of a distributive lattice (Stone algebra).

Kaiser [259] studied a correspondence between lattices and universal classes. For a universal class B of systems with relations and for the first-order language $F(A)$ that corresponds to the type of B with an additional set A of constants, he constructed a lattice L called a logic. The main result says that the pseudocomplemented extension L^* of L is not compact for a "closed" A.

§I.4. Orthocomplemented lattices and partially ordered sets

See II.6 for quantum logics.

In the monograph [105] Beran gave an algebraic theory of orthomodular lattices (ol, for short) and pointed out certain applications. We have already discussed the monograph by Kalmbach [266] in [OSL, II]. This book, besides an algebraic theory, also contains chapters devoted to geometric aspects of the theory of ol (dimension theory, continuous geometries) and representations of ol. The survey [265] by Kalmbach is devoted to the main results related to ol and their applications in algebra, geometry, measure theory, and logic.

4.1. Orthocomplemented partially ordered sets. Recall that an orthocomplemented poset is a poset $(P; \leq, ', 0, 1)$ with 0 and 1 and having a unary operation $'$ (orthocomplementation) with the properties: (1) $a \leq b \Rightarrow b' \leq a'$; (2) $(a')' = a$; (3) $a \wedge a' = 0$ and $a \vee a' = 1$ for all $a \in P$.

Elements $p, q \in P$ are called nonorthogonal whenever $p \not\leq q'$. An M-base is a maximal subset consisting of pairwise nonorthogonal elements. Let $M(P)$ denote the set of all M-bases of a poset P. Let $Z_p = \{B \in M(P) : p \in B\}$, $Z(M(P)) = \{Z_p : p \in P\}$. Katrnoška ([279], [280]) showed that every orthocomplemented poset is ortho-isomorphic to an orthocomplemented poset $(Z(M(P)); \subseteq, ', \varnothing, M(P))$ of open-closed subsets of a 0-dimensional completely regular T_1-space $(M(P), J)$, whose topology J is generated by the family $Z(M(P))$ that is its subbase. Flachsmayer ([190], [191]) established the following theorem: for an arbitrary ring R with identity 1 the set of all its idempotents is an orthocomplemented poset, in which $x \leq y$ if and only if $xy = yx = x$ and the orthocomplement is defined as follows: $x' = 1 - x$. He also proved that for every orthocomplemented poset one can construct an orthodouble, a new orthocomplemented poset.

Mayet [324] suggested a duality theory for orthocomplemented posets, which, in the Boolean case, does not coincide with the Stone duality.

Godowski [211] characterized the commutativity relation in orthocomplemented posets. Pták [367] defined three categories of orthocomplemented posets, in which the category of Boolean algebras is a reflexive and coreflexive subcategory. It is known (see Kalmbach [264]) that an ol is completely determined by its Boolean sublattices. Krausser [303] extended this result to an arbitrary orthocomplemented poset.

Rogalewicz [382] found conditions characterizing those elements of orthocomplemented posets that are incompatible with the states.

4.2. Algebraic operations and varieties. For an ol L Beran [103] studied operations of skew Boolean structure $(L; +, \cdot, ', 0, 1)$, if $x+y = (x \wedge y') \vee y$ and $xy = (x \vee y') \wedge y$ (see also Beran [OSL, II, 104]). Šimon ([417] and [418]) investigated properties of the operations $x/y = (x \vee y) \wedge (x' \vee y) \wedge (y' \vee (x \wedge y) \vee (x' \wedge y))$, $x \backslash y = y/x$, $x+y = (x \vee y) \wedge (x' \vee y')$, and $xy = (x \wedge y) \vee (x' \wedge y')$ on an ol. Beran [106] found new triples of identities (in the signature $(\vee, \wedge, ')$) that define ol and Boolean algebras.

Beran [104] introduced the concept of a $\wedge$-Frattini element (polynomial) p in a free ol $F = F[x_1, \ldots, x_n]$ as follows: for any

$$q, r \in F, \quad qC(r \wedge p) \Leftrightarrow qCr.$$

Dually, a concept of $\vee$-Frattini polynomials was introduced. He established certain properties of these polynomials.

Using states on ol, Mayet [324] studied certain varieties of ol. In [325] he showed that there exists a continuum of varieties of ol.

Dietz [159] showed that a lattice L with orthocomplementations is an ol if and only if every orthogonal subset $M \subseteq L$ (that is, if $a, b \in M$ and $0 \neq a \neq b \neq 0$, then a and b are orthogonal) is completely contained in a certain block of L. Roddy [380] studied the lattice of subvarieties of modular lattices with orthocomplementation.

A mapping $m: L \to [0, 1] \subseteq R$ from an ol L is called a state on L if $m(1) = 1$, $m(a \vee b) = m(a) + m(b)$ for orthogonal elements a and b. A set $\{m_t : t \in T\}$ of states is called complete (strongly complete) if $m_t(a) \le m_t(b)$ for all $t \in T$ implies $a \le b$ (respectively, $m_t(a) = 1 \Rightarrow m_t(b) = 1$ for all $t \in T$ implies $a \le b$). A state is called two-valued if it takes values 0 and 1 only. Godowski [212] proved that the class K of ol with a complete set of two-valued states is a variety. In a variety generated by a class of ol with a strongly complete set of states no subvariety that contains K has a finite basis for identities. Godowski [213] considered states on homomorphic images and direct products of ol.

A polarity lattice is a lattice with an involutory unary operation $x \to x'$ that satisfies De Morgan laws. (These lattices generalized lattices with orthocomplementation.) Kamara [269] described upper neighbors (in the lattice of varieties of polarity lattices) of the variety of distributive polarity lattices. Hermann [235] characterized distributive polarity lattices in the class of modular lattices. He proved that a modular polarity lattice is distributive if and only if it does not contain a certain finite "forbidden" modular polarity lattice as a subalgebra.

Nishimura [345] (see also Cutland and Gibbins [II, 108]) considered Gentzen formal systems that described lattices with orthocomplementations. Tamura [429] showed that every sequence of this system, which has a deduction with the cut rule, also has a deduction without this rule. On Goldblatt's work see [II. 6].

4.3. Congruences, blocks. Chevalier [136] introduced in ol four binary relations defined, respectively, by the conditions $x \wedge y = 0$, $x \wedge y' = 0$, $x' \wedge y = 0$, and $x' \wedge y' = 0$. He studied properties of these relations and their connection with perspectivity and congruences on ol.

It is known that every ol is representable in the form $\operatorname{Proj} S$ for a suitable Baer semigroup S, where $\operatorname{Proj} S$ is the set of projections from S. For every congruence θ of a Baer semigroup S, Chevalier [137] considered its restriction θ_p on the subset $\operatorname{Proj} S$. Then θ_p is a congruence on the ol $\operatorname{Proj} S$. Not every congruence of $\operatorname{Proj} S$ is of this form. Conditions are found, under which every $\overline{\theta} \in \operatorname{Con}(\operatorname{Proj} S)$ can be obtained as an extension of a suitable congruence $\theta \in \operatorname{Con}(S)$. A Baer semigroup with a one-element focal ideal is called a Foulis semigroup. Kuhn [304] showed that a mapping from the category of ol to the category of Foulis semigroups, under which ol is transformed into a subsemigroup of a Foulis semigroup constructed for this lattice and coordinatized by it, cannot be a functor. Länger [309] showed that Baer semigroups form a variety.

A projection of an element x onto an element y in an ol L is the element $x \wedge (x' \wedge y)$. Among the ideals of L orthomodular ideals are characterized by the property that, for each of their elements, they contain its projection onto all elements of the lattice. Carrega, Chevalier, and Mayet [129] proved that an ol L is representable in the form $B \times X_1 \times \cdots \times L_n$, where B is a Boolean algebra and $L_1, \ldots, L_n$ congruence-free ol, if and only if every orthomodular ideal of L generated by commutators is principal. Roddy [379] showed that a block-finite ol is isomorphic to a finite direct product of a Boolean algebra and simple ol's.

Let $C(x)$ be the set of all elements of an ol L that commute with x. Bruns and Greechie ([120] and [121]) considered the following conditions: (A_n) L has at most n blocks (that is, maximal Boolean subalgebras); (B_n) there exists a covering of L by at most n blocks; (C_n) the set $\{C(x): x \in L\}$ has n elements at most; (D_n) at least two of any $n+1$ elements of the ol L commute. Put $A = \exists n(A_n)$, $B = \exists n(B_n)$, $C = \exists n(C_n)$, $D = \exists n(D_n)$. If $X, Y \in \{A, B, C, D\}$, we say that X "uniformly" implies Y if for every natural number n there exists a natural number m such that every ol that satisfies X_n satisfies Y_m as well. The uniform equivalence of conditions is defined in an obvious way. The authors proved that conditions A, B, C, and D are uniformly equivalent.

Dichtl [156] considered problems of amalgamation of Boolean algebras. He found conditions when an amalgam of a finite set of Boolean algebras is an orthomodular poset or an ol.

4.4. Additions to quantum logics. (See [II. 6].) Mushits, Ovchinnikov, and Terekhin [37] studied operations of product and coproduct in quantum logics. They showed that a logic decomposes into a product if and only if its center is nontrivial. Then they proved that a logic can be represented as a logic of sets (that is, as a system of sets ordered by inclusion with the operation of set-theoretic complementation). They found a criterion for a given logic to be isomorphic to a logic of sets.

Pulmannová ([369] and [370]) showed that a logic L can be embedded in the lattice $L_f(V)$ of all f-closed subspaces of a linear space V with a Hermitian form. In the next paper [371] she proved that $L_f(V)$ has the Hilbert property $M + M^\perp = L$ for all $M \in L_f(V)$ if and only if $a \vee e$ exists in L for every $a \in L$ and every atom $e \in L$.

Mittelstaedt [331] suggested a modification of quantum logic for description of dynamics of quantum systems.

Let S_1 and S_2 be (physical) systems. Spaces of events of S_1 and S_2 are, respectively, the lattices $L(H_1)$ and $L(H_2)$ of closed linear subspaces of Hilbert spaces H_1 and H_2. Stairs [422] introduced a concept of logical product of ol $L(H_1)$ and $L(H_2)$ to describe the composite system $S_1 + S_2$.

Takeuti [426] suggested an operation $\coprod: 2^L \to L$ on ol L. Properties of this operation indicate that the value of $\coprod(A)$ shows to which extent elements of A behave "classically". Cooke and van Lambalgen [143] made these ideas more precise. They investigated connections between structural properties of an ol L and the existence of two-valued homomorphisms on L.

Mukherjee [337] considered a poset with orthocomplementation in which every pair of elements a and b such that $a \leq b'$ possesses a supremum. He gave three conditions that characterize quantum logic.

Sarymsakov and Chilin [58] defined and studied T_2-uniformity on logics. It is known that different uniformities may exist on an arbitrary logic L, but, if L is a uniform logic defined in that paper, it possesses only one uniformity compatible with its order.

4.5. Miscellany. A closure on a set X is an idempotent, isotone, and extensive mapping $A \to \overline{A}$ on 2^X, which preserves the empty set. A subset A is defined to be regularly closed if $A = \overline{(\overline{A'})'}$. Sekanina [407] showed that the lattice of all regular closed subsets is a complete ol and every complete ol has such a representation. Iturrioz [246] simplified the proof of the latter statement.

Let $L^{(k)}$ denote the class of all ol in which every block is of the form 2^k, and $L_n^{(k)}$ the class of all ol from $L^{(k)}$ that have precisely n blocks. The rank of an atom a of a lattice from $L^{(k)}$ is the cardinality of the set of the blocks in which a is an atom. Köhler [293] showed that, for any natural t and $k \geq 3$ there exists n such that in $L_n^{(k)}$ a lattice L exists, every atom of which has rank t. A lattice with these properties is called t-regular. Various estimates are obtained in [293].

Havrda [230] studied L-independence in set ol. In particular, he proved that there exist maximal L-independent sets and that every two maximal L-independent sets have the same cardinality.

Romanovich [50] found necessary and sufficient conditions under which the ordered set $\mathrm{Ort}_0 S$ of semi-orthogonality relations on a lattice S is a (distributive) lattice.

Let E and E' denote vector spaces over a field. Let L and L' be complete distributive sublattices of the lattices of subspaces of E and E', respectively. Gross and Keller [219] showed that if L and L' satisfy the descending chain condition then any rank-preserving isomorphism $\varphi: L \to L'$ is induced by a linear bijection $\tau: E \to E'$. Moreover, if there is a bilinear form on E that admits an orthogonal basis and satisfies certain assumptions, then a theorem on the existence of orthogonal complements is proved.

Gensheimer [202] showed the difference between the measure theory on general ol and the measure theory on Boolean algebras.

Kalmbach [267] showed that every group is isomorphic to the automorphism group of a suitable ol.

Savel'ev [53] showed that a lattice L with orthocomplements and continuous measure can be continuously embedded into the lattice of additive mappings of L in a topological group. Kalmár ([261] and [262]) and Cossinelli and Truini [146] generalized classical results on conditional probability or random variables for suitable ol.

Pulmannová [372] generalized the concept of a commutator introduced by Beran [OSL, II, 106]. Poguntke [359] studied the role of the commutator in a finitely generated lattice with orthocomplements. He described finite posets which generate congruence-free lattices with orthocomplements.

Iturrioz [247] and Strojewski [424] suggested representations of ol. In the former paper a generalization of the known Stone topological representation of Boolean algebras was obtained, while in the latter it was proved that an ol L is embeddable in the lattice $[0, 1]^S$ if and only if the set S of states on L is ordered.

Kalmár [263] introduced the operation $a * c = (a \vee c') \wedge c$ for ol L. He established certain properties of the algebra $(L; *,', 0, 1)$.

Bibliography

1. M. Ya. Antonovskiĭ, V. G. Boltyanskiĭ, and T. A. Sarymsakov, *Topological Boolean algebras*, Topological Semifields, no. 1, Izdat. Akad. Nauk Uzbek. SSR, Tashkent, 1963; English transl., Amer. Math. Soc. Transl. (2) **106** (1977).

2. A. A. Baĭzhumanov, *Certain estimates for solution of systems of linear Boolean equations*, Voprosy Kibernetiki, vol. 121, Tashkent, 1983, pp. 36–44. (Russian)

3. V. A. Baranskiĭ, *Algebraic systems whose elementary theory is compatible with an arbitrary group*, Algebra i Logika **22** (1983), no. 6, 599–607; English transl. in Algebra and Logic **22** (1983), no. 6.

4. F. D. Beznosikov, *External homomorphisms of Boolean algebras*, Ordered Spaces and Operator Equations, Perm', 1982, pp. 93–103. (Russian)

5. M. R. Bunyatov, *Theory of homotopy groups of distributive lattices*, Problems in Geometry and Algebraic Topology, Izdat. Azerbaidzhan Univ., Baku, 1985, pp. 37–55. (Russian)

6. M. R. Bunyatov and R. A. Baĭramov, *K-theory on a category of distributive lattices*, Dokl. Akad. Nauk AzSSR **39** (1983), no. 5, 7–11. (Russian)

7. M. R. Bunyatov and V. A. Kasimov, *The Eilenberg-Zil'ber theorem for Boolean algebras with closure*, Dokl. Akad. Nauk AzSSR **38** (1982), no. 2, 7–9. (Russian)

8. D. A. Vladimirov, *Boolean Algebras*, "Nauka", Moscow, 1969; German transl., Akademie-Verlag, Berlin, 1972.

9. V. A. Goncharov, *A recursively representable Boolean algebra*, Boundary Value Problems for Differential Equations and their Applications in Mechanics and Technology, "Nauka", Kazakh. SSR, Alma-Ata, 1983, pp. 43–46. (Russian)

10. S. S. Goncharov, *Universal recursively enumerable Boolean algebras*, Sibirsk. Mat. Zh. **24** (1983), no. 6, 36–43; English transl. in Siberian Math. J. **24** (1983), no. 6.

11. V. L. Dzgoev, *Constructivizations of direct products of algebraic systems*, Algebra i Logika **21** (1982), no. 2, 138–148; English transl. in Algebra and Logic **21** (1982), no. 2.

12. Z. A. Dulatova, *Extended theories of Boolean algebras*, Sibirsk. Mat. Zh. **25** (1984), no. 1, 201–204; English transl. in Siberian Math. J. **25** (1984), no. 1.

13. ———, *On extended theories of finite Boolean algebras with measure*, Voprosy Teorii Algebr. Sistem, Karaganda, 1981, pp. 43–48. (Russian)

14. D. P. Egorova, *Unipermanent matrices over distributive lattices*, Abstract of the XVIII All-Union Algebraic Conference (Kishinev, 1985), Part I, 1985, p. 183. (Russian)

15. Yu. L. Ershov, *Distributive lattices with relative complements*, Algebra i Logika **18** (1979), no. 6, 680–722; English transl. in Algebra and Logic **18** (1979), no. 6.

16. ———, *Existence of constructivizations*, Dokl. Akad. Nauk SSSR **204** (1972), no. 5, 1041–1044; English transl. in Soviet Math. Dokl. **13** (1972), no. 3.

17. V. V. Zharinov, *Distributive lattices and their applications in complex analysis*, Trudy Mat. Inst. Steklov **162** (1983), 3–80; English transl. in Proc. Steklov Inst. Math. **162** (1985).

18. M. V. Zakhar'yashchev, *On intermediate logics*, Dokl. Akad. Nauk SSSR **269** (1983), no. 1, 18–22; English transl. in Soviet Math. Dokl. **27** (1983), no. 2.

19. M. I. Kanovich, *Complexity and convergence of algorithmic mass problems*, Dokl. Akad. Nauk SSSR **272** (1983), no. 2, 289–293; English transl. in Soviet Math. Dokl. **28** (1983), no. 2.

20. C. C. Chang and H. J. Keisler, *Model Theory*, North-Holland, Amsterdam, 1973.

21. Yu. M. Kir'yakov, *Study of Certain Special Properties of Logical Formulas of the Proposition Calculus*, Rostov Civil Engineering Inst., Rostov-on-Don, 1983. (Russian)

22. A. I. Kokorin and A. G. Pinus, *Decision problems of extended theories*, Uspekhi Mat. Nauk **33** (1978), no. 2, 49–84; English transl. in Russian Math. Surveys **33** (1978), no. 2.

23. A. G. Kusraev and S. S. Kutateladze, *Subdifferentials in Boolean-valued models of set theory*, Sibirsk. Mat. Zh. **24** (1983), no. 5, 109–122; English transl. in Siberian Math. J. **24** (1983), no. 5.

24. A. G. Kusraev, *Order-continuous functionals in Boolean-valued models of set theory*, Sibirsk. Mat. Zh. **25** (1984), no. 1, 69–79; English transl. in Siberian Math. J. **25** (1984), no. 1.

25. P. I. Lazarev and I. I. Lazarev, *On complete Boolean algebras with an ergodic group of automorphisms*, Teoretich. i Prikladn. Issled. po Matematike i Tekhnike, Tashkent, 1983, pp. 52–58. (Russian)

26. A. G. Lutsenko, *Certain Problems of Topology in Boolean Algebras*, Tul. Gos. Ped. Inst., Tula, 1983. (Russian)

27. _____, *On Representation of Free Boolean Algebras as Union of Chains of their Subalgebras*, Izdat. Tul. Ped. Inst., Tula, 1983. (Russian)

28. V. P. Malaĭ and M. F. Ratsa, *On chain precomplete sets of operations of pseudo Boolean algebras*, Abstract of the XVIII All-Union Algebraic Conference (Kishinev, 1985), Part II, 1985. (Russian)

29. O. M. Mamedov, *On equationally compact algebras in certain classes*, Proc. Sem. Univ. Algebra (Baku, 1979), 1979, p. 17. (Russian)

30. S. I. Mardaev, *On finitely generated implicative semilattices*, Abstract of the XVIII All-Union Algebraic Conference (Kishinev, 1985), Part II, 1985. (Russian)

31. V. I. Mart'yanov, *Undecidability of the theory of Boolean algebras with an automorphism*, Sibirsk. Mat. Zh. **23** (1982), no. 3, 147–154; English transl. in Siberian Math. J. **23** (1982), no. 3.

32. A. S. Morozov, *Strong constructibility of countable saturated Boolean algebras*, Algebra i Logika **21** (1982), no. 2, 193–203; English transl. in Algebra and Logic **21** (1982), no. 2.

33. _____, *Countable homogeneous Boolean algebras*, Algebra i Logika **21** (1982), no. 3, 269–282; English transl. in Algebra and Logic **21** (1982), no. 3.

34. _____, *On constructive Boolean algebras with almost identity automorphisms*, Mat. Zametki **37** (1985), no. 4, 478–482; English transl. in Math. Notes **37** (1985).

35. _____, *Groups of recursive automorphisms of constructive Boolean algebras*, Algebra i Logika **22** (1983), no. 2, 138–158; English transl. in Algebra and Logic **22** (1983), no. 2.

36. _____, *Automorphisms of constructivizations of Boolean algebras*, Sibirsk. Mat. Zh. **26** (1985), no. 4, 98–110; English transl. in Siberian Math. J. **26** (1985), no. 4.

37. L. V. Mushits, P. G. Ovchinnikov, and O. B. Terekhin, *On Quantum Logics*, Kazan. Univ., Kazan, 1982. (Russian)

38. K. A. Nauryzbaev, *A criterion of equational compactness of distributive lattices*, Issledovaniya po Konstruktivnym Modelyam, Alma-Ata, 1982, pp. 46–53. (Russian)

39. _____, *Equational compactness of Boolean products in Mal'tsev's varieties and in the class of distributive lattices*, Abstract of the XVIII All-Union Algebraic Conference (Kishinev, 1985), Part II, 1985, p. 61. (Russian)

40. I. S. Negru, *Quasi distributivity of lattices*, Mat. Issled. (1982), no. 66, 113–127.

41. S. P. Odintsov, *Atomless ideals of constructive Boolean algebras*, Algebra i Logika **23** (1984), no. 3, 278–295; English transl. in Algebra and Logic **23** (1984), no. 3.

42. A. I. Omarov, *Elementary theory of D-degrees*, Algebra i Logika **23** (1984), no. 5, 530–537; English transl. in Algebra and Logic **23** (1984), no. 5.

43. A. I. Pavlovskiĭ, *Applied Problems of Boolean Algebra*, Izdat. Minsk. Ped. Inst., Minsk, 1982. (Russian)

44. V. V. Pashenkov, *Boolean algebras*, Ordered Sets and Lattices, No. 7, Izdat. Saratov Univ., Saratov, 1983, pp. 4–15. (Russian)

45. _____, *On a pair of adjoint functors*, Ordered Sets and Lattices, No. 8, Izdat. Saratov Univ., Saratov, 1982, pp. 70–81. (Russian)

46. A. G. Pinus, *On constructivizations of Boolean algebras*, Sibirsk. Mat. Zh. **22** (1981), no. 4, 169–175; English transl. in Siberian Math. J. **22** (1981), no. 4.

47. A. G. Poroshkin, *On a Condition of Horn-Tarski for Boolean Algebras*, Syktyvkar. Univ., Syktyvkar, 1982. (Russian)

48. _____, *On the problem of normability of Boolean algebras with a continuous outer measure*, Sibirsk. Mat. Zh. **21** (1980), no. 4, 216–220, 240; English transl. in Siberian Math. J. **21** (1980), no. 4.

49. V. A. Romanovich, *Dimension of an ordered product of lattices*, Abstract of the 7th Regional Conference Mathematics and Mechanics (Tomsk, 1981), Section of Algebra, Tomsk, 1981, pp. 37–38. (Russian)

50. _____, *On structural properties of the semi-orthogonality relation on lattices*, Ordered Sets and Lattices, No. 8, Izdat. Saratov Univ., Saratov, 1982, pp. 96–108.

51. V. V. Rybakov, *Elementary theories of free topo-Boolean and pseudo Boolean algebras*, Mat. Zametki **37** (1985), no. 6, 797–802; English transl. in Math. Notes **37** (1985).

52. _____, *Decidability of the problem of admissibility in finite-layer modal logics*, Algebra i Logika **23** (1984), no. 1, 100–116; English transl. in Algebra and Logic **23** (1984), no. 1.

53. L. A. Savel′ev, *Measures on ortholattices*, Dokl. Akad. Nauk SSSR **264** (1982), no. 5, 1091–1094; English transl. in Soviet Math. Dokl. **25** (1982), no. 3.

54. _____, *External measures and external topologies*, Sibirsk. Mat. Zh. **24** (1983), no. 2, 133–149; English transl. in Siberian Math. J. **24** (1983), no. 2.

55. V. N. Salii, *Lattices with unique complements*, Nauka, Moscow, 1984; English transl., Transl. Math. Monographs, vol. 69, Amer. Math. Soc., Providence, RI, 1988.

56. _____, *On quasi Boolean lattices and quasi Boolean matrices*, Abstract of the XVIII All-Union Algebraic Conference (Kishinev, 1985), Part II, 1985, p. 165. (Russian)

57. T. A. Sarymsakov and V. I. Chilin, *Measure on topological algebras*, Proc. Conf. Topology and Measure I, Greifswald, 1978. (Russian)

58. _____, *Uniformities and outer estimates on logics*, Proc. Conf. Topology and Measure, II, Part 2 (Rostock/Warnemunde, 1977), Ernst Moritz Arndt Univ., Greifswald, 1980, pp. 125–128. (Russian)

59. L. A. Skornyakov, *Eigenvectors of matrices·over a distributive lattice*, Vīsnik Kiïv. Unīv. Ser. Mat. Mekh., no. 27 (1985), 96–97. (Ukrainian)

60. _____, *Invertible matrices over distributive lattices*, Sibirsk. Mat. Zh. **27** (1986), no. 2, 182–185; English transl. in Siberian Math. J. **27** (1986), no. 2.

61. L. A. Skornyakov and D. P. Egorova, *Normal subgroups of the general linear group of degree 3 over a distributive lattice*, Algebra i Logika **23** (1984), no. 6, 670–683; English transl. in Algebra and Logic **23** (1984), no. 6.

62. Z. B. Todua, *Homology and cohomology groups of a distributive lattice*, Soobshch. Akad. Nauk Gruzin. SSR **102** (1981), no. 3, 557–560. (Russian)

63. _____, *Relative homology and cohomology groups of a distributive lattice over a pair of coefficient groups*, Soobshch. Akad. Nauk Gruzin. SSR **105** (1982), no. 3, 489–492. (Russian)

64. _____, *Some properties of homology groups of a distributive lattice*, Soobshch. Akad. Nauk Gruzin. SSR **113** (1984), no. 2, 277–280. (Russian)

65. M. P. Tropin, *Embeddability of a free lattice in the lattice of quasivarieties of pseudocomplemented distributive lattices*, Algebra i Logika **22** (1983), no. 2, 159–167; English transl. in Algebra and Logic **22** (1983), no. 2.

66. K. U. Uzakov, *On complexity of the set of atoms of a recursive Boolean algebra*, Issledovaniya po Konstruktivnym Modelyam, Alma-Ata, 1985, pp. 85–96. (Russian)

67. Yu. Flaksmaĭer [J. Flachsmayer], *Topological semifields and Boolean algebras corresponding to them*, Trudy Mat. Inst. Steklov **154** (1983), 252–264; English transl. in Proc. Steklov. Inst. Math. **154** (1984).

68. Ya. P. Tsirulis [J. P. Cirulis], *Relative pseudocomplements in semilattices*, Izv. Vyssh. Uchebn. Zaved. Mat. (1982), no. 2, 78–80; English transl. in Soviet Math. (Iz. VUZ) **26** (1982), no. 2.

69. V. M. Shiryaev, *The determinability of a finite semigroup by the specification of its action*, Vestnik Beloruss. Gos. Univ. Ser. I Fiz. Mat. Mekh. (1982). (Russian)

70. L. L. Esakia, *Algebras of logic*, Ordered Sets and Lattices, No. 7, Izdat. Saratov Univ., Saratov, 1983, pp. 15–26. (Russian)

71. A. Abian, *Two methods of construction of free Boolean algebras*, Studia Sci. Math. Hungar. **14** (1979), no. 1–3, 125–129.

72. M. E. Adams and D. M. Clark, *Universal terms for pseudo-complemented distributive lattices and Heyting algebras*, Lecture Notes in Math., vol. 1149, Springer-Verlag, New York and Berlin, 1985, pp. 1–16.

73. M. E. Adams and T. Katriňák, *A note on subdirectly irreducible distributive double p-algebras*, J. Austral. Math. Soc. Ser. A **35** (1983), no. 1, 46–58.

74. M. E. Adams, V. Koubek, and J. Sichler, *Pseudocomplemented distributive lattices with small endomorphism monoids*, Bull. Austral. Math. Soc. **28** (1983), no. 3, 305–318.

75. ____, *Homomorphisms and endomorphisms in varieties of pseudocomplemented distributive lattices (with applications to Heyting algebras)*, Trans. Amer. Math. Soc. **285** (1984), 57–79.

76. ____, *Homomorphisms and endomorphisms of distributive lattices*, Houston J. Math. **11** (1985), no. 2, 129–145.

77. ____, *Homomorphisms of pseudocomplemented distributive lattices with countably many universal prime ideals*, Preprint.

78. M. E. Adams and H. A. Priestley, *Kleene algebras are almost universal*, Bull. Austral. Math. Soc. **34** (1986), no. 3, 343–373.

79. ____, *De Morgan algebras are universal*, Discrete Math. **66** (1987), no. 1–2, 1–13.

80. M. E. Adams and V. Trnková, *Isomorphisms of sums of countable bounded distributive lattices*, Algebra Universalis **15** (1982), no. 2, 242–257.

81. I. H. Anellis, *Boolean groups*, Abh. Braunschweig. Wiss. Ges. **33** (1982), 85–97.

82. S. Argyros, *Boolean algebras without free families*, Algebra Universalis **14** (1982), no. 2, 244–256.

83. T. E. Armstrong and K. Prikry, *On the semimetric on a Boolean algebra induced by a finitely additive probability measure*, Pacific J. Math. **99** (1982), no. 2, 249–264.

84. R. Balbes, *On free pseudocomplemented and relatively pseudocomplemented semilattices*, Fund. Math. **78** (1973), 119–131.

85. B. Balcar and F. Franěk, *Independent families in complete Boolean algebras*, Trans. Amer. Math. Soc. **274** (1982), no. 2, 607–618.

86. B. Balcar and P. Simon, *Cardinal invariants in Boolean spaces*, General Topology and its Relations to Modern Analysis and Algebra, V (Prague, 1981), Sigma Ser. Pure Math., vol. 3, Heldermann, Berlin, 1983, pp. 39–47.

87. B. Balcar and P. Štěpánek, *Embedding theorems for Boolean algebras and consistency results on ordinal definable sets*, J. Symbolic Logic **42** (1977), no. 1, 64–76.

88. R. N. Ball, *Distributive Cauchy lattices*, Algebra Universalis **18** (1984), no. 2, 134–174.

89. B. Banaschewski, G. C. L. Brümmer, and K. A. Hardie, *Biframes and bispaces*, Proceedings of the Symposium on Categorical Algebra and Topology (Cape Town, 1981), Quaestiones Math. **6** (1983), no. 1–3, 13–25.

90. H.-J. Bandelt, *Coproducts of bounded (α, β)-distributive lattices*, Algebra Universalis **17** (1983), no. 1, 92–100.

91. H.-J. Bandelt and M. Erné, *Representations and embeddings of M-distributive lattices*, Houston J. Math. **10** (1984), no. 3, 315–324.

92. H.-J. Bandelt and M. Petrich, *Subdirect products of rings and distributive lattices*, Proc. Edinburgh Math. Soc. (2) **25** (1982), no. 2, 155–171.

93. J. E. Baumgartner, A. D. Taylor, and S. Wagon, *Structural properties of ideals*, Dissertationes Math. (Rozprawy Mat.) **197** (1982).

94. R. Beazer, *A characterization of complete bi-Brouwerian lattices*, Colloq. Math. **29** (1974), 55–59.

95. ____, *Pseudocomplemented algebras with Boolean congruence lattices*, J. Austral. Math. Soc. Ser. A **26** (1978), 163–168.

96. ____, *On congruence lattices of p-algebras and double p-algebras*, Algebra Universalis **13** (1981), no. 3, 379–388.

97. ____, *Affine complete Stone algebras*, Acta Math. Hungar. **39** (1982), no. 1, 169–174.

98. _____, *Affine complete double Stone algebras with bounded core*, Algebra Universalis **16** (1983), no. 2, 237–244.

99. _____, *Lattices whose ideal lattice is Stone*, Proc. Edinburgh Math. Soc. (2) **26** (1983), no. 1, 107–112.

100. A. M. Bell, M. R. Brown, and G. A. Fraser, *The tensor product of distributive lattices: structural results*, Proc. Edinburgh Math. Soc. (2) **27** (1984), no. 3, 237–245.

101. J. L. Bell, *On the strength of the Sikorski extension theorem for Boolean algebras*, J. Symbolic Logic **48** (1983), no. 3, 841–846.

102. M. Bell, *Two Boolean algebras with extreme cellular and compactness properties*, Canad. J. Math. **35** (1983), no. 5, 824–838.

103. L. Beran, *Some applications of Boolean skew-lattices*, Studia Sci. Math. Hungar. **14** (1979), no. 1–3, 183–188.

104. _____, *Boolean and orthomodular lattices—a short characterization via commutativity*, Acta Univ. Carolin.—Math. Phys. **23** (1982), no. 1, 25–27.

105. _____, *Orthomodular Lattices. Algebraic Approach*, Reidel, Dordrecht, 1985.

106. _____, *Special polynomials in orthomodular lattices*, Comment. Math. Univ. Carolin. **26** (1985), 641–650.

107. C. Berge, *A theorem related to the Chvátal conjecture*, Proc. 5th British Combinatorial Conf. (Aberdeen, 1975), Utilitas Math., Winnepeg, Manitoba, 1976, pp. 35–40.

108. C. Bernardi, *Lo spazio duale di un prodotto di algebra di Boole e le compatificationi di Stone*, Ann. Mat. Pura Appl. **126** (1980), 253–266.

109. L. Beznea, *A topological characterization of complete distributive lattices*, Discrete Math. **49** (1984), no. 2, 117–120.

110. A. Błaszczyk, *A note on rigid spaces and rigid Boolean algebras*, General Topology and its Relations to Modern Analysis and Algebra. V (Prague, 1981), Sigma Ser. Pure Math., vol. 3, Heldermann, Berlin, 1983, pp. 48–52.

111. _____, *Aspekty Topologiczne Algebr Boole'a* (1982), Uniwersytet Śląski, Katowice.

112. T. S. Blyth and J. C. Varlet, *Sur la construction de certaines MS-algebras*, Portugal. Math. **39** (1980), 489–496.

113. _____, *Corrigendum sur la construction de certaines MS-algebras*, Portugal. Math. **42** (1983/84), 469–471.

114. _____, *On a common abstraction of De Morgan algebras and Stone algebras*, Proc. Roy. Soc. Edinburgh Sect. A **94** (1983), no. 3–4, 301–308.

115. R. Bonnet, *On homomorphism types of superatomic interval Boolean algebras*, Model and Sets (Aachen, 1983), Lecture Notes in Math., vol. 1103, Springer-Verlag, New York and Berlin, 1984.

116. B. A. Bosbach, *A representation theorem for completely join distributive algebraic lattices*, Period. Math. Hungar. **13** (1982), 113–118.

117. G. Brenner, *A simple construction for rigid and weakly homogeneous Boolean algebras answering a question of Rubin*, Proc. Amer. Math. Soc. **87** (1983), no. 4, 601–606.

118. G. Brenner and D. Monk, *Tree algebras and chains*, Lecture Notes in Math., vol. 1004, Springer-Verlag, New York and Berlin, 1983.

119. G. Bruns, *Varieties of modular ortholattices*, Houston J. Math. **9** (1983), no. 1, 1–7.

120. G. Bruns and R. Greechie, *Orthomodular lattices which can be covered by finitely many blocks*, Canad. J. Math. **34** (1982), no. 3, 696–699.

121. _____, *Some finiteness conditions for orthomodular lattices*, Canad. J. Math. **34** (1982), no. 3, 535–549.

122. B. Budde, *Some Boolean Properties of Stone Lattices*, University of Wyoming, Laramie, 1980.

123. J. Bulatović, *On a random function defined on a pseudo Boolean algebra*, Publ. Inst. Math. (Beograd) **32** (1982), 33–36.

124. E. Burlacu, *On the inverses of Boolean matrices*, Bul. Ştiinţ. Tehn. Inst. Politehn. Timişoara **14** (1969), no. 1, 15–20. (Romanian)

125. S. Burris, *The first order theory of Boolean algebras with a distinguished group of automorphisms*, Algebra Universalis **15** (1982), no. 2, 151–161.

126. S. Burris and R. McKenzie, *Decidability and Boolean representations*, Mem. Amer. Math. Soc. **32** (1981), no. 246.

127. G. G. Călugăreanu, *Torsion in lattices*, Mathematica (cluj) **25(48)** (1983), no. 2, 127–129.

128. J.-C. Carrega, *Exclusion d'algèbres*, C. R. Acad. Sci. Paris Sér. I Math. **295** (1982), no. 2, 43–46.

129. J.-C. Carrega, G. Chevalier, and R. Mayet, *Une classe de treillis orthomodulaires en liaison avec un théorème de décomposition*, C. R. Acad. Sci. Paris Sér. I Math. **299** (1984), no. 14, 639–642.

130. J.-C. Carrega and M. Fort, *Un problème d'exclusion de treillis orthomodulaires*, C. R. Acad. Sci. Paris Sér. I Math. **296** (1983), no. 12, 485–488.

131. M. Chacron, *A note on matrices with entries in a distributive lattice*, Bull. Soc. Math. Belg. **22** (1970), no. 2, 143–145.

132. V. R. Chandran and V. Varalakshmi, *On Mal'cev varieties*, J. Math. Phys. Sci. **17** (1983), no. 2, 201–202.

133. Z. K. Charzyński and T. Prucnal, *On certain characteristics of the family of prime filters of distributive lattices*, Demonstratio Math. **17** (1984), no. 2, 495–497.

134. C. C. Chen and G. Grätzer, *Stone lattices* II. *Structure theorems*, Canad. J. Math. **21** (1969), 895–903.

135. C. C. Chen, K. M. Koh, and K. L. Teo, *On the length of the lattice of subalgebras of a finite Stone algebra*, Bull. Malaysian Math. Soc. (2) **5** (1982), no. 2, 101–104.

136. G. Chevalier, *Relations binaires et congruences dans un treillis orthomodulaire*, C. R. Acad. Sci. Paris Sér. I Math. **296** (1983), no. 19, 785–788.

137. _____, *Les congruences d'un treillis orthomodulaire de projections*, C. R. Acad. Sci. Paris Sér. I Math. **299** (1984), no. 15, 731–734.

138. W. Chromik, *The sum of a double system of graphs*, Demonstratio Math. **16** (1983), no. 2, 429–433.

139. J. Cichoń, *On the compactness of some Boolean algebras*, J. Symbolic Logic **49** (1984), no. 1, 63–67.

140. H. Cohen, R. J. Koch, and J. D. Lawson, *Semigroups defined by lattice polynomials*, Lecture Notes in Math., vol. 998, Springer-Verlag, New York and Berlin, 1983, pp. 50–56.

141. P. M. Cohn, *Ringe mit distributivem Faktorverband*, Abh. Braunschweig. Wiss. Ges. **33** (1982), 35–40.

142. M. Contessa, *Ultraproducts of PM-rings and MP-rings*, J. Pure Appl. Algebra **32** (1984), no. 1, 11–20.

143. R. M. Cooke and M. van Lambalgen, *The representation of Takeutis II-operator*, Studia Logica **42** (1983), no. 4, 407–415.

144. W. H. Cornish, *A ternary variety generated by lattices*, Comment. Math. Univ. Carolin. **22** (1981), no. 4, 773–784.

145. W. H. Cornish, T. Sturm, and T. Traczyk, *Embedding of commutative BCK-algebras into distributive lattice BCK-algebras*, Math. Japon. **29** (1984), no. 2, 309–320.

146. G. Cossinelli and P. Truini, *Conditional probabilities on orthomodular lattices*, Rep. Math. Phys. **20** (1984), no. 1, 41–52.

147. J. Coulon and J.-L. Coulon, *Fuzzy Boolean algebras*, J. Math. Anal. Appl. **99** (1984), no. 1, 248–256.

148. R. Crawley, *Lattices whose congruences form a Boolean algebra*, Pacific J. Math. **16** (1960), 787–795.

149. C. Crociani and M. Moscucci, *Una traduzione algebrica di un concetto logico di indipendenza*, Matematiche (Catania) **36** (1981), no. 2, 261–280.

150. R. A. Cuninghame-Green, *The characteristic maxpolynomial of a matrix*, J. Math. Anal. Appl. **95** (1983), no. 1, 110–116.

151. J. Czelakowski, *Partial Boolean algebras in a broader sense and Boolean embeddings*, Colloq. Math. **45** (1981), no. 2, 171–180.

152. B. A. Davey, *Dualities for Stone algebras, double Stone algebras, and relative Stone algebras*, Colloq. Math. **46** (1982), no. 1, 1–14.

153. A. Day, *Splitting algebras and a weak notion of projectivity*, Algebra Universalis **5** (1975), no. 2, 153–162.

154. D. E. Daykin, A. J. W. Hilton, and D. Miklós, *Pairings from down-sets and up-sets in distributive lattices*, J. Combin. Theory Ser. A **34** (1983), no. 2, 215-230.

155. J. Demel, *Fast algorithms for finding a subdirect decomposition and interesting congruences on finite algebras*, Kybernetika (Prague) **18** (1982), 121-130.

156. M. Dichtl, *Astroids and pastings*, Algebra Universalis **18** (1984), no. 3, 380-385.

157. Y. Diers, *Un critère de représentabilité par sections continues de faisceaux*, Category Theory (Gummersbach, 1981), Lecture Notes in Math., vol. 962, Springer-Verlag, New York and Berlin, 1982, pp. 51-61.

158. _____, *Une description axiomatique des catégories de faisceaux de structures algébriques sur les espaces topologiques booléens*, Adv. Math. **47** (1983), no. 3, 258-299.

159. U. Dietz, *A characterization of orthomodular lattices among ortholattices*, Contributions to General Algebra, 3 (Vienna, 1984), Hölder-Pichler-Tempsky, Vienna, 1985, pp. 99-101.

160. A. Di Nola and S. Sessa, *A representation of n-fuzzy sets by means of Boolean matrices*, Ricerca (4) (Napoli) **30** (1979), no. 2, 15-24.

161. G. Dimov, *On the stone duality*, General Topology and its Relations to Modern Analysis and Algebra, V (Prague, 1981), Sigma Ser. Pure Math., vol. 3, Heldermann, Berlin, 1983, pp. 145-146.

162. H. Dobbertin, *On Vaught's criterion for isomorphisms of countable Boolean algebras*, Algebra Universalis **15** (1982), no. 1, 95-114.

163. _____, *Refinement monoids, Vaught monoids, and Boolean algebras*, Math. Ann. **265** (1983), no. 4, 473-487.

164. _____, *Verfeinerungsmonoide, Vaught Monoide und Boolesche Algebren*, Diss. Dokt. Naturwiss. Fachbereich. Math., Univ. of Hanover (1983).

165. E. K. van Douwen, *Cardinal functions on compact F-spaces and on weakly countably complete Boolean algebras*, Fund. Math. **114** (1981), no. 3, 235-256.

166. _____, *Nonsupercompactness and the reduced measure algebra*, Comment. Math. Univ. Carolin. **21** (1980), no. 3, 507-512.

167. E. K. van Douwen, J. D. Monk, and M. Rubin, *Some questions about Boolean algebras*, Algebra Universalis **11** (1980), no. 2, 220-243.

168. C. H. Dowker and D. Strauss, T_1 *and* T_2 *axioms for frames. Introduction*, London Math. Soc. Lecture Note Ser., vol. 93, Cambridge Univ. Press, Cambridge, 1985, pp. 325-335.

169. R. G. Downey, *Abstract dependence, recursion theory, and the lattice of recursively enumerable filters*, Bull. Austral. Math. Soc. **27** (1983), no. 3, 461-464.

170. J. M. Dunn, *A relational representation of quasi-Boolean algebras*, Notre Dame J. Formal Logic **23** (1982), no. 4, 353-357.

171. I. Düntsch, *A class of projective Stone algebras*, Bull. Austral. Math. Soc. **24** (1981), no. 1, 133-147.

172. _____, *Projectivity, prime ideals, and chain conditions of Stone algebras*, Algebra Universalis **14** (1982), no. 2, 167-180.

173. _____, *On a problem of Chen and Grätzer*, Algebra Universalis **14** (1982), 401.

174. _____, *A description of the projective Stone algebras*, Glasgow Math. J. **24** (1983), no. 1, 75-82.

175. _____, *A note on atomic Stone algebras*, Algebra Universalis **16** (1983), no. 3, 398-399.

176. _____, *On free or projective regular double Stone algebras*, Houston J. Math. **9** (1983), no. 4, 455-463.

177. _____, *Finite projective double Stone algebras*, Proc. Edinburgh Math. Soc. (2) **29** (1984), no. 2, 65-67.

178. I. Düntsch and S. Koppelberg, *Complements and quasicomplements in the lattice of subalgebras of* $P(\omega)$, Discrete Math. **53** (1985), 63-78.

179. Ph. Dwinger, *Completeness of Boolean powers of Boolean algebras*, Universal Algebra (Esztergom, 1977), Colloq. Math. Soc. János Bolyai, vol. 29, North-Holland, Amsterdam and New York, 1982, pp. 209-217.

180. _____, *Characterization of the complete homomorphic images of a completely distributive complete lattice* I, Nederl. Akad. Wetensch. Indag. Math. **44** (1982), no. 4, 403-414.

181. _____, *Characterization of the complete homomorphic images of a completely distributive complete lattice* II, Nederl. Akad. Wetensch. Indag. Math. **45** (1983), no. 1, 43-49.

182. _____, *Unary operations on completely distributive complete lattices*, Lecture Notes in Math., vol. 1149, Springer-Verlag, New York and Berlin, 1985, pp. 46–81.

183. W. Dziobiak, *Concerning axiomatizability of the quasivariety generated by a finite Heyting or topological Boolean algebra*, Studia Logica **41** (1982), no. 4, 415–428.

184. _____, *Concerning axiomatizability of the quasivariety generated by a finite Heyting or topological Boolean algebra*, Polish Acad. Sci. Inst. Philos. Sociol. Bull. Sect. Logic **10** (1981), no. 4, 177–180.

185. _____, *On subquasivariety of some varieties related with distributive p-algebras*, Algebra Universalis **21** (1985), 62–67.

186. _____, *The subvariety lattice of the variety of distributive double p-algebras*, Bull. Austral. Math. Soc. **31** (1985), 377–381.

187. G. Epstein, *On Rine's view of Boolean algebras*, Polish Acad. Sci. Inst. Philos. Sociol. Bull. Sect. Logic **10** (1981), no. 2, 91–92.

188. P. Erdös, *The Scottish Book*, Birkhäuser, Boston, 1981.

189. L. Español, *Constructive Krull dimension of lattices*, Rev. Acad. Cienc. Zaragoza (2) **37** (1982), 5–9.

190. J. Flachsmayer, *Note on orthocomplemented posets*, Proc. Conf. Topology and Measure III, Part 1 (Vitte/Hiddensee, 1980), Ernst Moritz Arndt Univ., Greifswald, 1982, pp. 65–73.

191. _____, *Note on orthocomplemented posets* II, Rend. Circ. Mat. Palermo (2) Suppl. **31** (1982), no. 2, 67–74.

192. I. Fleischer, *One-variable equationally compact distributive lattices*, Math. Slovaca **34** (1984), no. 4, 385–386.

193. J. Flum, *Modelltheorie—topologische Modelltheorie*, Jahresber. Deutsch. Math.-Verein. **86** (1984), no. 2, 69–82.

194. M. Foreman, *Games played on Boolean algebras*, J. Symbolic Logic **48** (1983), no. 3, 714–723.

195. G. A. Fraser, *The semilattice tensor product of distributive lattices*, Trans. Amer. Math. Soc. **217** (1976), 183–194.

196. F. J. Freniche, *The number of nonisomorphic Boolean subalgebras of a power set*, Proc. Amer. Math. Soc. **91** (1984), no. 2, 199–201.

197. S. Fujishige and N. Tomizawa, *A note on submodular functions on distributive lattices*, J. Oper. Res. Soc. Japan **26** (1983), no. 4, 309–318.

198. S. Fujiwara, *On the parallelism in some left complemented lattice*, Res. Bull. Fac. Liberal Arts Oita Univ. **2** (1964), no. 4, 1–8.

199. E. R. Gansner, *Parenthesizations of finite distributive lattices*, Algebra Universalis **16** (1983), no. 3, 287–303.

200. M. Gavalec, *Dimensions of distributive lattices*, Mat. Časopis Sloven. Akad. Vied **21** (1971), no. 3, 177–190.

201. _____, *Independent complete subalgebras of collapsing algebras*, Colloq. Mat **45** (1981), no. 2, 181–189.

202. H. Gensheimer, *Measures on orthomodular lattices*, Contributions to General Algebra, 2 (Klagenfurt, 1982), Hölder-Pichler-Tempsky, Vienna, 1983, pp. 115–121.

203. H. Gerstmann, *n-Distributivgesetze*, Acta Sci. Math. (Szeged) **46** (1983), no. 1–4, 99–113.

204. _____, *n-Distributivgesetze*, Diss. Dokt. Naturwiss. Fachbereich. Math., Univ. of Hanover (1983).

205. G. Gierz and J. D. Lawson, *Generalized continuous and hypercontinuous lattices*, Rocky Mountain J. Math. **11** (1981), no. 2, 271–296.

206. G. Gierz, J. D. Lawson, and A. D. Stralka, *Metrizability conditions for completely distributive lattices*, Canad. Math. Bull. **26** (1983), no. 4, 446–453.

207. G. Gierz and A. Stralka, *Distributive lattices with sufficiently many chain intervals*, Algebra Universalis **20** (1985), no. 1, 77–89.

208. K. Gilezan, *Some fixed point theorems in Boolean algebras*, Publ. Inst. Math. (Beograd) **28** (1980), 77–82.

209. Y. Give'on, *Lattice matrices*, Inform. and Control (Shenyang) **7** (1965), no. 4, 477–484.

210. K. Głazek and T. Katriňák, *Weak homomorphisms of distributive p-algebras*, Universal Algebra and Applications (Warsaw, 1978), Banach Center Publ., 9, PWN, Warsaw, 1982, pp. 383–390.

211. R. M. Godowski, *Commutativity in orthomodular posets*, Rep. Math. Phys. 18 (1980), no. 3, 347–351.

212. _____, *States on orthomodular lattices*, Demonstratio Math. 15 (1982), no. 3, 817–822.

213. _____, *Varieties of orthomodular lattices with a strongly full set of states*, Demonstratio Math. 14 (1981), no. 3, 725–733.

214. H. Göktas, *On lattice ideals in a translation lattice*, Hacettepe Bull. Natur. Sci. and Eng. 11 (1982), 207–215.

215. M. S. Goldberg, *Distributive double p-algebras whose congruence lattices are chains*, Algebra Universalis 17 (1983), no. 2, 208–215.

216. G. Grätzer, *Lattice Theory: First Concepts and Distributive Lattices*, Freeman, San Francisco, 1971.

217. G. Grätzer and E. T. Schmidt, *Ideals and congruence relations in lattices*, Acta Math. Hungar. 9 (1958), 137–175.

218. J. R. Griggs, *Maximum antichains in the product of chains*, Order 1 (1984), no. 1, 21–28.

219. H. Gross and H. A. Keller, *On the problem of classifying infinite chains in projective and orthogonal geometry*, Ann. Acad. Sci. Fenn. Ser. A I Math. 8 (1983), no. 1, 67–86.

220. H.-P. Gumm and W. Poguntke, *Boolesche Algebra*, Bibliographisches Institut, Mannheim, 1981.

221. Y. Gurevich, *Decision problem for separated distributive lattices*, J. Symbolic Logic 48 (1983), no. 1, 193–196.

222. J. G. Hagendorf, *Extensions immédiates de chaînes*, Z. Math. Logik Grundlag. Math. 28 (1982), no. 1, 15–44.

223. W. Hanf, *Primitive Boolean algebras* (Proc. Sympos. Pure Math., Univ. of Calif., Berkeley, CA, 1971), vol. XXV, Amer. Math. Soc., Providence, RI, 1974, pp. 75–90.

224. G. Hansoul, *Algèbres de Boole primitives*, Discrete Math. 53 (1985), 103–116.

225. _____, *A duality for Boolean algebras with operators*, Algebra Universalis 17 (1983), no. 1, 34–49.

226. G. Hansoul and L. Vrancken-Mawet, *Décompositions Booléennes de lattis distributifs bornés*, Bull. Soc. Roy. Sci. Liège 53 (1984), no. 2, 88–92.

227. H. Hashimoto, *Subuniverses of fuzzy matrices*, Fuzzy Sets and Systems 12 (1984), no. 2, 155–158.

228. J. Hashimoto, *Direct subdirect decompositions and congruence relations*, Osaka J. Math. 9 (1957), 87–112.

229. M. Havier and T. Katriňák, *Lattices whose congruence lattice is relative Stone*, Acta Sci. Math. (Szeged) 51 (1987), no. 1–2, 81–91.

230. J. Havrda, *Independence in a set with orthogonality*, Časopis. Pěst. Mat. 107 (1982), no. 3, 267–272.

231. J. Hawranek and J. Zygmunt, *Some elementary properties of conditionally distributive lattices*, Polish Acad. Sci. Inst. Philos. Sociol. Bull. Sect. Logic 12 (1983), no. 3, 117–121.

232. L. Heindorf, *Regular ideals and Boolean pairs*, Z. Math. Logik Grundlag. Math. 30 (1984), no. 6, 547–560.

233. _____, *Beiträge zur Modelltheorie der Booleschen Algebren*, Seminarberichte, vol. 53, Humboldt Universität, Berlin, 1984.

234. H. E. Hendry, *Complete extensions of the calculus of individuals*, Noûs 16 (1982), no. 3, 453–460.

235. C. Herrmann, *A characterization of distributivity for modular polarity lattices*, Contributions to Lattice Theory (Szeged, 1980), Colloq. Math. Soc. János Bolyai, vol. 33, North-Holland, Amsterdam and New York, 1983, pp. 473–490.

236. U. Höhle, *Compact g-fuzzy topological spaces*, Fuzzy Sets and Systems 13 (1984), no. 1, 39–63.

237. C. S. Hoo, *Pseudocomplemented and implicative semilattices*, Canad. J. Math. 34 (1982), no. 2, 423–437.

238. _____, *Atoms, primes and implicative lattices*, Canad. Math. Bull. **27** (1984), no. 3, 279–285.

239. C. S. Hoo and P. V. Ramana Murty, *Modular and admissible semilattices*, Canad. J. Math. **36** (1984), no. 5, 795–799.

240. A. P. Huhn, *A reduced free product of distributive lattices*, Acta Math. Hungar. **42** (1983), no. 3–4, 349–354.

241. B. Hutton, *Uniformities on fuzzy topological spaces*, J. Math. Anal. Appl. **58** (1977), no. 3, 559–571.

242. Iqbalunnisa, *On lattices whose lattice of congruences are Stone lattices*, Fund. Math. **79** (1971), 315–318.

243. J. Isbell, *A frame with no admissible topology*, Math. Proc. Cambridge Philos. Soc. **94** (1983), no. 3, 447–448.

244. _____, *Graduation and dimension in locales*, Aspects of Topology, London Math. Soc. Lecture Note Ser., vol. 93, Cambridge University Press, Cambridge, 1985, pp. 195–210.

245. _____, *Review on "Stone spaces", by P. T. Johnstone*, Bull. Amer. Math. Soc. **11** (1984), 389–392.

246. L. Iturrioz, *A simple proof of a characterization of complete orthocomplemented lattices*, Bull. London Math. Soc. **14** (1982), no. 6, 542–544.

247. _____, *A topological representation theory for orthomodular lattices*, Contributions to Lattice Theory (Szeged, 1980), Colloq. Math. Soc. János Bolyai, vol. 33, North-Holland, Amsterdam and New York, 1983, pp. 503–524.

248. M. F. Janowitz, *Complemented congruences on complemented lattices*, Pacific J. Math. **73** (1977), 87–90.

249. C. Jayaram, *Quasicomplemented semilattices*, Acta Math. Hungar. **39** (1982), no. 1–3, 39–47.

250. _____, *Semilattices and finite Boolean algebras*, Algebra Universalis **16** (1983), no. 3, 390–394.

251. C. G. Jockusch and J. Mohrherr, *Embedding the diamond lattice in the recursively enumerable truth-table degrees*, Proc. Amer. Math. Soc. **94** (1985), no. 1, 123–128.

252. P. T. Johnstone, *Stone Spaces*, Cambridge Studies in Advanced Math., vol. 3, Cambridge Univ. Press, New York, 1983.

253. _____, *The point of pointless topology*, Bull. Amer. Math. Soc. (N. S.) **8** (1983), no. 1, 41–53.

254. G. T. Jones, *Projective pseudocomplemented semilattices*, Pacific J. Math. **52** (1974), 442–456.

255. B. Jónsson and A. Tarski, *Boolean algebras with operators*, Amer. J. Math. **73** (1951), 891–939.

256. A. Joyal and M. Tierney, *An extension of the Galois theory of Grothendieck*, Mem. Amer. Math. Soc. **51** (1984), no. 309.

257. P.-F. Jurie, A. Touraille, and R. Garnier, *Idéaux elémentairement équivalents dans une algèbre Booléenne*, C. R. Acad. Sci. Paris Sér. I Math. **299** (1984), no. 10, 415–418.

258. W. Just and A. Krawczyk, *On certain Boolean algebras $\mathscr{P}(\omega)$ I*, Trans. Amer. Math. Soc. **285** (1984), no. 1, 411–429.

259. K. Kaiser, *Lattices acting on universal classes*, Z. Math. Logik Grundlag. Math. **27** (1981), no. 2, 127–130.

260. _____, *On complementedly normal lattices. II: extensions*, Z. Math. Logik Grundlag. Math. **30** (1984), no. 6, 567–573.

261. I. G. Kalmár, *Conditional probability measures on propositional systems*, Publ. Math. Debrecen **30** (1983), no. 1–2, 101–115.

262. _____, *On random variables defined on the atom space of an orthomodular atomistic σ-lattice*, Publ. Math. Debrecen **31** (1984), no. 1–2, 85–93.

263. _____, **-structures and orthomodular lattices*, Publ. Math. Debrecen **32** (1985), no. 1–2, 1–5.

264. G. Kalmbach, *Orthomodular lattices do not satisfy any special lattice equation*, Arch. Math. (Basel) **28** (1977), 7–8.

265. _____, *Orthomodulare Verbände*, Jahresber. Deutsch. Math. Verein. **85** (1983), no. 1, 33–49.

266. _____, *Orthomodular Lattices*, Academic Press, London, 1983.

267. _____, *Automorphism groups of orthomodular lattices*, Bull. Austral. Math. Soc. **29** (1984), no. 3, 309–313.

268. M. Kamara, *Nichtdistributive, modulare Polaritätsverbände*, Arch. Math. (Basel) **39** (1982), no. 2, 126–133.

269. _____, *Spaltende modulare Polaritätsverbände*, Contributions to General Algebra, 2 (Klagenfurt, 1982), Hölder-Pichler-Tempsky, Vienna, 1983, pp. 179–190.

270. T. Katriňák, *Notes on Stone lattices*, II, Mat. Časopis Sloven. Akad. Vied **17** (1967), 20–37.

271. _____, *Free p-algebras*, Algebra Universalis **15** (1982), no. 2, 176–186.

272. _____, *Complemented p-algebras*, Acta Math. Univ. Comenian. **44/45** (1984), 37–38.

273. _____, *Über eine Konstruktion der distributiven pseudokomplementären Verbände*, Math. Nachr. **53** (1972), 85–99.

274. _____, *Splitting p-algebras*, Algebra Universalis **18** (1984), no. 2, 199–224.

275. T. Katriňák and S. El-Assar, *Algebras with Boolean and Stonean congruence lattices*, Acta Math. Hungar. **48** (1986), no. 3-4, 301–316.

276. _____, *p-algebras with Stone congruence lattices*, Acta Sci. Math. (Szeged) **51** (1987), no. 3-4, 371–387.

277. T. Katriňák and P. Mederly, *Constructions of p-algebras*, Algebra Universalis **17** (1983), no. 3, 288–316.

278. T. Katriňák and K. Mikula, *On a construction of MS-algebras*, Portugal. Math. **45** (1988), no. 2, 157–163.

279. F. Katrnoška, *On the representation of orthocomplemented posets*, Comment. Math. Univ. Carolin. **23** (1982), no. 3, 489–498.

280. _____, *M-base and the representation of orthocomplemented posets*, General Topology and its Relations to Modern Analysis and Algebra, V (Prague, 1981), Sigma Ser. Pure Math., vol. 3, Heldermann, Berlin, 1983, p. 434.

281. M. Katz, *Deduction systems and valuation spaces*, Logique et Anal. (N.S.) **26** (1983), no. 102, 157–175.

282. D. Kelly, *A note on equationally compact lattices*, Algebra Universalis **2** (1972), 80–84.

283. D. C. Kent and C. R. Atherton, *The order topology in a bicompactly generated lattice*, J. Austral. Math. Soc. **8** (1968), no. 2, 345–349.

284. J. B. Kim, *A certain matrix semigroup*, Math. Japon. **22** (1978), no. 5, 519–522.

285. _____, *On circulant fuzzy matrices*, Math. Japon. **24** (1979), no. 1, 35–40.

286. _____, *Note on the semigroup of fuzzy matrices*, J. Korean Math. Soc. **16** (1979/80), no. 1, 1–7.

287. _____, *On the semigroup of the circulant fuzzy matrices*, Bull. Malaysian Math. Soc. (2) **4** (1981), no. 1, 9–15.

288. J. B. Kim and Y. M. Lee, *Combinatorial properties of a fuzzy matrix semigroup*, Proc. 10th Southeastern Conf. Combinatorics, Graph Theory and Computing (Boca Raton, FL, 1979), Utilitas Math. (1979), Winnipeg, Manitoba, 569–576.

289. K. H. Kim, *Boolean Matrix Theory and Applications*, M. Dekker, New York, 1982.

290. K. H. Kim and F. W. Roush, *Generalized fuzzy matrices*, Fuzzy Sets and Systems **4** (1980), no. 3, 293–315.

291. E. W. Kiss, *Finitely Boolean representable varieties*, Proc. Amer. Math. Soc. **89** (1983), no. 4, 579–582.

292. E. W. Kiss, L. Márki, P. Pröhle, and W. Tholen, *Categorical algebraic properties. A compendium on amalgamation, congruence extensions, epimorphisms, residual smallness and injectivity*, Studia Sci. Math. Hungar. **18** (1983), 79–141.

293. E. Köhler, *Orthomodulare Verbände mit Regularitätsbedingungen*, J. Geom. **19** (1982), no. 2, 130–145.

294. P. Köhler, *Brouwerian semilattices: The lattice of total subalgebras*, Universal Algebra and Applications (Warsaw, 1978), Banach Center Publ., 9, PWN, Warsaw, 1982, pp. 47–56.

295. M. Kolibiar, *Weak homomorphisms in some classes of algebras*, Studia Sci. Math. Hungar. **19** (1984), 413–420.

296. R. König, *Theorie der formalen Sprachen*, Arbeitsber. Inst. Math. Masch. Datenverarb. Inform. **16** (1983), no. 2.

297. S. Koppelberg, *On Boolean algebras with distinguished subalgebras*, Enseign. Math. (2) **28** (1982), no. 3-4, 233–252.

298. _____, *On Boolean algebras with distinguished subalgebras*, Monograph Enseign. Math. (1982), no. 30, 297–316.

299. _____, *Groups of permutations with few fixed points*, Algebra Universalis **17** (1983), no. 1, 50–64.

300. S. Koppelberg, R. McKenzie, and J. D. Monk, *Cardinality and cofinality of homomorphs of products of Boolean algebras*, Algebra Universalis **19** (1984), no. 1, 38–44.

301. S. Koppelberg and J. D. Monk, *Homogeneous Boolean algebras with very nonsymmetric subalgebras*, Notre Dame J. Formal Logic **24** (1983), no. 3, 353–356.

302. V. Koubek and J. Sichler, *Universal varieties of distributive double p-algebras*, Glasgow Math. J. **26** (1985), 121–131.

303. D. Krausser, *On orthomodular amalgamations of Boolean algebras*, Arch. Math. (Basel) **39** (1982), no. 1, 92–96.

304. K. P. Kuhn, *Extending homomorphisms from orthomodular lattices to Foulis-semigroups*, Contributions to General Algebra, 2 (Klagenfurt, 1982), Hölder-Pichler-Tempsky, Vienna, 1983, pp. 229–232.

305. R. Kumar, *R.e. presented linear orders*, J. Symbolic Logic **48** (1983), no. 2, 369–376.

306. F. Lacava, *Alcune proprietà delle algebre di Boole principali*, Atti Accad. Naz. Lincei Rend. Cl. Sci. Fis. Mat. Natur. (8) **74** (1983), no. 3, 131–135.

307. A. H. Lachlan, *The elementary theory of recursively enumerable sets*, Duke Math. J. **35** (1968), no. 1, 123–146.

308. H. Lakser, *The semilattice tensor product of projective distributive lattices*, Algebra Universalis **13** (1981), no. 1, 78–81.

309. H. Länger, *Klassen von Baer ∗-Halbgruppen und orthomodularen Verbänden*, Österreich. Akad. Wiss. Math.-Natur. Kl. Sitzungsber. II **192** (1983), no. 1–3, 17–24.

310. R. Lidl and G. Pilz, *Applied Abstract Algebra*, Springer-Verlag, New York and Berlin, 1984.

311. P. Lipparini, *Existentially complete closure algebras*, Boll. Un. Mat. Ital. D (6) **1** (1982), no. 1, 13–19.

312. F. Ch. Liu, *Embedding distributive lattices in vector lattices*, Bull. Inst. Math. Acad. Sinica **11** (1983), no. 3, 459–462.

313. W. J. Liu, *Fuzzy invariant subgroups and fuzzy ideals*, Fuzzy Sets and Systems **8** (1982), no. 2, 133–139.

314. Y. M. Liu, *A formula for the intersection operation on union-preserving mappings in lattices*, Fuzzy Math. **3** (1983), no. 2, 43–46.

315. R. D. Luce, *A note on Boolean matrix theory*, Proc. Amer. Math. Soc. **3** (1952), no. 3, 383–388.

316. E. W. Madison, *The existence of countable totally nonconstructive extensions of the countable atomless Boolean algebra*, J. Symbolic Logic **48** (1983), no. 1, 167–170.

317. K. D. Magill, *The semigroups of endomorphisms of a Boolean ring*, J. Austral. Math. Soc. **11** (1970), 411–416.

318. P. Mangani and A. Marcja, *Shela rank for Boolean algebras and some applications to elementary theories* I, Algebra Universalis **10** (1980), no. 2, 247–257.

319. A. Marcja and C. Toffalori, *On Cantor-Bendixson spectra containing* $(1, 1)$. I, Lecture Notes in Math., vol. 1103, Springer-Verlag, New York and Berlin, 1984, pp. 331–350.

320. _____, *On Cantor-Bendixson spectra containing* $(1, 1)$. II, J. Symbolic Logic **50** (1985), no. 3, 611–618.

321. _____, *On pseudo* $\aleph_0$ *categorical theories*, Z. Math. Logik Grundlag. Math. **30** (1984), no. 6, 533–540.

322. C. J. Maxson, *On semigroups of Boolean ring endomorphisms*, Semigroup Forum **4** (1972), 79–82.

323. R. Mayet, *Une dualité pour les ensembles ordonnés orthocomplémentés*, C. R. Acad. Sci. Paris Sér. I Math. **294** (1982), no. 2, 63–65.

324. _____, *Classes équationnelles de treillis orthomodulaires liées aux états*, Publ. Dép. Math. (Lyon) (N. S.) **2/A** (1983).

325. _____, *Varieties of orthomodular lattices related to states*, Algebra Universalis **20** (1985), 368–396.

326. R. McKenzie, *Equational bases and nonmodular lattice varieties*, Trans. Amer. Math. Soc. **174** (1972), 1–42.

327. R. McKenzie and D. Monk, *Chains in Boolean algebras*, Ann. Math. Logic **22** (1982), no. 2, 137–175.

328. S. Meskhi, *Complete Boolean powers of quasi-primal algebras with lattice reducts*, Algebra Universalis **14** (1982), no. 3, 388–390.

329. Ž. Mijajlović, *Two remarks on Boolean algebras*, Algebraic Conference (Skopje, 1980), Univ. Kiril et Metodij, Skopje, 1980, pp. 35–41.

330. A. W. Miller and K. Prikry, *When the continuum has cofinality* ω_1, Pacific J. Math. **115** (1984), no. 2, 399–407.

331. P. Mittelstaedt, *Relativistic quantum logic*, Internat. J. Theoret. Phys. **22** (1983), no. 4, 293–314.

332. B. Molzan, *The theory of superatomic Boolean algebras in the logic with the binary Ramsey quantifier*, Z. Math. Logik Grundlag. Math. **28** (1982), no. 4, 365–376.

333. J. D. Monk, *Independence in Boolean algebras*, Period. Math. Hungar. **14** (1983), no. 3–4, 269–308.

334. F. Montagna, *The wellfounded algebras*, Algebra Universalis **16** (1983), no. 1, 38–46.

335. _____, *A completeness result for fixed-point algebras*, Z. Math. Logik Grundlag. Math. **30** (1984), no. 6, 525–532.

336. T. Mori, *Results on a compact space which has certain properties and their application to superatomic Boolean algebras*, Rep. Fac. Sci. Engrg. Saga Univ. Math. (1983), no. 11, 1–8.

337. M. K. Mukherjee, *A generalized characterization theorem for quantum logic*, Lett. Nuovo Cimento (2) **40** (1984), no. 15, 453–456.

338. Ch. J. Mulvey and J. W. Pelletier, *On the points of focales in a De Morgan topos*, Categorical Topology (Toledo, Ohio, 1983), Sigma Ser. Pure Math., vol. 5, Heldermann, Berlin, 1984, pp. 392–407.

339. P. V. R. Murty, *On the triple construction of distributive p-algebras*, Algebra Universalis **22** (1985), 229–234.

340. P. V. R. Murty and T. Engelbert, *On "Constructions of p-algebras"*, Algebra Universalis **22** (1985), 215–228.

341. P. V. R. Murty and V. Raman, *Triple construction of semilattices with 1 admitting neutral p-closure operators*, Math. Slovaca **32** (1982), no. 4, 367–378.

342. P. Năslău, E. Burlacu, and S. Matei, *The semiderivatives of Boolean functions*, Inst. Politehn. "Traian Vuia" Timişoara. Lucrăr. Sem. Mat. Fiz. **1982**, 23–24.

343. W. Nemitz, *Extension of Brouwerian semilattices*, Houston J. Math. **10** (1984), 545–558.

344. A. Nerode and J. Remmel, *A survey of lattices of r. e. substructures*, Recursion Theory (Ithaca, NY, 1982), Proc. Sympos. Pure Math., vol. 42, Amer. Math. Soc., Providence, RI, 1985, pp. 323–375.

345. H. Nishimura, *Sequential method in quantum logic*, J. Symbolic Logic **45** (1980), no. 2, 339–352.

346. T. Nordahl, *Lattice representation in exchange transfusion*, Semigroup Forum **23** (1981), no. 3, 275.

347. D. Normann, *R. e. degrees of continuous functionals*, Arch. Math. Logik Grundlag. **23** (1983), no. 1–2, 79–98.

348. U. Oberst, *Anwendungen des chinesischen Restsatzes*, Exposition Math. **3** (1985), no. 2, 97–148.

349. A. J. F. de Oliveira, *A note on cyclic Boolean algebras*, Proc. 8th Portuguese-Spanish Conf. Mathematics, vol. I (Coimbra, 1981), Univ. Coimbra, Coimbra, 1981, pp. 153–161.

350. R. Padmanabhan, *A first order proof of a theorem of Frink*, Algebra Universalis **13** (1981), no. 3, 397–400.

351. _____, *A self-dual equational basis for Boolean algebras*, Canad. Math. Bull. **26** (1983), no. 1, 9–12.

352. G. H. Pakyh and Y. S. Park, *Ordered topological representations of distributive lattices*, Kyungpook Math. J. **23** (1983), no. 2, 109–113.

353. W. Peters, *On the (0)-topology on ABS-Boolean algebras*, Proc. Conf. Topology and Measure III, Part 1, 2 (Vitte/Hiddensee, 1980), Ernst Moritz Arndt Univ., Greifswald, 1982, pp. 193–199.

354. L. Pezzoli, *On D-complementation*, Adv. in Math. **51** (1984), no. 3, 226–239.

355. C. I. Picu, *Characterization of distributive lattices by equivalences and implications*, Stud. Cerc. Mat. **34** (1982), no. 4, 367–369.

356. R. S. Pierce, *Compact zero-dimensional metric spaces of finite type*, Mem. Amer. Math. Soc. (1972), no. 130.

357. _____, *Tensor products of Boolean algebras*, Lecture Notes in Math., vol. 1004, Springer-Verlag, New York and Berlin, 1983, pp. 232–239.

358. J. Płonka, *On mixed product of Boolean algebras*, Colloq. Math. **23** (1971), no. 1, 29–32.

359. W. Poguntke, *Finitely generated ortholattices: The commutator and some applications*, Contributions to Lattice Theory (Szeged, 1980), Colloq. Math. Soc. János Bolyai, vol. 33, North-Holland, Amsterdam and New York, 1983, pp. 651–666.

360. W. B. Powell and C. Tsinakis, *The distributive lattice free product as a sublattice of the Abelian l-group free product*, J. Austral. Math. Soc. Ser. A **34** (1983), no. 1, 92–100.

361. _____, *Free products of lattice ordered groups*, Algebra Universalis **18** (1984), no. 2, 178–198.

362. R. Prata dos Santos, *On certain sets of congruences on the pseudocomplemented algebra*, Bull. Soc. Roy. Sci. Liège **54** (1985), 119–126.

363. _____, *Congruences of pseudocomplemented algebras*, Portugal. Math. **42** (1983).

364. H. A. Priestley, *Ordered sets and duality for distributive lattices*, Ann. Discrete Math. **23** (1984), 39–60.

365. _____, *Catalytic distributive lattices and compact zero-dimensional topological lattices*, Algebra Universalis **19** (1984), no. 3, 322–329.

366. H. J. Prömel and B. Voigt, *Canonical partition theorems for finite distributive lattices*, Rend. Circ. Mat. Palermo (2) Suppl. **31** (1982), no. 2, 223–237.

367. P. Pták, *Categories of orthomodular posets*, Math. Slovaca **35** (1985), no. 1, 59–65.

368. Ch. Pühringer, *Ein Vollständigkeitsbeweis für die Theorie der atomlosen $\cap^{-c}$-partiellen Booleschen Algebren*, Conceptus **16** (1982), no. 38, 81–88.

369. S. Pulmannová, *Superpositions of states and a representation theorem*, Ann. Inst. H. Poincaré Sect. A **32** (1980), no. 4, 351–360.

370. _____, *Compatibility and partial compatibility in quantum logics*, Ann. Inst. H. Poincaré Sect. A **34** (1981), no. 4, 391–403.

371. _____, *On representations of logics*, Math. Slovaca **33** (1983), no. 4, 357–362.

372. _____, *Commutators in orthomodular lattices*, Demonstratio Math. **18** (1985), 187–208.

373. A. Pultr, *Remarks on metrizable locales*, Rend. Circ. Mat. Palermo (2) Suppl. **33** (1984), no. 6, 247–258.

374. _____, *Pointless uniformities I. Complete regularity*, Comment. Math. Univ. Carolin. **25** (1984), no. 1, 91–104.

375. _____, *Pointless uniformities II. Dia-metrization*, Comment. Math. Univ. Carolin. **25** (1984), no. 1, 105–120.

376. M. Ramalho, *Radicals in pseudocomplemented lattices*, Algebra Universalis **20** (1985), 243–253.

377. W. Rehder, *Von Wright's "and next" versus a sequential tense-logic*, Logique et Anal. (N.S.) **25** (1982), no. 97, 33–46.

378. Ch. Reutenauer and H. Straubing, *Inversion of matrices over commutative semiring*, J. Algebra **88** (1984), no. 2, 350–360.

379. M. Roddy, *An orthomodular analogue of the Birkhoff-Menger theorem*, Algebra Universalis **19** (1984), no. 1, 55–60.

380. _____, *Varieties of modular ortholattices*, Order **3** (1987), no. 4, 405–426.

381. V. Rödl, *A note on finite Boolean algebras*, Acta Polytech. Práce ČVUT Praze Ser. IV Tech. Teoret. **15** (1982), no. 1, 47–50.

382. V. Rogalewicz, *Remarks about measures on orthomodular posets*, Časopis Pěst. Mat. **109** (1984), no. 1, 93–99.

383. A. Romanowska, *Algebras of functions from partially ordered sets into distributive lattices*, Universal Algebra and Lattice Theory (Puebla, 1982), Lecture Notes in Math., vol. 1004, Springer-Verlag, New York and Berlin, 1983, pp. 245–256.

384. M. Rubin, *The theory of Boolean algebras with a distinguished subalgebra is undecidable*, Ann. Sci. Univ. Clermont II Math. **60** (1976), no. 13, 129–134.

385. _____, *On the reconstruction of Boolean algebras from their automorphism groups*, Arch. Math. Logik Grundlag. **20** (1980), no. 3-4, 125–146.

386. _____, *A Boolean algebra with few subalgebras, interval Boolean algebras and retractiveness*, Trans. Amer. Math. Soc. **278** (1983), no. 1, 65–89.

387. D. E. Rutherford, *Inverses of Boolean matrices*, Proc. Glasgow Math. Assoc. **6** (1983), no. 1, 49–53.

388. _____, *The Cayley-Hamilton theorem for semi-rings*, Proc. Roy. Soc. Edinburgh Sect. A **66** (1963), no. 1, 49–53.

389. _____, *The eigenvalue problem for Boolean matrices*, Proc. Roy. Soc. Edinburgh Sect. A **67** (1964), no. 1, 25–38.

390. _____, *Orthogonal Boolean matrices*, Proc. Roy. Soc. Edinburgh Sect. A **67** (1964), no. 2, 126–135.

391. H. P. Sankappanavar, *On pseudocomplemented semilattices with Stone congruence lattices*, Math. Slovaca **29** (1979), 381–395.

392. _____, *Congruence-semimodular and congruence-distributive pseudocomplemented semilattices*, Algebra Universalis **14** (1982), no. 1, 68–81.

393. _____, *Heyting algebras with dual pseudocomplementation*, Pacific J. Math. **117** (1985), 405–415.

394. _____, *Pseudocomplemented Ockham and De Morgan algebras*, Z. Math. Logik Grundlag. Math. **32** (1986), no. 5, 385–394.

395. B. M. Schein, *Ordered sets, semilattices, distributive lattices and Boolean algebras with homomorphic endomorphisms semigroups*, Fund. Math. **68** (1970), 31–50.

396. J. H. Schmerl, $\aleph_0$ *categorical distributive lattices of finite breadth*, Proc. Amer. Math. Soc. **87** (1983), no. 4, 707–713.

397. J. Schmid, *Model companions of distributive p-algebras*, J. Symbolic Logic **47** (1982), no. 3, 680–688.

398. _____, *Algebraically closed distributive p-algebras*, Algebra Universalis **15** (1982), no. 1, 126–141.

399. _____, *On the structure of free p-semilattices*, Houston J. Math. **16** (1990), no. 1, 71–85.

400. _____, *Algebraically closed p-semiliattices*, Arch. Math. (Basel) **45** (1985), 501–510.

401. _____, *Lee classes and sentences for p-semilattices*, Preprint.

402. _____, *Fast subdirect decomposition of pseudocomplemented semilattices*, Contributions to General Algebra, 5 (Salzburg, 1986), Hölder-Pichler-Tempsky, Vienna, 1987, pp. 321–353.

403. M. G. Schwarze and R. Chuaqui, *Axiomatizations for σ-additive measurement structures*, Mathematical Logic in Latin America (Proc. IV Latin Amer. Sympos. Math. Logic, Santiago, 1978), North-Holland, Amsterdam, 1980, pp. 351–364.

404. D. Schweigert and M. Szymańska, *On distributive correlation lattices*, Contributions to Lattice Theory (Szeged, 1980), Colloq. Math. Soc. János Bolyai, vol. 33, North-Holland, Amsterdam and New York, 1983, pp. 697–721.

405. D. S. Scott, *Background to formalization*, Truth, Syntax and Modality (Proc. Conf. on Alternative Semantics, Temple Univ., Philadelphia, PA, 1970), Studies in Logic and the Foundations of Math., vol. 68, North-Holland, Amsterdam, 1973, pp. 244–273.

406. D. Seese, H.-P. Tuschik, and M. Weese, *Undecidable theories in stationary logic*, Proc. Amer. Math. Soc. **84** (1982), no. 4, 563–567.

407. M. Sekanina, *On a characterization of the system of all regularly closed sets in general closure spaces*, Math. Nachr. **38** (1968), 61–66.

408. A. Sendlewski, *Pretabular varieties of N-lattices*, Polish Acad. Sci. Inst. Philos. Sociol. Bull. Sect. Logic **12** (1983), no. 1, 17-20.

409. S. S. Sengupta, *Lattices, bargaining and group decisions*, Theory and Decision **16** (1984), no. 2, 111-134.

410. B. Šešelja and G. Vojvodić, *Fuzzy generalized equivalence relations and partitions*, Univ. u Novom Sadu Zb. Rad. Prirod.-Mat. Fak. Ser. Mat. **11** (1981), 267-273.

411. S. Shelah, *Classification theory and the number of nonisomorphic models*, Studies in Logic and the Foundations of Mathematics, vol. 92, North-Holland, Amsterdam and New York, 1978.

412. _____, *Remarks on Boolean algebras*, Algebra Universalis **11** (1980), no. 1, 77-89.

413. _____, *Models with second order properties IV. A general method and eliminating diamonds*, Ann. Pure Appl. Logic **325** (1983), no. 2, 183-212.

414. _____, *Constructions of many complicated uncountable structures and Boolean algebras*, Israel J. Math. **45** (1983), no. 2-3, 100-146.

415. N. D. Shi, *Creative pairs of the subalgebras of recursively enumerable Boolean algebras*, Acta Math. Sinica **25** (1982), no. 6, 737-745.

416. H. Simmons, *A couple of triples*, Topology Appl. **13** (1982), no. 2, 201-223.

417. J. Šimon, *Opérations dérivées des treillis orthomodulaires (Part I)*, Acta Univ. Carolin.—Math. Phys. **22** (1981), no. 2, 7-14.

418. _____, *Opérations dérivées des treillis orthomodulaires (Part II)*, Acta Univ. Carolin.—Math. Phys. **23** (1982), no. 1, 29-36.

419. P. Simon and M. Weese, *Nonisomorphic thin-tall superatomic Boolean algebras*, Comment. Math. Univ. Carolin. **26** (1985), no. 2, 241-252.

420. C. Smoryński, *Fixed point algebras*, Bull. Amer. Math. Soc. (N.S.) **6** (1982), no. 3, 317-356.

421. L. Sobolska, *On mixed product of n Boolean algebras*, Demonstratio Math. **17** (1984), no. 1, 85-96.

422. A. Stairs, *On the logic of pairs of quantum systems*, Synthese **56** (1983), no. 1, 47-60.

423. P. Štěpánek, *Boolean algebras with no rigid or homogeneous factors*, Trans. Amer. Math. Soc. **270** (1982), no. 1, 131-147.

424. D. Strojewski, *Numerical representations of orthomodular lattices and Boolean algebras with infinite operations*, Bull. Polish Acad. Sci. Math. **33** (1985), 341-348.

425. K. L. N. Swamy and V. Swaminathan, *On compactly packed lattices and autometrized algebras*, Math. Sem. Notes Kobe Univ. **10** (1982), no. 2, 543-549.

426. G. Takeuti, *Quantum set theory*, Current Issues in Quantum Logic (Erice, 1979), Ettore Majorana Internat. Sci. Ser. Phys. Sci., vol. 8, Plenum, New York and London, 1981, pp. 303-322.

427. _____, *Boolean completion and m-convergence*, Lecture Notes in Math., vol. 915, Springer-Verlag, New York and Berlin, 1982, pp. 333-350.

428. R. Talamo, *On σ-saturations of a Boolean algebra*, Rend. Circ. Mat. Palermo (2) **33** (1984), no. 2, 201-210.

429. S. Tamura, *Decision procedure for pseudocomplemented lattices*, Proceedings of the 8th Symposium on Semigroups (Matsue, 1984), Shimane Univ., Matsue, 1985, pp. 36-39.

430. A. Gentzen, *Formulation without the cut rule for ortholattices*, Kobe J. Math. **5** (1988), no. 1, 133-150.

431. T. Tanaka, *Canonical subdirect factorization of lattices*, J. Sci. Hiroshima Univ. Ser. A **16** (1952), 239-246.

432. B. Tembrowski, *On some class of Boolean algebras with an additional binary operation*, Demonstratio Math. **15** (1982), no. 1, 189-206.

433. _____, *Normal ultrafilters in B_3-algebras*, Demonstratio Math. **16** (1983), no. 3, 601-617.

434. N. K. Thakare and Y. S. Pawar, *Minimal prime ideals in 0-distributive semilattices*, Period. Math. Hungar. **13** (1982), no. 3, 237-246.

435. J. Tiller, *Augmented compact spaces and continuous lattices*, Houston J. Math. **7** (1981), no. 3, 411-453.

436. S. Todorčević, *On minimal separating Boolean algebras*, Publ. Inst. Math. (Beograd) (N.S.) **29(43)** (1981), 241-247.

437. R. Tošić, *Constant-preserving Boolean functions over the finite Boolean algebras*, Math. Balkanica **8** (1978), 227–234.

438. A. Touraille, *Élimination des quantificateurs dans la théorie élémentaire des algèbres de Boole munies d'une famille d'idéaux distingués*, C. R. Acad. Sci. Paris Sér. I Math. **300** (1985), no. 5, 125–128.

439. S. Tulipani, *Model-completions of the theories of finitely additive measures with values in an ordered field*, Z. Math. Logik Grundlag. Math. **27** (1981), no. 6, 481–488.

440. _____, *A use of the method of interpretations for decidability or undecidability of measure spaces*, Algebra Universalis **15** (1982), no. 2, 228–232.

441. A. Urquhart, *Distributive lattices with a dual homomorphic operation* II, Studia Logica **46** (1981), 391–404.

442. _____, *Equational classes of distributive double p-algebras*, Algebra Universalis **14** (1982), no. 2, 235–243.

443. J. Varlet, *Distributive semilattices and Boolean lattices*, Bull. Soc. Roy. Sci. Liège **41** (1972), 5–10.

444. _____, *Regularity in p-algebras and p-semilattices*, Universal Algebra and Applications (Warsaw, 1978), Banach Center Publ., 9, PWN, Warsaw, 1982, pp. 369–278.

445. _____, *Regularity in double p-algebras*, Algebra Universalis **18** (1984), no. 1, 95–105.

446. M. van de Vel, *Binary convexities and distributive lattices*, Proc. London Math. Soc. **48** (1984), no. 1, 1–33.

447. _____, *Dimension of binary convex structures*, Proc. London Math. Soc. **48** (1984), no. 1, 34–54.

448. K. Venkataraman, *The Boolean algebras $A(G)$, $F(G)$*, J. Indian Math. Soc. **44** (1980), no. 1-4, 275–280.

449. P. Vojtáš, *A transfinite Boolean game and a generalization of Kripke's embedding theorem*, General Topology and its Relations to Modern Analysis and Algebra, V (Prague, 1981), Sigma Ser. Pure Math., vol. 3, Heldermann, Berlin, 1983, pp. 657–662.

450. _____, *Game properties of Boolean algebras*, Comment. Math. Univ. Carolin. **24** (1983), no. 2, 349–369.

451. L. Vrancken-Mawet, *Sur des congruences d'un ensemble ordonné. Application à l'étude du lattis des sous-algèbres d'un demi-lattis de Brouwer fini*, Bull. Soc. Roy. Sci. Liège **51** (1982), no. 5-8, 174–187.

452. _____, *Le lattis des sous-algèbres d'une algèbre de Heyting finie*, Bull. Soc. Roy. Sci. Liège **51** (1982), no. 1-2, 82–94.

453. L. Vrancken-Mawet and G. Hansoul, *The subalgebra lattice of a Heyting algebra*, Czechoslovak Math. J. **37** (1987), no. 1, 34–41.

454. Guo Juń Wang, *Order-homomorphisms on fuzzes*, Fuzzy Sets and Systems **12** (1984), no. 3, 281–288.

455. J. H. M. Wedderburn, *Boolean linear associative algebra*, Ann. Math. **35** (1934), no. 1, 185–194.

456. M. Weese, *Undecidable extensions of the theory of Boolean algebras*, Handbook of Boolean Algebras, vol. 3, North-Holland, Amsterdam and New York, 1989, pp. 1067–1096.

457. _____, *The theory of Boolean algebras extended by a group of automorphisms*, Seminarberichte, vol. 60, Humboldt Universität, Berlin, 1984, pp. 218–222.

458. M. Weese and H.-J. Goltz, *Boolean algebras*, Seminarberichte, vol. 62, Humboldt Universität, Berlin, 1984.

459. B. Wolniewicz, *The Boolean algebra of objectives*, Polish Acad. Sci. Inst. Philos. Sociol. Bull. Sect. Logic **10** (1981), no. 1, 17–23.

460. _____, *A formal ontology of situations*, Studia Logica **41** (1982), no. 4, 381–413.

461. F. M. Yaqub, *α-representable coproducts of distributive lattices*, Proc. Edinburgh Math. Soc. (2) **25** (1982), no. 3, 229–235.

462. J. W. Zhang, *Fuzzy set structure with strong implication*, Advances in Fuzzy Sets, Possibility Theory, and Applications, Plenum, New York and London, 1983, pp. 107–136.

463. U. Zimmermann and P. Köhler, *Products of finitely based varieties of Brouwerian semilattices*, Algebra Universalis **18** (1984), 110–116.

464. T. Katriňák, *Corrigendum to "Constructions of p-algebras"*, Preprint.

465. D. M. Clark, *The structure of algebraically and existentially closed Stone and double Stone algebras*, J. Symbolic Logic **54** (1989), no. 2, 363–375.

466. M. E. Adams, *Principal congruences in De Morgan algebras*, Proc. Edinburgh Math. Soc. (2) **30** (1987), no. 3, 415–421.

467. M. E. Adams and R. Beazer, *The intersection of principal congruences on De Morgan algebras*, Houston J. Math. **16** (1990), no. 1, 59–70.

Translated by BORIS M. SCHEIN

Chapter II
Algebras of Logic

N. YA. KOMARNITSKIĬ

The present survey is devoted to those algebras of logic whose properties and structure explicitly or implicitly depend on the existence of a partial order. It is a natural continuation of a paper by Esakia [51] with the same title, where papers of the period 1973–1977 were reflected. Some of the works on algebras of logic, reviewed from 1977 through 1982, are reflected in surveys published in [276].

The author is sincerely grateful to Professor L. A. Skornyakov, who has kindly suggested that the author write this paper and who wrote a part of the second section on relatives. The author appreciates the help and advice of T. S. Fofanova and H. Draškovičová, who made the work of selection of the material essentially easier.

§II.1. Cylindric and polyadic algebras

Henkin, Monk, and Tarski published a voluminous work [175], which is Part I of the book [109]. (See also [176].) It is devoted to a systematic study of cylindric set algebras (csa). For cylindric algebras such concepts and constructions as subalgebras, homomorphisms, products, and ultraproducts are considered. The processes of change of basis and relativization of cylindric algebras (ca) are described. Analogous results are presented in the paper by Andréka and Németi [59], which forms Part II of that book. Here the attention was focused on relativized csa and the exposition is more detailed. As in the monograph of Henkin, Monk, and Tarski [174], both papers contain a wealth of open problems.

A number of results on isomorphisms of csa were announded by Biró [73]. Larson [196] considered diagonal-free csa, model examples of which

---

1991 *Mathematics Subject Classification*. Primary 03Gxx; Secondary 06Axx, 06Bxx, 06Cxx, 06Dxx, 06Exx.

are Boolean algebras of subsets of a Cartesian product of two sets endowed with two additional operations. He established that, up to isomorphism, there exist only seven types of infinite diagonal-free csa of dimension 2 with a single generator. He produced an explicit description for all seven. He showed that lifting any of these restrictions leads to the existence of a continuum of types of such algebras. Further, Larson [197] proved that, for every $n \geq 3$, there exists a continuum of types of single-generated diagonal-free csa of dimension n. Bergman [69] introduced the concept of rank of a diagonal-free csa of dimension 2 and proved that, for every natural n, there exists a diagonal-free csa of rank n isomorphic to a diagonal-free csa of rank 1. He also characterized a sufficiently large class of csa of finite rank that are not isomorphic to csa of lower rank. These characterizations permitted the author to answer Larson's question in the positive: does there exist a 2-dimensional csa generated by n elements that cannot be isomorphically embedded in a csa generated by fewer elements?

An algebra A of sets is called compact if $\bigcap S \neq \varnothing$ holds for every $S \subset A$ with $\bigcap F \neq \varnothing$ for every finite $F \subset S$. It was proved in [260] that the compactness property for csa corresponds to the universality property of models in logic; a brief description of an algebraic study of this correspondence can also be found there.

Németi [223] solved Problem 2.II from [174], which asked whether the class $Nr_\alpha CA_\beta$ of all α-dimensional reducts of β-dimensional cylindric algebras is closed under homomorphic images and passage to subalgebras, where α and β are ordinal numbers and $\alpha < \beta$. The answer is positive for homomorphic images and negative for subalgebras. He conjectured that the class $Nr_\alpha CA_\beta$ is elementary. Andréka and Némati [58] studied dimension-complemented ca. They established that every such algebra is elementarily equivalent to some locally finite-dimensional ca. This served as a foundation of the claim that the theories of locally finite-dimensional and dimension-complemented ca coincide, and the former is axiomatized by a simple finite scheme of axioms. Zlatoš [288] gave a negative answer to the question on the axiomatizability of the class of locally finite ca. Nevertheless, the category of locally finite ca of dimension α turned out to be isomorphic to the variety of heterogeneous algebras.

A new system of axioms for strict ca was suggested by Sudkamp [266]. He proved that the theory of strict ca of dimension 2 is polynomially equivalent to the theory of MacKinsey algebras. Pinter [227] studied ca satisfying the Rasiowa-Sikorski condition and formulated several other equivalent conditions. Larson [195] computed the cardinality of single-generated ca of dimension at least 3. Erdös, Faber, and Larson [130] applied their results on density of sets of natural numbers to construction of csa of dimension 2 which possess specific properties. The example they constructed answers a series of questions from a known book by Ulam [48] concerning projective set algebras. Ferencsi [136] studied measures on locally compact ca, established properties of these measures, considered the problem of extension of measures, and applied the general results obtained to csa. A theory of ca with an additional operation which is an algebraic analog of the quantifier "for almost all" was developed by Schwarts [252] who proved that such algebras could be represented as set algebras.

Imieliński and Lipski [177] considered applications of ca. It turned out that ca are convenient for the Codd relational model of data, because such a model could be embedded in a ca. This made it possible to raise problems of axiomatizability, decision problem, and duality for the objects connected with the Codd relational model of data.

A cycle of Georgescu papers contained research of polyadic algebras (pa) and their generalizations. For example, in [149], starting from the topological logic of Makowski and Ziegler $L(I)$, he defined $L(I)$-polyadic algebras and proved a representation theorem for such algebras. He transferred to the case of $L(I)$-polyadic algebras a theorem on omitting of types he proved earlier [147] for locally finite pa. Quellet [244] suggested a definition of polyadic algebras which does not coincide with the classic one. He listed properties of such pa and, in particular, investigated locally finite pa of this type. A connection between the logic $L(Q)_{mon}$ with a monotone quantifier and pa was traced in [148], where Georgescu proved that the so-called monotone pa are adequate algebraic structures for this logic. Besides, he proved a representation theorem for monotone pa, which is an algebraic version of the completeness theorem for $L(Q)_{mon}$. A representation theorem for the polyadic Heyting algebra in a specially constructed Heyting pa, which consists of mappings into a two-element Boolean algebra, was given in [152]. In [153] Georgescu gave a definition of a pa, which uses a logic with the quantifier "there exist uncountably many." Such a definition made it possible to prove that every countable pa of countable degree is representable by pa functions on a suitable set, which take values in $\{0, 1\}$. Georgescu's results on the Popper-Carnap probabilistic functions defined on pa [150] stand somewhat apart from his general direction, but he used them to obtain a generalization of Leblanc's theorem on weak completeness of the Lindelöf-Tarski algebras of the first-order predicate calculus.

§II.2. Other Boolean algebras with additional structure

Lindenbaum algebras are important among algebras occurring in logic. Interesting considerations of sublattices of Lindenbaum algebras are due to Cherlin [96]. He singled out sublattices consisting of classes that contain the formulas in prenex normal form whose quantifier prefix begins with a certain quantifier and contains at most a given number of changes of quantifiers. It turned out that many important model-theoretic concepts could be stated in the language of these sublattices. Among them are the properties of amalgamation, model completeness, and the existence of a model companion. Moreover, using this approach Cherlin identified a new property of theories, which is stronger than the existence of a model companion, but weaker than the existence of a model companion with amalgamation property. Interesting open questions are raised at the end of that paper.

Stepién [264] says that a consistent set of sentences has Tarski's property if it has exactly one Lindenbaum extension (that is, a complete consistent extension). Using the concept of extensional completeness for a set of formulas that he introduced, he characterized Tarski's property. Lindenbaum extensions and their connection with Tarski's property are also studied in [72].

Goltz [162] proved that the Lindenbaum algebra of the theory of linearly ordered sets with respect to the logic $L(Q_\alpha^m)$ with Malitz quantifier, where α is a regular cardinal number, is an atomic Boolean algebra. Molzan [216] established the undecidability of the theory of Boolean algebras with quantifier Q_0^2, leaving open the case of quantifiers Q_0^n for $n > 2$. Results of [14] are connected with Lindenbaum algebras as well.

Lindenbaum algebras of formal theories, which contain a sufficient segment of arithmetic, are examples of diagonalizable algebras (da). Diagonalizable algebras are Boolean algebras with an additional unary operation $x \mapsto p(x)$ that possesses the properties $p(1) = 1$; $p(x \wedge y) = p(x) \wedge p(y)$; $p(x \vee \neg p(x)) = p(x)$. Magari [210] wrote a paper of retrospective character, in which the principal moments of the development of algebraization of diagonal constructions in the theory of da, as well as historic information on the entire algebraic logic, was contained. Smoryński [258] generalized to da the Stone theorem on representation of Boolean algebras in the space of ultrafilters and explained a connection between these algebras and Kripke's semantics. Also, he established that the theories of da and finite da are undecidable. The hereditary undecidability of the variety of da containing an algebra A, in which $p(0) \neq 1$, was proved in [275]. The same result holds for intuitionist da, which had been studied earlier in [274]. Smoryński [259] proved the finite inseparability of the elementary theory of da in the sense that the set of formulas that hold on all finite da is effectively inseparable from the set of formulas whose negation holds on all finite da. Janicka-Zuk [183] added to that, that the class of da possesses the amalgamation property. An interpretation of the equivalence of formulas of Peano's arithmetic in terms of da was given in [70].

Prucnal [233] called an algebra codiagonalizable (cda) if it has an additional operation δ with the properties dual to those of the operation p on da. He pointed out that cda are isomorphic to subalgebras of algebras of subsets of topological spaces, where δx is the closure of the complement of x. Buczkowski and Prucnal [93] proved a representation theorem for cda, which says that, for every cda A, there exists an embedding h of A into the field of all subsets of a topological space X such that $\delta(h(x))$ is Cantor's derivative of the set $h(x)$.

Using as a starting point an example based on Lindenbaum algebra, Montagna [217] studied fixed-point algebras. These are pairs (A, B), where A and B are Boolean algebras such that

(1) each element of B is a mapping of A into A;
(2) every constant map $\lambda: A \to A$ belongs to B;
(3) Boolean operations of B act pointwise on A; and
(4) for every $\lambda \in B$ there exists $a \in A$ such that $\lambda(a) = a$.

For example, if A_T is the Lindenbaum algebra of a sufficiently strong theory and B_T is the Boolean algebra of operations defined on A by extensional formulas $F(x)$, then (A_T, B_T) is a fixed-point algebra. The main results of Montagna are about fixed-point algebras with an additional structure. Solovay [262] proved that fixed-point algebras of any two reasonable theories are isomorphic. Such theories are, for example, Peano's arithmetic, ZF set theory, and Gödel-Bernays set theory. This result confirms a conjecture of Smoryński.

Some authors continued studying Magari algebras. Recall that a Boolean algebra A with an additional unary operation d is called a Magari algebra if $d(0) = 0$, $d(a) \leq d(a - d(a))$, and $d(a \vee b) = d(a) \vee d(b)$. Grigoliya [12] described free Magari algebras of finite nonzero rank and, as a corollary, obtained a criterion for such an algebra to be finitely generated. Abashidze [1] calls the height of a Magari algebra the least integer $n > 0$ such that $d^n(1) = 0$. Using this concept, he studied in [1] an interconnection of algebras which split the variety of Magari algebras and residual finiteness. He defined a bijective functor from the category of Magari algebras to the category of Grzegorczyk algebras which transforms each algebra into its reduct. Muravitskiĭ [31] considered the class of Magari algebras whose logic is an extension of the provability logic; he proved that this class is not axiomatizable.

Researchers' attention was drawn to algebras that could facilitate the study of modal logics. Tembrowski [270] renewed studies of B-algebras introduced by Huntington in 1937. They are Boolean algebras with an additional binary operation subject to certain properties. It turned out that some properties of topological Boolean algebras were naturally reflected in special classes of B-algebras. For example, the class of monadic algebras is polynomially equivalent to a certain class of B-algebras. Here Tembrowski obtained information on systems of filter closures in the class of B-algebras. A B-algebra A is called a QB-algebra (a normal B-algebra) if it possesses a Q-filter (respectively, a normal ultrafilter). In the note [271] necesesary and sufficient conditions for a B-algebra to be a QB-algebra (a normal B-algebra) are given.

A Boolean algebra with two unary operations $x \mapsto \bar{x}$ and $x \mapsto \Box x$ is called a modal algebra if $(\Box x \to x) = 1$; $\Box 1 = 1$; $\Box(x \to y) \to (\Box x \to \Box y) = 1$, where $x \to y = \bar{x} \vee y$. Rybakov [39] proved that a subdirectly irreducible finite modal algebra has a finite basis for quasi-identities and, contrary to that, there exists a finite modal algebra which has no basis for quasi-identities of finitely many variables. Loureiro [201] considered free tetravalent modal algebras. She proved that such an algebra of finite rank is finite and computed its order. Maksimova [25] proved that modal companions of Dummett logic lack the interpolation property. She also proved that every variety of topo-Boolean algebras, which correspond to modal companions of Dummett logic, does not have the amalgamation property. Certain other algebras are called modal as well. For example, Schweigert and Szymańska ([253], [254]) considered modal algebras which are closer to diagonalizable algebras than to modal algebras in the above sense. They proved that a modal algebra in their sense which satisfies the condition $\overline{x} \leq x$ has no nontrivial congruences if and only if it has no nontrivial fixed points. They characterized subdirectly irreducible modal algebras of this type.

A relative or a relation algebra is a universal algebra R of signature $\{+, \cdot, ^-, 0, 1, \circ, *, E\}$, where $+, \cdot,$ and $\circ$ are binary operations, $^-$ and * are unary and $0, 1,$ and E nullary operations such that $\{R|+, \cdot, ^-, 0, 1\}$ is a Boolean algebra, $\{R|\circ, ^*, E\}$ a monoid with involution, $(x+y) \circ z = x \circ z + y \circ z$, $(x+y)^* = x^* + y^*$, and $x^* \circ (\overline{x \circ y}) \leq \overline{y}$. The history of research of relatives, which began essentially at the end of the last century, is reflected in Martin's paper [212]. A number of properties of relatives established in the fifties is collected in Jónsson's paper [185]. A certain refinement of axiomatics was suggested in [OSL, I, 99] and [83]. The main result of Jónsson's paper we have just mentioned is a description of atoms and co-atoms in the lattice of varieties of relatives. He obtained a description of certain other varieties in the lower strata of this lattice. The starting point of the theory of relatives is the relative $\operatorname{Rel} X$ of all binary relations on a set X. Schönfeld [251] proved that the theory of all finite relatives of such a form is undecidable (the undecidability of the theory of all relatives of the form $\operatorname{Rel} X$ had been known earlier). Jónsson [186] established an anti-isomorphism between the subgroup lattice of the group $\operatorname{Aut}(\operatorname{Rel} X)$ (observe that it is isomorphic to the group of all permutations of X) and a certain lattice of subrelatives of $\operatorname{Rel} X$. This led to a description of certain maximal subrelatives of the above relative. Representations of relatives by relatives of binary relations were investigated by Schmidt and Ströhein [250]. The paper [85] belongs to this direction too. The set $\operatorname{Mat} X$ of $(X \times X)$-matrices over a Boolean algebra is a natural generalization of $\operatorname{Rel} X$. Skornyakov ([43] and [44]) suggested a characterization of such relatives. A relative $\{2^G | \cup, \cap, G \backslash -, \cdot, -1, \varnothing, G, \{e\}\}$ is associated with each group G. An analogous construction is possible if a group is replaced by a polygroup, that is, a set with a multivalued associative binary operation that satisfies natural conditions. It turns out that the category of polygroups is equivalent to the complete atomic relatives (this means the completeness and atomicity of the corresponding Boolean algebra) without zero divisors (that is, $x, y \neq 0$ imply $x \circ y \neq 0$) ([101], [102]). A generalization of relatives when the associativity of $\circ$ is given up was considered by Maddux ([208], [209]), and was studied by Skornyakov [44] when the Boolean algebra is replaced by the distributive lattice. A new characterization of the algebra $\{\operatorname{Rel} X | \cap, \circ, *\}$ was suggested by Bredikhin [3]. Bednarek and Ulam ([66], [67]) considered a generalization of relatives connected with projective algebra, i.e., a Boolean algebra with two additional unary operations and a partial binary operation. These algebras are connected with algebraization of the predicate logic. Brink [84] defined a module over a relative R as a Boolean algebra A for which a mapping $R \times A \to A$ is defined and the axioms are chosen in such a way that each relative is a module over itself if the above product coincides with $\circ$.

§II.3. Post and Łukasiewicz algebras

I. A. Mal'tsev [27] studied invariants of quasi-cells of iterative Post algebras (Pa). Using them he characterized algebras belonging to certain varieties. Skordev [42] proved that every iterative Pa P_A of all operations on a nonempty set A with a positive number of arguments, considered as a monoid with respect to a certain special operation, could be embedded in a semigroup in which the theorem on the normal form of products holds. Negru [35] studied lattices SP_k of Post classes closed with respect to superposition of functions of k-valued logic. He established that no identities or quasi-identities hold on SP_k for $k > 3$. For $k = 2$ he found such identities, while the problem remains open for $k = 3$.

Yaqub published several papers related to the theory of Post L-algebras (PLa). A definition of PLa was obtained from the usual one after the chain of constants was replaced by an arbitrary lattice L. He proved in the first paper [284] that the class of PLa is a variety if and only if L is finite. In the second paper [283] a structure of a commutative ring was defined in a unique way on each PLa; this ring turned out to be isomorphic to the ring of continuous functions defined on the Stone space S of the center of the algebra L (here L is treated as a discrete topological space). In the third paper [285] Post L-spaces over PLa were defined, and a theorem proved by Priestley for distributive lattices generalized.

Ribeiro [245] investigated connections between Pa of order $m > 1$ and fuzzy structures which permitted construction of fuzzy sets in the sense of Zadeh. Pigozzi [240] reduced the study of subvarieties of the variety of Post pseudoalgebras to analogous problems for Heyting algebras. This was possible due to a strong equivalence between the categories of Post pseudoalgebras and Heyting algebras.

For every finite Pa with m elements, Sett [257] constructed a category whose functors into the category of m-valued r-place relations were in one-to-one correspondence with the classes of logically equivalent formulas with exactly r free variables of the first-order language with equality over the algebra L.

θ-valued Łukasiewicz algebras (Ła) and their analogs belong to generalizations of Pa. Glas [155] continued investigating θ-valued Ła, where θ is an ordinal type. He was interested in atoms of the lattice CA of all those elements of a θ-valued Łukasiewicz algebra A which are complements of other elements from A. He deduced from his general results that the atomicity of CA is equivalent to the atomicity of A. Cignoli [98] modified a definition of n-valued Ła and called these new algebras proper Łukasiewicz algebras. He showed that proper Ła are adequate for Łukasiewicz logics for $n > 4$. Karpenko [191] constructed another adequate model of an $(n + 1)$-valued propositional Łukasiewicz logic based on Boolean algebras. His paper [18] is devoted to analogous considerations. Komori's paper [192] belongs to this direction too. Cignoli [97] proved an analog of the completeness theorem for an n-valued predicate Łukasiewicz logic, it was based on a study of proper Ła. His paper [99], co-authored by de Gallego, contains a characterization of 5-valued Ła by means of De Morgan algebras. It follows from this characterization that the structure of 5-valued Ła on a bounded distributive lattice, if it exists, is uniquely determined. The same authors [100] applied to Ła their theorem on duality between the category of De Morgan algebras with a certain additional operation and the category of totally disconnected topological spaces which possess an order-preserving involutive homomorphism.

Hatvany [170] proved that every α-ideal of a Ła A over a De Morgan algebra M, where $\alpha \in M$, is represented as an intersection of α-prime ideals. Kabziński [187] found that each equivalential Ła possesses the regularity property.

Iorgulescu's research is devoted to Łukasiewicz-Moisil monadic algebras (LMA). She defined an equivalence relation on an n-valued LMA, factorization over which leads to 3-valued such algebras [179]. Her other paper [178] contains a construction for special 3-valued LMA in which every 3-valued LMA is embeddable. Palasiński [225] produced a method permitting one to obtain all BCK-algebras from Ła using amalgamations. In analogy with the classical case, Georgescu [151] defined Chang's algebra as an algebraic model of Chang's modal logic and proved a representation theorem for such algebras. Mundici [221] used Chang's algebras to investigate the elementary equivalence between residually finite-dimensional C^*-algebras. For each C^*-algebra A of this kind he produced a certain set $\Theta(A)$ of theories in a Łukasiewicz $\aleph_0$-valued calculus. Then A is elementarily equivalent to B if and only if $\Theta(A) = \Theta(B)$.

Ayissi Eteme [132] discovered interesting connections between θ-valued Ła and certain algebraic structures called θ-valued Chrysippian rings. He observed that any θ-valued Ła is embeddable in a θ-valued Chrysippian ring [133]. Filipou [138] studied m-complete Chrysippians, that is, m-valued Ła, possessing certain additional properties; he proved a theorem on the equivalence of the m-representability of m-complete Chrysippians and the m-representability of their Boolean subalgebras of fixed points.

Generalized three-valued La and their connection with De Morgan algebras were studied in Fowler's dissertation [143]. Other algebras connected with La were studied in [137].

§II.4. Heyting algebras. Heyting-valued sets and related topics

The monograph [52] of Esakia is devoted specifically to Heyting algebras (Ha) and closure algebras which are tightly connected with Ha. It contains elementary information from the theory of Heyting algebras and considers more specialized topics. In particular, it contains an exposition of the theory of stencil algebras. It discusses in detail the concept of hybrid that appeared as a result of uniting the Stone duality with Kripke models. The duality between the category of Ha and the category of strict hybrids is proved; it is subsequently applied to representations of Ha by rings of cones of suitable ordered sets. In its conclusion the book contains a brief description of the contents of the forthcoming second part of the monograph. The monograph contains a most detailed exposition of its author's results.

Jankowski [184] characterized complete Ha by means of closure spaces. Grigoliya and Meskhi [165] constructed a countable family of critical varieties of biclosure algebras. Grigoliya [11] described free and projective algebras with conjugate closures. Sankappanavar [248] investigated Ha with dual pseudocomplements. The congruence lattice of such an algebra is isomorphic to the lattice of its normal filters. He characterized subdirectly irreducible, prime, and those dually pseudocomplemented Ha, whose lattice of normal filters is Boolean.

Meskhi ([28] and [30]) studied varieties of Ha with involution whose restriction to regular elements coincides with the Boolean complementation. He characterized subdirectly irreducible and injective algebras in these varieties and in [29] constructed an example of the continuum property of pairwise disjoint varieties of Ha with regular involution, each of which is generated by its finite elements. Rybakov [41] showed that a free Ha had no basis of quasi-identities of finitely many variables. Wroński [282] investigated quasivarieties of Ha generated by their finite elements. Dziobiak [126] formulated a distributivity criterion for the lattice of subquasivarieties of an arbitrary variety of Ha. In another paper [127] he showed that there exists a finite Ha that generates a nonfinitely axiomatizable quasivariety. Also, a union of two finitely axiomatizable quasivarieties of Ha need not be fiintely axiomatizable. Dziobiak observed that analogous results hold for topological Ha. A topological Ha H is an ordinary Ha with two additional operators: a closure operator K and an interior operator I, which, besides the ordinary properties of these operators, satisfy the condition $K(I(x) \to K(y)) = I(x) \to K(y)$. Topological Ha were studied by

Luppi ([203] and [204]). For a complete topological Ha he found necessary and sufficient conditions for the isomorphicity of such an algebra and the algebra of all open subsets of a topological space. We also note that he succeeded in obtaining certain results on the embeddability of topological Ha in special topological Ha. Font [141] obtained useful information on monodic and semisimple topological Ha.

It was clarified in [88] when every prime filter of a Ha is contained in no more than n noncomparable filters. Such Ha are called strongly n-normal. It was proved there that a projective Ha is strongly n-normal if and only if it has finite breadth. Wrancken-Mawet [281] investigated finite Ha. Among his results we note a description of finite Ha with the distributive, dually atomic, or with Boolean subalgebra lattice. He investigated the structure of the Frattini subalgebra of a finite Ha.

0-endomorphic rigid Ha were defined in [55] as Ha with exactly two endomorphisms. The authors proved that the classes of 0-endomorphic rigid Ha form a proper class. Zakhar'yashchev [17] represented Ha as a system S of subsets of a set K and called the pair (K, S) a model structure. He found nontrivial necessary and sufficient conditions of refutability of an arbitrary formula in a given model structure.

A Heyting algebra is called double if the dual (in the lattice sense) algebra is a Heyting algebra too. The implication operations in a Heyting algebra and its dual are denoted by $\rightarrow$ and $\leftarrow$, respectively. An element c of a double Ha H is called a core point if there exists $x \in H$ such that $x \rightarrow c = 0$ and $x \leftarrow c = 0$ simultaneously. Epstein and Horn [129] considered the core of a double Ha, that is, the set of its core points, and established that a subset $Q \subseteq H$ contains the core if and only if for any $x, y \in H$ there exists $q \in Q$ such that $x \rightarrow y = y \vee q$. They defined dissectable lattices, that is, bounded lattices with a chain $0 = e_0 \leq e_1 \leq \cdots \leq e_n = 1$, in which every interval $[e_i, e_{i+1}]$ is a Boolean lattice. Dissectable lattices are interesting because each of their elements has a handy monotonous representation in terms of core points. This representation possesses many useful properties. Beazer [62] investigated cores of double Stone algebras and gave several equivalent conditions for every proper interval of a double Stone algebra to be a Boolean lattice. Düntsch [120] showed that, under certain restrictions, a double Ha could be characterized as a projective Stone algebra. Further research of free and projective double Stone algebras was undertaken in [121] and [122]. Congruences of double Ha were studied in [248]. Congruence extension properties of double Ha and connection of congruences with the injective hull of a distributive lattice were found in [156].

Pitts [228] preferred a categorial research of Ha. He is interested in stability of monomorphisms with respect to co-universal squares, the interpolation property for universal squares which stems from Craig's interpolation theorem, and the amalgamation property in the category of Ha. Completions of products of Ha and completions of Ha in extensions were considered in [128]. Lópes-Escobar [198] used categorial analysis to explain why Beth models are equivalent to Ha, while Ha are not equivalent to Beth models. Adams, Koubek, and Sichler [55] proved that the category C of nonoriented graphs is embeddable in the category H of Ha in such a way that between any two objects of C considered as objects of H there exists exactly one homomorphism φ in H such that $\operatorname{Im} \varphi = \{0, 1\}$.

Maksimova [26] proved that the amalgamation property of the variety of pseudo-Boolean algebras (implicative lattices) is equivalent to Craig's interpolation property for superintuitionistic (respectively, positive) logic. Also, she proved the algorithmic undecidability of the amalgamation property for finitely axiomatizable varieties of pseudo-Boolean algebras, implicative lattices, and closure algebras. Rybakov [40] proved the hereditary undecidability of the theory of free algebras of countable rank of the variety of Ha.

Now we consider results related to generalizations of Ha and Ha with additional structure. Grishin [13] suggested a generalization of Heyting-Brouwer algebras. This generalization is interesting because it comprises both Ha and ordered groups. Iturrioz [181] suggested another direction for generalizations. She defined SH_k-algebras and called them symmetric Ha of order k. In the process of research she clarified when such an algebra is a Łukasiewicz algebra, a Kleene algebra, or a Post algebra. In the previous paper [180] she justified the necessity of studying these algebras. Analogously to the case of the Lindenbaum algebras, Porta [232] constructed an algebra for intuitionistic logic in which not all logical connectives are reflected. This algebra is a Ha and possesses certain specific properties. Zarębski [286] suggested a characterization of certain Ha in the language of dual spaces.

Kuznetsov [20] considered a calculus I^Δ whose algebraic interpretations are Δ-pseudo-Boolean algebras. It follows from his main logical result that every pseudo-Boolean algebra can be embedded as a subalgebra in an algebra from the variety it generates, which can be turned into a Δ-pseudo-Boolean algebra. Shum [50] introduced a concept of structurally correct Ha and proved that propositional superintuitionistic logic had a structurally correct model. In [145] a definition of a pseudo-Boolean algebra was given, which differs from the classical one, and a new semantics for the intuitionistic propositional calculus was suggested on that base.

Goldblatt [159] called a Ha H local if it has an additional unary operation j that satisfies the following conditions:

$$x \leq j(x); \qquad jj(x) = j(x); \qquad j(x \wedge y) = j(x) \to j(y),$$

for all $x, y \in H$. Any local Ha gives rise to various semantics of the propositional languages that contain intuitionistic propositional connectives and a modal operator. The author constructed four such semantics and observed that the algebra of global elements of the classificator of subobjects of any Gröthendieck topos is a local Ha. Incidentally, a fine introduction to the results of this paper can be found in Goldblatt's book [161] whose second edition with a new chapter "Logical geometry" appeared in 1984 (see also its Russian translation [8]). This book contains the basics of the theory of Heyting-valued sets, whose particular case are Boolean-valued sets.

It is well known that the algebra of subobjects of any object of an elementary topos is a Heyting algebra. As proved earlier by Isbell (MR 50 #11184) and now reproved by Borceaux and Van den Bossche [89], the partially ordered set of topologies of an elementary topos is also a Heyting algebra. But Borceaux and Van den Bossche added that in a Grothendieck topos this algebra is complete. A more systematic investigation of the lattice of topologies of a Grothendieck topos was undertaken by Patryshev [226]. Borceaux and Van den Bossche [90] gave much attention to Kripke scales, Heyting subobjects, and applications of their results to Gel'fand rings. Fourman [144] considered categories of generalized spaces. These are the categories whose duals are isomorphic to full subcategories of the category of complete Ha with morphisms preserving the operations $\bigwedge$ and $\bigvee_{i \in I}$ and the constant t. He proved that a category of generalized spaces is a Grothendieck topos and that the principles of continuity, local choice, and local compactness hold in that topos.

Takeuti and Titani published the paper [268] on Heyting-valued sets and their connections with intuitionistic logic. Higgs [171] gave a complete description of injective objects in the category of Heyting-valued sets. Other problems of the theory of Heyting-valued sets were considered in [163], [154], [269], [202], and [87]. Kusraev [21] suggested a method for studying abstract normed spaces and modules based on the apparatus of Boolean-valued models of the set theory. He observed that this approach is very effective in solving many problems of convex analysis and the theory of vector lattices. Lyubetskiĭ ([22] and [24]), investigating truthfulness of various (including Horn) types of formulas on sheaves defined on a Ha, obtained interesting corollaries. For example, if a strictly Horn formula holds in the class of all quasiprime rings or skew fields, then it holds in the class of all biregular (strictly regular) rings. He continued this research in [23] obtaining analogous results for other classes of rings and modules. Unfortunately, he did not note a connection of his results with those of other researchers on this subject. Analogous results are contained in Gordon [10].

In analogy with Heyting-valued sets, Eytan [134] tried to prove that the category Fuz(H) of fuzzy sets was a topos for any Heyting algebra H. Recall that objects of Fuz(H) are pairs (X, φ), where X is a set and φ a mapping of X into a Heyting algebra H. Morphisms of this category are mappings $\Theta \colon X \times Y \to H$ satisfying three special conditions. The error was corrected in [135], where Eytan proved that if Fuz(H) is a topos then H is a Boolean algebra. The same result was independently obtained by Pitts [229] (cf. also [222] and [231]). Connections between the category Fuz(H) and the topos of sheaves over a complete Ha were analyzed by Stout [265].

Dzik ([124] and [125]) studied distributive lattices of logics. Among his results we note a representation of these lattices as algebras of open sets of a topological space. Also, he characterized the content of the lattices of logics.

Sendlewski [255] continued studying Nelson algebras (see Rasiowa's monograph, MR 56 #5285) and determined those congruences on them for which the factor algebras are Ha. He succeeded in describing all Nelson algebras for which the corresponding factor algebra is isomorphic to a fixed Ha. In [256] he studied N-lattices, which correspond to the intuitionistic logic with a strong Markov-Nelson negation, similarly to Ha corresponding to the ordinary intuitionistic logic. The main results characterize subdirectly irreducible N-lattices and describe pre-primitive varieties of N-lattices.

Some information on Ha can be found in [15] and [117].

§II.5. De Morgan algebras and other related distributive lattices

A De Morgan lattice (or, equivalently, a De Morgan algebra, DMa) is a bounded distributive lattice with a unary operation of complementation $x \to \overline{x}$ that satisfies the identities $\overline{x \vee y} = \overline{x} \wedge \overline{y}$ and $\overline{\overline{x}} = x$. A bounded distributive lattice in which each (closed) interval is a De Morgan lattice is called a relative De Morgan lattice. Trillas and Valverde [273] considered DMa $(P(X), \cup, \cap, \overline{}^n, X, \varnothing)$, where $P(X) = [0, 1]^X$, $\cup$ and $\cap$ are defined pointwise, and the complementation operation $\overline{}^n$ is defined by means of a continuous involution n on $[0, 1]$ with the property $n(1) = 1 - n(0) = 0$ by the rule $\overline{A^n}(x) = nA(x)$.

Cornish [103] demonstrated the possibility of singling out DMa as subalgebras of BCK-algebras. He showed that if a BCK-algebra A is an upper semilattice, then for every $a \in A$ the subset $O_a = \{b \in A | a(ab) = b\}$ is a DMa. Here O_a is a Boolean algebra if and only if A is positively implicative. Varlet [279] described relative De Morgan lattices in the class of finite distributive lattices. These are exactly the lattices that do not contain intervals isomorphic to the lattices $2^2 + 1$ and $1 + 2^2$. In another paper [277] he showed that there exist precisely two nonisomorphic minimal DMa whose subalgebra lattices are isomorphic to 2^k. On a finite lattice that is a direct product of chains one can define many various structures of DMa. Their number was computed in [278]. Among these structures only one produces a Kleene algebra. In a more general situation, there exist $[\frac{1}{2}(n + 2)]$ structures of DMa on a free distributive algebra $FD(n)$ of rank n. Certain results on dualities on $FD(n)$ were obtained. De Morgan algebras satisfying the identity $x \wedge \overline{x} \leq y \vee \overline{y}$ are called Kleene algebras. Varlet [277] proved that a DMa is a Kleene algebra if and only if the subset of the elements of the form $x \vee \overline{x}$ is a filter. (For complete DMa this result was contained in [53].) Also, he found necessary and sufficient conditions under which a DMa is Boolean. The condition is that each maximal ideal of the algebra must be the kernel of a congruence. Varlet [278] also answered the question of which DMa can be epimorphically mapped onto a Boolean algebra. Božić [86] studied T-filters of DMa, i.e., the filters that do not contain an element of a lattice and its complement simultaneously, but contain either the element or its complement. He gave a criterion for a fixed filter to be contained in a T-filter.

Hon-Fei [172] studied the subcategory A of the category K of Kleene algebras, whose objects are Kleene algebras corresponding to Post algebras of order 3. He proved A is a reflexive subcategory of K. This result is useful in the study of injective and projective objects of A. As a corollary to general propositions he proved that finite Kleene algebras are injective in A.

Sankappanavar [249] characterized principal congruences of De Morgan algebras and showed that the variety of DMa has definable principal congruences. He proved also that the congruence lattice of a DMa is Boolean if and only if the algebra is finite. It is interesting that compact elements of the congruence lattice form a Boolean sublattice. Subdirectly irreducible DMa were described.

The lattice of subvarieties of the variety of DMa was described by Cignoli and de Gallego [100]. This description is based on a duality, proved in the paper, between the category of DMa and a category of special topological spaces endowed with an additional structure.

A mapping $f: A \to B$, where A and B are DMa, is called completely additive if $f(\bigvee a_k) = \bigvee f(a_k)$ for every family $\{a_k | k \in K\} \subseteq A$. Hamburg [168] defined a De Morgan mapping of a DMa as a completely additive symmetric mapping and found necessary and sufficient conditions for a function $f: [0, 1] \to [0, 1]$, where $[0, 1]$ is the interval of the real line considered as a DMa, to be De Morgan.

Esteva [131] considered representations of DMa by fuzzy sets, analogously to representations of Boolean algebras as algebras of ordinary sets. Connections between DMa and compact totally disconnected ordered topological spaces were pointed out in Fowler's dissertation [143]. Verbal congruences on DMa corresponding to subvarieties of Kleene and Boolean algebras were described in the same dissertation. Loureiro [199] introduced an additional operation $x \mapsto \nabla x$ on a DMa and called the resulting algebra a tetravalent modal algebra if the following conditions hold: $\overline{x} \vee \nabla(x) = 1$ and $x \wedge \overline{x} = \overline{x} \wedge \nabla x$. She described completely the DMa on which there exists a structure of a tetravalent modal algebra and proved in [200] that a tetravalent modal algebra that contains more than a single element is isomorphic to a subdirect product of its factor algebras modulo all maximal deductive subsystems. The proof of the finiteness of tetravalent modal algebras of finite rank and computation of their order is contained in [201].

Romanowska [246] considered p-algebras which are DMa and observed that each such algebra is a double p-algebra. Also, she described certain connections between finite subdirectly irreducible algebras of this class and prime DMa. Kleene M-algebras were considered in [218]. It was proved there that a Kleene M-algebra is embeddable in the lattice of subsets of a set.

A bounded distributive lattice A is called an existential algebra if it is endowed with a closure operator K that satisfies the following condition: for any $x, y \in A$ $K(x \wedge Ky) = Kx \wedge Ky$. Servi [261] established that every finite existential algebra is embeddable in a monadic Boolean algebra. Using this fact, he concluded that if an open formula that has occurrences of K, connectives $\cap$ and $\cup$, and constant 0 and 1 holds on every monadic Boolean algebra, then it holds on every existential algebra.

A bounded distributive lattice is called a distributive Ockham algebra if it has a unary operation $\circ$ that is a dual endomorphism of the lattice. Various researchers have studied varieties of Ockham algebras. For example, Beazer [63] considered the variety of distributive Ockham algebras defined by the identity $x^\circ = x^{\circ\circ\circ}$ and described all subdirectly irreducible algebras in it. There are 12 of them. Also, he observed that the subalgebra $A^{\circ\circ} = \{x \in A | x = x^{\circ\circ}\}$ of a distributive subdirectly irreducible Ockham algebra is a prime DMa. Another variety of Ockham algebras defined by the identity $x = x \wedge x^{\circ\circ}$ was considered in [64]. It was established there that the lattice of subvarieties of this variety is finite. Also, subvarieties of this variety were pointed out that are rich in injective algebras. Goldberg [157] studied subvarieties of the variety of Ockham algebras generated by their finite subdirectly irreducible distributive Ockham algebras. He gave a detailed description of the identities of these varieties and of free and injective algebras in them. The variety of Ockham algebras defined by the identity $(\ldots(x^\circ)^\circ \ldots)^\circ = x$, where $\circ$ is applied $2n$ times, was studied in [158]. Subdirectly irreducible algebras of this variety were described there, and it was noted that all of them are prime. Some results on algebras from this variety can be found in [280]. For example, the congruence lattice of each finite algebra is Boolean, and if $n = 2$ then the lattice of subvarieties consists of seven elements. Varieties of distributive Ockham algebras were investigated in the last chapter of Fowler's dissertation [143]. Goldberg [158] investigated the problem of representability of distributive Ockham algebras by means of hom-functors.

Blyth and Varlet considered MS-algebras in a series of papers ([75]–[81]). A bounded distributive lattice A with a unary operation $\circ$ is called an MS-algebra if, for every $x, y \in A$, the following conditions hold:

$$x \leq x^{\circ\circ}; \quad (x \wedge y)^\circ = x^\circ \vee y^\circ; \quad 1^\circ = 0.$$

The subset $A^{\circ\circ} = \{x \in A | x = x^{\circ\circ}\}$ of an MS-algebra A is, in fact, a DMa, and, as they proved in [74], for a subdirectly irreducible MS-algebra A, the algebra $A^{\circ\circ}$ is prime. They proved that there exist exactly nine subdirectly irreducible MS-algebras (up to isomorphism), and described their structure completely. Congruences of MS-algebras were studied in the same paper. In [75] the lattice of subvarieties of the variety of MS-algebras was found and identities defining these subvarieties given. The lattice consists of 20 elements and was represented as a union of a filter F and an ideal J. The filter F is generated by the variety of DMa, $\operatorname{Card} F = 9$, and $\operatorname{Card} J = 11$. Let $\operatorname{Fix}(A) = \{x \in A | x = x^\circ\}$. It was established in [81] that $\operatorname{Card}\operatorname{Fix}(A) \leq 1$ for every algebra A from the variety that is an element of J. For $V \in F$ and any $n \geq 2$ there exists a countable MS-algebra $A \in V$ such that $\operatorname{Card}\operatorname{Fix}(A) = n$. Also, [74] contains information on verbal congruences corresponding to subvarieties of the variety of MS-algebras. A list of varieties of MS-algebras possessing a somewhat strengthened amalgamation property was given in [79]. Blyth and Varlet [77] obtained a characterization of MS-algebras by means of principal congruences and studied connections between principal congruences and their structural reducts. Certain propositions from [249] were obtained as corollaries. Among other results of this paper we note the following: (1) the congruence lattice of an MS-algebra A is Boolean if and only if A is a finite De Morgan algebra; and (2) a Stone algebra A has a Boolean congruence lattice if and only if A is a finite Boolean algebra. Every algebra A in the variety of MS-algebras generated by the four-element chain $b^\circ < a = a^\circ < b < c$, besides the sublattice $A^{\circ\circ}$, contains another sublattice $A^\vee$ such that every congruence on A is completely determined by its restrictions to $A^{\circ\circ}$ and $A^\vee$ [65].

An MS-algebra A is called double if A has an additional operation $x \mapsto x^+$ with the properties $x^{+\circ} = x^{++}$ and $x^{\circ+} = x^{\circ\circ}$ and if the lattice A^* dual to A is an MS-algebra with respect to the operation $+$. Blyth and Varlet [76] characterized MS-algebras on which the structure of a double MS-algebra exists and described all varieties K of MS-algebras such that for every double MS-algebra A the condition $(A, \circ) \in K$ implies $(A^*, +) \in K$. In [78] they found all 22 types of subdirectly irreducible double MS-algebras, of which 11 are prime. It follows that there exist only four types of subdirectly irreducible double Stone algebras. The core of double Stone algebras was studied in [62]. Duality properties for Stone algebras and double Stone algebras were established in [113].

Necessary and sufficient conditions for the existence of a structure of an MS-algebra on a bounded distributive lattice were found in [80]. The exact number of such structures was found in certain cases.

Another type of algebras appearing in logic was considered in [256]. These are Belknap algebras, that is, lattices L with a given unary operation $x \mapsto \overline{x}$ and a singled out truth filter T that satisfies the condition $a \in \Gamma \Leftrightarrow \overline{a} \notin T$, and, in addition, $(L, \leq, ^-)$ is a De Morgan lattice. The author considered implicative extensions of Belknap algebras by means of a two-element Boolean algebra. In these extensions the implication was defined, although Belknap algebras do not have it. Certain results on properties of Belknap algebras preserved under implicative extensions were obtained.

Quasi-Boolean algebras were studied in [119], where a possibility of representation of these algebras by relatives was explored.

§II.6. Quantum logic

As in previous periods, a great number of papers were devoted to quantum logic (ql) in which, as is known, an essential role belongs to the theory of orthomodular lattices. First of all we note the monographs of Kalmbach [189] and Beran [68] specifically devoted to orthomodular lattices. New results on ql obtained from 1980 to 1982 were reflected in Kalmbach's survey [190]. Among books we note also [107], which contains a great majority of papers on ql reviewed in that period (it contains materials of a school on ql that took place on December 2–9, 1979 in the city of Erice). Some of these papers have critical direction which is explained by a dissatisfaction in the existing approaches to the foundations of ql.

Takeuti [267] used the thesis that ql describes the real world more adequately than classical logic does, and suggested that all of mathematics be built on ql. Here he started a construction of a quantum theory of sets as follows. Let L be a quantum logic, that is, a complete orthomodular lattice with certain additional properties. Ordinary sets S are taken as sets, and mappings $S \times S \to L$ with certain additional properties are taken as relations. Special relations $\in_S$ and $=_S$ are defined. To interpret formulas containing $\in_S$ and $=_S$ as well as logical connectives, the author defined expressions $a \to b$ as $\neg a \vee (a \wedge b)$, $(\forall x) Ax$ as $\bigwedge_x Ax$, and $(\exists x) Ax$ as $\bigvee_x Ax$. The axioms of the quantum theory of sets are the axioms of ZF written in terms of $\in_S$ and $=_S$, where $\cap$, $\cup$, and other set-theoretical operations are replaced by the corresponding quantum logical operations. Takeuti's main result proved the consistence of the ZF quantum theory of sets; the proof was given by a standard method of construction of a corresponding model. Certain problems of calculus based on the quantum theory of sets were touched upon in [166].

Foulis and Randall [142] gave a brief survey of empirical logic and stated five general requirements that every ql had to satisfy in the interests of practice. Finkelstein [139] thinks that one of the main drawbacks of the ql is nonlocality of its atoms, while physical processes described by the relativistic theory are local. This contradicts the distributivity of space-time logic, characteristic for the relativistic theory, which does not correspond to quantum processes. He suggested avoiding this contradiction by means of quantum sets, assemblies, and plexi he introduced.

New approaches to ql were suggested in papers of other authors. For example, Cook and Hilgevoord [104] tried to formulate ql in a way that permitted discussion of hidden variables. To this end they defined an equivalence relation τ on a lattice L and another equivalence relation on the set M of measurable procedures. The factor lattice L/τ is not orthomodular. Although this is not needed for the theory of hidden variables, they succeeded in finding conditions securing the orthomodularity of L/τ. Research of Pták [235] is connected with the theory of hidden variables as well. He showed that an extension of the class of queries to the class of all mappings of the form $L \times L \to [0, 1]$ ensures the possibility of hidden variables. Brabec and Pták [82] determined which subsets of a ql L are embeddable in a subalgebra of L that is a Boolean algebra.

Pták and Rogalewicz [238] gave a positive solution to the unicity problem for observables. Possibilities for introducing linear and topological structures on certain sets of observables were discussed in [236].

Mączyński, Dorninger, and Länger [118] continued studying σ-continuous logics, that is, a σ-complete orthomodular lattice L. Let $D(L)$ denote the orthomodular lattice of all observables of a quantum mechanical system with logic L. It was established that L and $D(L)$ generate the same variety. Possibilities of extension of homomorphisms from L to $D(L)$ were studied. Pykacz [243] observed that Mączyński's quantum logic has additional structures and introduced affine Mączyński logics on compact convex sets of states. He determined when this logic is Boolean. Thieffine [272] investigated ql that satisfy Piron's axioms. He translated this system to a modal language and observed that certain drawbacks, for which Piron's system was criticized, disappear in that interpretation. In [57] pure states in an algebraic approach to the foundations of quantum mechanics were considered and a hypothesis discussed that pure states are exhausted by vectors of a certain nonstandard extension of the space of states. Deliannis [116] gave an axiomatization of ql in terms of the set of pure states.

Cutland and Gibbins [108] suggested a new regular sequential calculus, in which $\wedge$ and $\vee$ are dual, that represents ql. Earlier Morgan and LeBlanc constructed correct and complete semantics for various logical systems in terms of functions $P_S(A, B)$ that give the probability of truthfulness of a formula A given a condition B. In [219] Morgan constructed analogous systems for orthologic and quantum logic in terms of $P_S(A, \Gamma)$, where Γ runs over sets of formulas. He proved an analog of the completeness theorem for these systems.

Deliannis [115] found a geometric interpretation of ql with conditional probabilities as sets of linear operators preserving a cone in a suitable linear space. He pointed out conditions ensuring restoration of ql by the set of operators and the cone.

Pták [234] proved that for an arbitrary Boolean algebra B and a convex compact subset K of a locally compact topological linear space there exists an orthomodular lattice L whose center is Boolean isomorphic to B and the set of states of the ql L is affinely homeomorphic to K. Thus, for every Boolean algebra B there exists a pure (without states) ql L_1 and a rigid (with a single state) ql L_2 with the centers of each of them isomorphic to B.

Dalla [111] investigated a minimal ql and an orthomodular ql to see whether they had the Lindenbaum property and discovered interesting situations. Philosophic and physical interpretations of the results obtained are curious.

The problem of a reasonable definition of implications in ql continued to attract attention. Bernini [71] considered a minimal ql QL whose language L_0 contains the logical symbols $\neg, \cap, \cup, \forall, \exists$ and defined an interpretation V_L of QL in the complete orthocomplemented lattice L. Further, he considered extensions of QL containing implication. Here he obtained necessary and sufficient conditions, expressed in the language of interpretation, for deducibility of the formula $A \to B$ in the extension. An axiomatic approach to the introduction of the implication was suggested in [193].

Ovchinnikov studied logics of functions on a set and proved Wigner's theorem on automorphisms of the logic of projectors on that basis. In the same paper [36] he stated the Gleason theorem for that logic. In [38] he proved a theorem on atoms—preserving embeddings of ql in ql such that every finitely additive function on them is constant on atoms. Similar results for logics of continuous functions on a topological space were published in [37].

Tsirulis (Cirulis) [49] noticed an analogy between certain structures appearing in the theory of nondeterministic automata and structures studied in ql. He connected an ortho-ordered set L and an algebra E of events with each automaton. The main result is that, in reality, L could be embedded in E and a filter in L corresponds to each state of the automaton.

In many papers on ql the emphasis is on orthomodular lattices and, in particular, on the problem of embedding of an orthomodular lattice in a Hilbert space. Gudder [167] gave necessary and sufficient conditions for such embeddability in terms of the existence of a Baer *-semigroup which was interpreted as a set of operations over quantum systems. Cook and Lambalgen [105] investigated connections beween structural properties of an orthomodular lattice L and the existence of two-valued homomorphisms of L. Various measures on the lattice of events of a quantum-mechanical system were studied by Pták and Rogalewicz [238]. Goldblatt [160] discovered that the orthomodularity property of the lattice of closed subspaces of a pre-Hilbert space could not be expressed in the first-order language with variables over vectors and the only relation, the orthogonality relation. He concluded that the property of being a Hilbert space is not elementary in that language. Bazhanov [2] considered the first-order language of the complete atomic lattices in which elementary formulas are two-term equalities and inequalities consisting of variables connected by $\cup, \cap$, and $\perp$. In this language one obtains a theory of ql which turns out to be undecidable. Another axiomatics of orthomodular lattices of 13 axioms and one deduction rule was produced in [188].

Cegla [95] constructed a new orthomodular lattice based on a study of Borel sets in a Minkowski space; he used it to describe a logic of causal events.

Daniel [112] generalized the definition of entropy to arbitrary quantum logics and gave arguments in favor of such a generalization.

An alternative ql approach to formalization of quantum physics was suggested by Mielnik [215]. It is based on a set Φ of states, a set Ξ of alternatives, and a mapping

$$g(\xi, t, t'): \Phi \to \Phi, \quad \xi \in \Xi,$$

which describes the evolution of the system from moment t to moment t'. Bugajski [92] described in detail a procedure of finding an inner language and inner logic of physical theories.

Lock and Hardegree ([205] and [206]) suggested the most general concept of orthoalgebra. They showed that the logic of events constructed earlier by Foulis and Randall could be deduced from their theory.

Zecca [287] studied tensor products of complete atomic orthomodular lattices. Pulmanová [241] suggested a new definition of a tensor product of ql. Its usefulness is in the fact that now both the tensor product of Hilbert logics and that of a Hilbert and classical logics are already defined.

Certain special problems of ql were considered in [110], [123], [224], [230], and [242].

§II.7. Menger algebras. Functionally complete algebras. Miscellaneous

Burtman [4] introduced notions of a suv-ideal, s-ideal, and a representative ideal of a Menger algebra. He succeeded in characterizing Menger algebras of multiplace linear selfmaps of a linear space over a division ring in terms of dense embeddings of these ideals. He proved that such a characterization does not hold for semigroups. Further, he proved in [6] that, for a finite-dimensional linear space A over a field F, the Menger algebra L of all multiplace linear mappings $f: A^n \to A$ contains an n-generated subalgebra whose linear envelope coincides with L. If A is countable-dimensional, such a subalgebra may have only two generators. A description of all congruences of the algebra of n-place linear self-maps of a linear space over a division ring was given in [5]. Here, for every cardinal number, there exists a unique congruence. Trokhimenko [45] found an elementary characterization of the connection relation on Menger algebras of multiplace functions ordered by the relation of extension of functions and quasi-ordered by the relation of inclusion of their domains (two functions are called connected if their graphs have a nonempty intersection). In [46] he found a criterion for a subset of a fundamentally ordered Menger algebra to be stationary. In [47] he gave a characterization of Menger algebras of multiplace functions bijective in each of the variables separately, with certain restrictions imposed on the superposition and their domains.

Gadzhiev [7] studied congruences on Menger algebras defined over a topological space. He clarified whether a topological space is determined by the Menger algebras over it. A joint paper by Gadzhiev and Mal'tsev [146] contains interesting results on the existence of dense subalgebras generated by certain sets of mappings in Menger algebras of continuous mappings of metrizable compacta.

Golunkov and Savel'ev [9] studied properties of the subalgebra lattices of algebras of single-place partially recursive functions (of predicates and systems of algorithmic algebras). In particular, it turned out that the lattice of systems of algorithmic algebras is not modular.

An algebra A is called functionally complete if every function $f\colon A^n \to A$ can be represented by a polynomial. Schweigert and Szymańska [253] showed that a prime modal algebra A that satisfies certain additional conditions is functionally complete. In another paper [254] they give analogous results and produce a great number of examples of lattice-polynomially complete algebras. Demetrovics and Hannák [114] established the existence of a continuum of nonequivalent types (signatures) of functionally complete algebras on any finite set A with $\mathrm{Card}(A) > 3$ (the equivalence of types is understood in the sense of the theory of clones of operations). Muter [32] studied simple, from the point of view of their realization by technical means, functionally complete bases formed by logic functions. His paper [33] belongs to this direction. In [34] he suggested new nondistributive algebras which, in his opinion, are more convenient for representing multivalued logic functions as two-level forms.

Beazer [61] generalized to Stone algebras with the least dense element a known result of Grätzer on the affine completeness of Boolean algebras. It turned out that such a Stone algebra A is affine complete if and only if the filter $D(A)$ of all its dense elements possesses no proper Boolean intervals. The latter condition is equivalent to the affine completeness of the lattice $D(L)$.

Algebras and Sugihara matrices were investigated in [220]. Drugush [16] introduced a notion of a forest logic, which models partially ordered sets with a tree order, and proved the decidability of the calculus so obtained. Also, he established that such a logic is residually finite.

A survey of BCK-algebras is contained in [167].

Bibliography

1. M. A. Abashidze, *Some properties of Magari algebras*, Studies in Logic and Semantics, "Metsniereba", Tbilisi, 1981, pp. 111–127. (Russian)

2. V. A. Bazhanov, *Logic of quantum mechanics and its decidability problems*, Logic, Cognition, and Reflection, Sverdlovsk, 1984, pp. 58–65. (Russian)

3. D. A. Bredikhin, *On Jónsson relation algebras*, Problems of Theoretical and Applied Mathematics, Tatru, 1985, pp. 35–36. (Russian)

4. M. I. Burtman, *Dense embeddings in the Menger algebra of linear mappings*, Inst. of Math. and Mechanics, Baku, 1983. (Russian)

5. ____, *Congruences on the Menger algebra of linear mappings*, Izv. Akad. Nauk Azerbaĭdzhan SSR Ser. Fiz.-Tekhn. Mat. Nauk 5 (1984), no. 3, 7–14. (Russian)

6. ____, *Finitely generated subalgebras of the Menger algebra of linear mappings*, Izv. Akad. Nauk Azerbaĭdzhan SSR Ser. Fiz.-Tekhn. Mat. Nauk 5 (1984), no. 2, 3–9. (Russian)

7. F. A. Gadzhiev, *Menger algebras*, Trudy Mat. Inst. Steklov. **163** (1984), 78–80; English transl. in Proc. Steklov Inst. Math. **163** (1985).

8. R. Goldblatt, *Topoi: the categorial analysis of logic*, North-Holland, Amsterdam and New York, 1979; 2nd ed., 1984.

9. B. V. Golunkov and A. A. Savelev, *A lattice of systems of algorithmic algebras of partial recursive functions and predicates*, Izv. Vyssh. Uchebn. Zaved. Mat. (1984), no. 11, 57–59; English transl. in Soviet Math. (Iz. VUZ) **28** (1984), no. 11.

10. E. I. Gordon, *Strongly unitary injective modules as linear spaces in Boolean-valued models of the set theory*, Gorkov. Gos. Univ., Gorki, 1984. (Russian)

11. R. Sh. Grigoliya, *Projective algebras with conjugate closures, Projective Skolem algebras*, Studies in Logic and Semantics, "Metsniereba", Tbilisi, 1981, pp. 103–110. (Russian)

12. ___, *Free Magari algebras with a finite number of generators*, Studies in Logic and Semantics, "Metsniereba", Tbilisi, 1983, pp. 135–149. (Russian)

13. V. N. Grishin, *A generalization of a system of Ajdukiewicz and Lambek*, Studies in Nonclassical Logics and Formal Systems, "Nauka", Moscow, 1983, pp. 315–334. (Russian)

14. V. A. Gudovshchikov, *Completeness theorem for propositional calculi characterized by nondistributive lattices*, Complexity Problems of Mathematical Logic, Kalinin Gos. Univ., Kalinin, 1985, pp. 33–36. (Russian)

15. E. Z. Dyment, *On an interpretation of the intuitionist predicate calculus by means of a calculus of problems with parameters*, Manuscript No. 1716-85, deposited at VINITI by the Editors of Izv. Vyssh. Uchebn. Zaved. Mat., 1985. (Russian)

16. Ya. M. Drugush, *Finite approximability of forest superintuitionist logics*, Mat. Zametki **36** (1984), no. 5, 755–764; English transl. in Math. Notes **36** (1984), no. 5–6.

17. M. V. Zakhar'yashchev, *On intermediate logics*, Dokl. Akad. Nauk SSSR **269** (1983), 18–22; English transl. in Soviet Math. Dokl. **27** (1983).

18. A. S. Karpenko, *Factor semantics for n-valued logics*, Studia Logica **42** (1983), no. 2–3, 179–184 (1984).

19. Yu. M. Kir'yakov, *A study of certain special properties of logic formulas of the calculus of propositions*, Rostov Civil Engineering Institute, Rostov-on-Don, 1983. (Russian)

20. A. V. Kuznetsov, *The proof-intuitionistic propositional calculus*, Dokl. Akad. Nauk SSSR **283** (1985), 27–30; English transl. in Soviet Math. Dokl. **32** (1985).

21. A. G. Kusraev, *Boolean-valued analysis of the duality of extended models*, Dokl. Akad. Nauk SSSR **267** (1982), 1049–1052; English transl. in Soviet Math. Dokl. **26** (1982).

22. V. A. Lyubetskiĭ, *Algebraic aspects of nonstandard analysis*, Moscow State Pedagogical Institute, Moscow, 1983. (Russian)

23. ___, *Sheaves on Heyting algebras: the case of rings*, Moscow State Pedagogical Institute, Moscow, 1984. (Russian)

24. ___, *Some algebraic questions of nonstandard analysis*, Dokl. Akad. Nauk SSSR **280** (1985), 38–41; English transl. in Soviet Math. Dokl. **31** (1985).

25. L. L. Maksimova, *Absence of the interpolation property at model companions of the Dummett logic*, Algebra i Logika **21** (1982), no. 6, 690–694; English transl. in Algebra and Logic **21** (1982), no. 6.

26. ___, *Interpolation properties of superintuitionistic, positive and modal logics*, Acta Philos. Fenn. **35** (1982), 70–78.

27. I. A. Mal'tsev, *Invariants of quasi cells of iterative Post algebras*, Sibirsk. Mat. Zh. **26** (1985), 220–223; English transl. in Siberian Math. J. **26** (1985).

28. V. Yu. Meskhi, *A variety of Heyting algebras with regular involution*, Studies in Logic and Semantics, "Metsniereba", Tbilisi, 1981, pp. 90–102. (Russian)

29. ___, *Varieties of Heyting algebras with involution which are not generated by finite elements*, Studies in Logic and Semantics, "Metsniereba", Tbilisi, 1984, pp. 42–46. (Russian)

30. ___, *On a discriminant variety of Heyting algebras with involution*, Algebra i Logika **21** (1982), no. 5, 537–552; English transl. in Algebra and Logic **21** (1982), no. 5.

31. A. Yu. Muravitskiĭ, *On extensions of the logic of provability*, Mat. Zametki **33** (1983), no. 6, 915–927; English transl. in Math. Notes **33** (1983), no. 5–6

32. V. M. Muter, *A representation of functions of many-valued logic in nondistributive logic-algebraic bases*, Northwestern Correspondence Polytechnic Institute, Leningrad, 1982. (Russian)

33. ____, *Nondistributive abstract algebras for analytic representation of functions of many-valued logic*, Northwestern Correspondence Polytechnic Institute, Leningrad, 1982. (Russian)

34. ____, *A few-level representations of functions of many-valued logic*, Northwestern Correspondence Polytechnic Institute, Leningrad, 1982.

35. I. S. Negru, *Algebraic properties of the lattice of Post classes and their many-valued generalizations*, Studies in Nonclassical Logics and Formal Systems, "Nauka", Moscow, 1983, pp. 300–315. (Russian)

36. P. G. Ovchinnikov, *On functional representations of quantum logics*, Izv. Vyssh. Uchebn. Zaved. Mat. (1984), no. 6, 75–78; English transl. in Soviet Math. (Iz. VUZ) **28** (1984), no. 6.

37. ____, *A finitely generated infinite quantum logic*, Funktsional′nyĭ Analiz. Spektral′naya teoriya, Ul′yanovsk, 1984, pp. 67–78. (Russian)

38. ____, *Finitely additive functions on extensions of quantum logics*, Izv. Vyssh. Uchebn. Zaved. Mat. (1985), no. 5, 74–75; English transl. in Soviet Math. (Iz. VUZ) **29** (1985).

39. V. V. Ryakov, *Bases of quasi identities of modal algebras*, Algebra i Logika **21** (1982), no. 2, 219–227; English transl. in Algebra and Logic **21** (1982), no. 2.

40. ____, *Elementary theories of free topo-Boolean and pseudo-Boolean algebras*, Mat. Zametki **37** (1985), no. 6, 797–802; English transl. in Math. Notes **37** (1985), no. 5–6.

41. ____, *Bases of admissible rules of logics* S4 *and* Int., Algebra i Logika **24** (1985), no. 1, 87–107; English transl. in Algebra and Logic **24** (1985), no. 1.

42. D. G. Skordev, *On an immersion of iterative Post algebras*, Algebra i Logika **21** (1982), no. 2, 228–241; English transl. in Algebra and Logic **21** (1982), no. 2.

43. L. A. Skornyakov, *Relation algebra of infinite matrices*, Problems of Theoretical and Applied Mathematics, Tartu, 1985, pp. 160–162. (Russian)

44. ____, *Matrix relational algebras*, Mat. Zametki **41** (1987), no. 2, 129–137; English transl. in Math. Notes **41** (1987), no. 1–2.

45. V. S. Trokhimenko, *A connection relation on ordered Menger algebras*, Ukrain. Mat. Zh. **35** (1983), no. 4, 461–466; English transl. in Ukrainian Math. J. **35** (1983), no. 4.

46. ____, *Stationary subsets of ordered Menger algebras*, Izv. Vyssh. Uchebn. Zaved. Mat. (1983), no. 8, 82–84; English transl. in Soviet Math. (Iz. VUZ) **27** (1983), no. 8.

47. ____, *On the theory of restrictive Menger algebras*, Ukrain. Mat. Zh. **36** (1984), 82–87; English transl. in Ukrainian Math. J. **36** (1984).

48. S. Ulam, *A collection of mathematical problems*, Interscience, New York, 1960.

49. Ya. P. Tsirulis (J. P. Cirulis), *Variations on the theme of quantum logic*, Algebra and Discrete Mathematics, Latv. Gos. Univ., Riga, 1984, pp. 146–158. (Russian)

50. A. A. Shum, *Structurally regular models of intuitionistic propositional calculus*, Complexity Problems of Mathematical Logic, Kalinin Gos. Univ., Kalinin, 1985, pp. 97–101. (Russian)

51. L. L. Esakia, *Algebras of logic*, Ordered Sets and Lattices, Saratov Gos. Univ., Saratov (1983), no. 7, 15–26. (Russian)

52. ____, *Heyting algebras*. 1. *Duality theory*, "Metsniereba", Tbilisi, 1985. (Russian)

53. M. Abad and L. Monteiro, *On the three-valued Moisil algebra*, Logique et Anal. **27** (1984), no. 108, 407–414.

54. M. C. Abbati and A. Mania, *The quantum logical and the operational description for physical systems*, Current Issues in Quantum Logic (Proc. Workshop on Quantum Logic; Erice, 1979), Plenum Press, New York, 1981, pp. 119–127.

55. M. E. Adams, V. Koubek and J. Sichler, *Endomorphisms and homomorphisms of Heyting algebras*, Algebra Universalis **20** (1985), no. 2, 167–178.

56. D. Aerts, *Description of compound physical systems and logical interaction of physical systems*, Current Issues in Quantum Logic (Proc. Workshop on Quantum Logic; Erice, 1979), Plenum Press, New York, 1981, pp. 381–404.

57. J. Anderson, *A conjecture concerning the pure states of* $B(H)$ *and a related theorem*, Topics in Modern Operator Theory (Timisoara/Herculane; 1980), Birkhauser, Basel and Boston, 1981, pp. 27–43.

58. H. Andréka and I. Németi, *Dimension-complemented and locally finite dimensional cylindric algebras are elementary equivalent*, Algebra Universalis **13** (1981), no. 2, 157–163.

59. ____, *Cylindric relativised set algebras*, Lecture Notes in Math. no. 883, Springer-Verlag, Berlin and New York, 1981, pp. 131–323.

60. M. Banai, *Propositional systems in field theories and lattice-valued quantum logic*, Current Issues in Quantum Logic (Proc. Workshop on Quantum Logic; Erice, 1979), Plenum Press, New York, 1981, pp. 425–435.

61. R. Beazer, *Affine complete Stone algebras*, Acta Math. Hungar. **39** (1982), no. 1–3, 169–174.

62. ___, *Affine complete double Stone algebras with bounded core*, Algebra Universalis **16** (1983), no. 2, 237–244.

63. ___, *On some small varieties of distributive Ockham algebras*, Glasgow Math. J. **25** (1984), no. 2, 175–181.

64. ___, *Injectives in some small varieties of Ockham algebras*, Glasgow Math. J. **25** (1984), no. 2, 183–191.

65. ___, *Congruence pairs for algebras abstracting Kleene and Stone algebras*, Czechoslovak Math. J. **35(110)** (1985), no. 1–2, 260–268.

66. A. R. Bednarek and S. M. Ulam, *Generators for algebras of relations*, Bull. Amer. Math. Soc. **82** (1976), no. 5, 781–782.

67. ___, *Projective algebra and the calculus of relations*, J. Symbolic Logic **43** (1978), no. 1, 56–64.

68. L. Beran, *Orthomodular lattices. Algebraic approach*, Mathematics and its Applications, Reidel, Dordrecht, Boston, 1985.

69. G. M. Bergman, *The rank of two-dimensional cylindric set algebras*, Colloq. Math. Soc. János Bolyai **29** (1982), 95–105.

70. C. Bernardi and F. Montagna, *Equivalence relations induced by extensional formulae: classification by means of a new fixed point property*, Fund. Math. **124** (1984), no. 3, 221–233.

71. S. Bernini, *Quantum logic as an extension of classical logic*, Current Issues in Quantum Logic (Proc. Workshop on Quantum Logic; Erice, 1979), Plenum Press, New York, 1981, pp. 161–171.

72. A. Biele and T. Stepién, *Lindenbaum's extensions*, Polish Acad. Sci. Inst. Philos. Sociol. Bull. Sect. Logic **10** (1981), no. 1, 42–47.

73. B. Biro, *On isomorphic but not lower-base isomorphic cylindric set algebras*, Polish Acad. Sci. Inst. Philos. Sociol. Bull. Sect. Logic **13** (1984), no. 4, 230–232.

74. T. S. Blyth and J. C. Varlet, *On a common abstraction of De Morgan algebras and Stone algebras*, Proc. Roy. Soc. Edinburgh. Sect. A **94** (1983), no. 3–4, 301–308.

75. ___, *Subvarieties of the class of MS-algebras*, Proc. Roy. Soc. Edinburgh Sect. A **95** (1983), no. 1–2, 157–169.

76. ___, *Double MS-algebras*, Proc. Roy. Soc. Edinburgh Sect. A **98** (1984), no. 1–2, 37–47.

77. ___, *Congruences on MS-algebras*, Bull. Soc. Roy. Sci. Liège **53** (1984), no. 6, 341–362.

78. ___, *Subdirectly irreducibile MS-algebras*, Proc. Roy. Soc. Edinburgh Sect. A **98** (1984) no. 3–4, 241–247.

79. ___, *Amalgamation properties in the class of MS-algebra*, Proc. Roy. Soc. Edinburgh Sect. A **100** (1985), 139–149.

80. ___, *MS-algebras definable on a distributive lattice*, Bull. Soc. Roy. Sci. Liège **54** (1985), no. 3, 167–182.

81. ___, *Fixed point in MS-algebras*, Bull. Soc. Roy. Sci. Liège **53** (1984), no. 1, 3–8.

82. J. Brabec and P. Pták, *On compatibility in quantum logics*, Fund. Theories Phys. **12** (1982), no. 2, 207–212.

83. Ch. Brink, *On Birkhoff's postulates for a relation algebra*, J. London Math. Soc. **15** (1977), no. 3, 391–394.

84. ___, *Boolean modules*, J. Algebra **71** (1981), no. 2, 291–313.

85. Ch. Brink and C. Hall, *The algebra of relatives*, Notre Dame J. Formal Logic **20** (1979), no. 4, 900–908.

86. M. Božić, *p-mappings in De Morgan lattices and their applications onto truth-filters*, Algebraic Conference (Skopje, 1980), Univ. "Kiril et Metodij", Skopje, 1981, pp. 43–51.

87. S. Bozzi and G. C. Meloni, *Representations of Heyting algebras with covering and propositional intuitionistic logic with local operator*, Bol. Un. Mat. Ital. A **17** (1980), no. 3, 436–442.

88. G. Bordalo, *Strongly N-normal lattices*, Nederl. Akad. Wetensh. Proc. Ser. A. **87** (1984), no. 2, 113–125.

89. F. Borceaux and G. Van den Bossche, Structure des topologies d'un topos, Cahiers Topologie Geom. Différentielle **25** (1984), no. 1, 37–39.

90. ____, *Algebra in a localic Topos with applications to ring theory*, Lecture Notes in Math. no. 1038, Springer-Verlag, Berlin and New York, 1983, pp. 1–240.

91. J. Bub, *What does quantum logic explain?* Current Issues in Quantum Logic (Proc. Workshop on Quantum Logic; Erice, 1979), Plenum Press, New York, 1981, pp. 89–100.

92. S. Bugajski, *What is quantum logic*, Studia Logica **41** (1982), no. 4, 311–316.

93. W. Buczkowski and T. Prucnal, *Topological representation of co-diagonalisable algebras*, Math. Res. **20** (1984), 63–65.

94. D. Busneag, *A note on deductive systems of a Hilbert algebra*, Kobe J. Math. **2** (1985), no. 1, 29–35.

95. W. Cegla, *Causal logic of Minkowski space*, Current Issues in Quantum Logic (Proc. Workshop on Quantum Logic; Erice, 1979), Plenum Press, New York, 1981, pp. 419–424.

96. G. L. Cherlin, *Lindenbaum algebras and model companions*, Fund. Math. **104** (1979), no. 3, 213–219.

97. R. Cignoli, *An algebraic approach to elementary theory based on n-valued Lukasiewicz logics*, Z. Math. Logik Grundlag. Math. **30** (1984), no. 1.

98. ____, *Proper n-valued Lukawiewicz algebras as S-algebras of Lukasiewicz n-valued propositional calculi*, Studia Logica **41** (1982), no. 1, 3–16.

99. R. Cignoli and M. S. de Gallego, *The lattice structure of some Lukasiewicz algebras*, Algebra Universalis **13** (1981), no. 3, 315–328.

100. ____, *Dualities for some De Morgan algebras with operators and Lukasiewicz algebras*, J. Austral. Math. Soc. Ser. A **34** (1983), no. 3, 377–393.

101. S. D. Comer, *A new foundation for the theory of relations*, Notre Dame J. Formal Logic **24** (1983), no. 2, 181–197.

102. ____, *Combinatorial aspects of relations*, Algebra Universalis **18** (1984), no. 1, 77–94.

103. W. H. Cornish, *BCK-algebras with a supremum.* II *Distributivity and interpolation.* Math. Japon **29** (1984), no. 3, 439–447.

104. R. M. Cook and J. Hilgevoord, *A new approach to equivalence in quantum logic*, Current Issues in Quantum Logic (Proc. Workshop on Quantum Logic; Erice, 1979), Plenum Press, New York, 1981, pp. 101–113.

105. R. M. Cook and M. Lambalgen, *The representation of Takeuti's operator*, Studia Logica **42** (1983), no. 4, 407–415.

106. J. Coulon and J.-L. Coulon, *Fuzzy Boolean algebras*, J. Math. Anal. Appl. **99** (1984), no. 1, 248–256.

107. *Current issues in quantum logic* (Proc. Workshop on Quantum Logic; Erice, 1979), Plenum Press, New York, 1981.

108. N. J. Cutland and P. F. Gibbins, *A regular sequant calculus for quantum logic in which $\wedge$ and $\vee$ are dual*, Logique et Anal. **25** (1982), no. 99, 221–248.

109. L. Henkin, J. D. Monk, A. Tarski, H. Andréka, and I. Németi, *Cylindric set algebras*, Lecture Notes in Math. vol. 883, Springer-Verlag, New York and Berlin, 1981.

110. J. Czelakowski, *Partial referential matrices for quantum logics*, Current Issues in Quantum Logic (Proc. Workshop on Quantum Logic; Erice, 1979), Plenum Press, New York, 1981, pp. 131–146.

111. M. L. Dalla Chiara, *Some metalogical pathologies of quantum logic*, Current Issues in Quantum Logic (Proc. Workshop on Quantum Logic; Erice, 1979), Plenum Press, New York, 1981, pp. 147–159.

112. W. Daniel, *The entropy of observables on quantum logic*, Rep. Math. Phys. **19** (1984), no. 3, 325–334.

113. B. A. Davey, *Dualities for Stone algebras, double Stone algebras and relative Stone algebras*, Coll. Mat. **46** (1982), no. 1, 1–14.

114. J. Demetrovics and L. Hannák, *On the number of functionally complete algebras*, (Proc. 12th Internat. Sympos. Multiple-Valued Logic, Paris, May 25–27, 1982), New York, 1982, pp. 329–330.

115. P. S. Deliannis, *Geometrical models for quantum logics with conditioning*, J. Math. Phys. **25** (1984), no. 10, 2939–2949.

116. ___, *Quantum logics derived from asymmetric Mielnik forms*, Internat. J. Theoret. Phys. (1984), no. 3, 217–226.

117. H. Dobbertin, *Measurable refinement monoids and applications to distributive semilattices, Heyting algebras and Stone algebras*, Math. Z. **187** (1984), no. 1, 13–21.

118. D. Dorninger, H. Länger, and M. Mączyński, *Zur Darstellung von Observablen auf σ-stetigen Quantenlogiken*, Österreich. Akad. Wiss. Math.-Natur. Kl. Sitzungsber. II **192** (1983), ABT. 2, no. 4–7, 169–176.

119. J. M. Dunn, *A relational representation of quasi-Boolean algebras*, Notre Dame J. Formal Logic **23** (1982), no. 4, 353–357.

120. I. Düntsch, *A description of the projective Stone algebras*, Glasgow Math. J. **24** (1983), no. 1, 75–82.

121. ___, *On free or projective regular double Stone algebras*, Houston J. Math. **9** (1983), no. 4, 455–463.

122. ___, *Projective, prime ideals and chain conditions of Stone algebras*, Algebra Universalis **14** (1982), no. 2, 167–180.

123. A. Dvurečenskij and S. Pulmanová, *Connection between joint distribution and compatibility*, Rep. Math. Phys. **19** (1984), no. 3, 349–359.

124. W. Dzik, *On the content of lattices of logics. Part I, The representation theorem for lattices of logics*, Rep. Math. Logic (1981), no. 13, 17–27.

125. ___, *On the content of lattices of logics. Part II*, Rep. Math. Logic (1982), no. 14, 29–47.

126. W. Dziobiak, *On distributivity of the lattice of subquasivarieties of variety of Heyting algebras*, Polish Acad. Sci. Inst. Philos. Sociol. Bull. Sect. Logic **12** (1983), no. 1, 37–40.

127. ___, *Concerning axiomatisability of the quasivariety generated by a finite Heyting or topological Boolean algebra*, Studia Logica **41** (1982), no. 4, 415–428.

128. K. Eda, *Completions and co-products of Heyting algebras*, Tsukuba J. Math. **5** (1981), no. 2, 195–222.

129. G. Epstein and A. Horn, *Core points in double Heyting algebras and dissectable lattices*, Algebra Universalis **16** (1983), no. 2, 204–218.

130. F. Erdös, V. Faber, and J. Larson, *Sets of natural numbers of positive density and cylindric set algebras of dimension 2*, J. Math. Anal. Appl. **100** (1984), no. 2, 463–469.

131. F. Esteva, *Some representable De Morgan algebras*, J. Math. Anal. Appl. **100** (1984), no. 2, 463–469.

132. F. Ayissi Eteme, *Complétion chrysippienne d'un algèbre de Lukasiewicz θ-valent*, C. R. Acad. Sci. Paris Sér. I Math. **299** (1984), no. 3, 69–72.

133. ___, *Anneau chrysippiens θ-valents*, C. R. Acad. Sci. Paris Sér. I Math. **299** (1984), no. 1, 1–4.

134. M. Eytan, *Fuzzy sets: a topos-logical point of view*, Fuzzy Sets and Systems **5** (1981), no. 1, 47–67.

135. ___, *Some remarks on the category "Ponnase D."*, Fuzzy Sets and Systems **9** (1983), no. 2, 199–204.

136. M. Ferencsi, *Measures on cylindric algebras*, Polish Acad. Sci. Inst. Philos. Sociol. Bull. Sect. Logic **11** (1984), no. 3–4, 142–153.

137. A. V. Figallo, I_3-∇-*algebras*, Rev. Colombana Mat. **17** (1983), no. 3–4, 105–116.

138. A. Filipou, *Some remarks on the representation theorem of Moisil*, Discrete Math. **33** (1981), no. 2, 163–170.

139. D. Finkelstein, *Quantum sets, assemblies and plexi*, Current Issues in Quantum Logic (Proc. Workshop on Quantum Logic; Erice, 1979), Plenum Press, New York, 1981, pp. 323–331.

140. B. C. van Fraasen, *A model interpretation of quantum mechanics*, Current Issues in Quantum Logic (Proc. Workshop on Quantum Logic; Erice, 1979), Plenum Press, New York, 1981, pp. 229–258.

141. J. M. Font, *Monadicity in topological pseudo-Boolean algebras*, Lecture Notes in Math. no. 1103, Springer-Verlag, Berlin and New York, 1984, pp. 169–192.

142. P. J. Foulis and C. H. Randall, *What are quantum logics and what ought they to be?*, Current Issues in Quantum Logic (Proc. Workshop on Quantum Logic; Erice, 1979), Plenum Press, New York, 1981, pp. 35–52.

143. P. Fowler, *De Morgan algebras*, Bull. Austral. Math. Soc. **25** (1982), no. 2, 305–307.

144. M. P. Fourman, *Continuous Truth.* 1. *Nonconstructive objects*, Logic Colloquium 82 (Florence, 1982), North-Holland, Amsterdam, 1984, pp. 161–180.

145. P. Gärdenfors, *Propositional logic on the dynamics of belief*, J. Symbolic Logic **50** (1985), no. 2, 390–394.

146. F. A. Gadzhiev and A. A. Mal'tsev, *On dense subalgebras of Post algebras and Menger algebras of continuous functions*, Lecture Notes in Math. no. 1060, Springer-Verlag, New York and Berlin, 1984, pp. 258–266.

147. G. Georgescu, *On a polyadic version of the omitting types theorem*, Stud. Cerc. Mat. **32** (1980), no. 5, 505–515.

148. ____, *Monotone quantifiers on polyadic algebras*, Stud. Cerc. Mat. **34** (1982), no. 2, 125–145.

149. ____, *Algebraic analysis of the topological logic $L(I)$*, Z. Math. Logik Grundlag. Math. **28** (1982), no. 5, 447–454.

150. ____, *On the Popper-Carnap probability functions in polyadic algebras*, Rev. Roumaine Math. Pures Appl. **28** (1983), no. 7, 583–585.

151. ____, *Chang's modal operators in algebraic logic*, Studia Logica **42** (1983), no. 1, 43–48.

152. ____, *A representation theorem for polyadic Heyting algebras*, Algebra Universalis **14** (1982), no. 2, 197–209.

153. ____, *Algebraic analysis of the logic with the quantifier "there exist uncountably many"*, Algebra Universalis **19** (1984), no. 1, 99–105.

154. G. Georgescu and I. Voiculescu, *Eastern model theory for Boolean-valued theories*, Z. Math. Logik Grundlag. Math. **31** (1985), no. 1, 79–88.

155. M. Glas, *Sur l'atomicité des algèbres de Lucasiewicz*, Rev. Roumaine Math. Pures Appl. **29** (1984), no. 3, 231–233.

156. G. Gierz and A. Stralka, *The injective hull of a distributive lattice and congruence extension*, Quart. J. Math. Oxford Ser. (2) **35** (1984), no. 137, 25–36.

157. M. S. Goldberg, *Distributive Ockham algebras: Free algebras and injectivity*, Bull. Austral. Math. Soc. **24** (1981), no. 2, 161–203.

158. ____, *Topological duality for distributive Ockham algebras*, Studia Logica **42** (1983), no. 1, 21–31.

159. R. Goldblatt, *Grothendieck topology as geometric modality*, Z. Math. Logik Grundlag. Math. **27** (1981), no. 6, 495–529.

160. ____, *Orthomodularity is not elementary*, J. Symbolic Logic **49** (1984), no. 2, 401–404.

161. ____, *Topoi: the categorial analysis of logic*, North-Holland, Amsterdam, 1984.

162. H. J. Goltz, *The Boolean sentence algebra of the theory of linear ordering is atomic with respect to logics with Malitz quantifier*, Z. Math. Logik Grundlag Math. **31** (1985), no. 2, 131–162.

163. R. J. Grayson, *Heyting-valued semantics*, Logic Colloquium '82. (Florence, 1982), North-Holland, Amsterdam, 1984, pp. 181–208.

164. R. J. Greechie, *A non-standard quantum logic with a strong set of states*, Current Issues in Quantum Logic, (Proc. Workshop on Quantum Logic; Erice, 1979), Plenum Press, New York, 1981, pp. 375–380.

165. R. Sh. Grigoliya and S. Meskhi, *Critical varieties of biclosure algebras*, Polish Acad. Sci. Inst. Philos. Sociol. Bull. Sect. Logic **10** (1981), no. 1, 2–8.

166. S. P. Gudder, *Measure and integration in quantum set theory*, Current Issues in Quantum Logic (Proc. Workshop on Quantum Logic; Erice, 1979), Plenum Press, New York, 1981, pp. 341–352.

167. ____, *Representations of Baer ∗-semigroups and quantum logics in Hilbert space*, Current Issues in Quantum Logic (Proc. Workshop on Quantum Logic; Erice, 1979), Plenum Press, New York, 1981.

168. P. Hamburg, *Applications entre des treillis morganiens*, Itinerant Seminar on Functional Equations, Approximation and Convexity (Cluj-Napoca, 1984), Preprint No. 84-6, Univ. "Babes-Bolyai", Cluj-Napoca, 1984, pp. 65–72.

169. G. M. Hardegre, *Some problems and methods in formal quantum logic*, Current Issues in Quantum Logic (Proc. Workshop on Quantum Logic; Erice, 1979), Plenum Press, New York, 1981, pp. 209–225.

170. C. Hatvany, *Ideals in Lukasiewicz algebras*, Inst. Politehn. Timisoara Lucrar. Sem. Mat. Fiz. (1984), May, 75–77.

171. D. Higgs, *Injectivity in the topos of compact Heyting-valued sets*, Canad. J. Math. **36** (1984), no. 3, 550–568.

172. L. Hon-Fei, *On Kleene algebras*, Algebre Universalis **11** (1980), no. 1, 117–126.

173. R. I. G. Hughes, *Realism and quantum logic*, Current Issues in Quantum Logic (Proc. Workshop on Quantum Logic; Erice, 1979), Plenum Press, New York, 1981, pp. 77–87.

174. L. Henkin, J. D. Monk, and A. Tarski, *Cylindric algebras. Part I*, North-Holland, Amsterdam, 1971.

175. ____, *Cylindric set algebras and related structures*, Lecture Notes in Math. no. 883, Springer-Verlag, Berlin and New York, 1981, pp. 1–130.

176. ____, *Cylindric algebras. Part II*, North-Holland, Amsterdam, 1985.

177. T. Imieliński and W. J. Lipski, *The relational model of data and cylindric algebras*, J. Comput. System Sci. **28** (1984), no. 1, 80–102.

178. A. Iorgulescu, *Functors between categories of three-valued Lukasiewicz-Moisil algebras. I*, Discrete Math. **49** (1984), no. 2, 121–131.

179. ____, *On the constructions of three-valued Lukasiewicz-Moisil algebra*, Discrete Math. **48** (1984), no. 2–3, 213–227.

180. L. Iturrioz, *Modal operators on symmetrical Heyting algebras*, Universal Algebra and Applications (Warsaw, 1978), PWN, Warsaw, 1982, pp. 289–303.

181. ____, *Symmetrical Heyting algebras with operators*, Z. Math. Logik Grundlag. Math. **29** (1983), no. 1, 33–79.

182. K. Iséki, *Some problems of BCK-algebras and Griss type algebras*, Universal Algebra and Applications (Warsaw 1978), PWN, Warsaw, 1982, pp. 423–430.

183. I. Janicka-Zuk, *Strong amalgamation properties of diagonalisable algebras*, Polish Acad. Sci. Inst. Philos. Sociol. Bull. Sect. Logic **12** (1983), no. 3, 105–110.

184. A. W. Jankowski, *A conjunction in closure spaces*, Studia Logica **43** (1984), no. 4, 341–351.

185. B. Jónsson, *Varieties of relation algebras*, Algebra Universalis **15** (1982), no. 3, 273–298.

186. ____, *Maximal algebras of binary relations*, Contemp. Math. **33** (1984), 299–307.

187. J. K. Kabziński, *Basic properties of the equivalence*, Studia Logica **41** (1982), no. 1, 17–40.

188. G. Kalmbach, *Omologic as a Hilbert type calculus*, Current Issues in Quantum Logic (Proc. Workshop on Quantum Logic; Erice, 1979), Plenum Press, New York, 1981, pp. 333–340.

189. ____, *Orthomodular lattices*, Academic Press, London, 1983.

190. ____, *1982 news about orthomodular lattices*, Discrete Math. **53** (1985), 125–135.

191. A. S. Karpénko, *Factor semantics for n-valued logics*, Studia Logica **42** (1983), no. 2–3, 179–185.

192. Y. Komori, *Super-Lukasiewicz propositional logics*, Nagoya Math. J. **84** (1981), no. 1, 119–133.

193. A. Kron, Z. Morić, and S. Vujošević, *Entailment and quantum logic*, Current Issues in Quantum Logic (Proc. Workshop on Quantum Logic; Erice, 1979), Plenum Press, New York, 1981, pp. 193–207.

194. Hon Fei Lai, *On Kleene algebras*, Algebra Universalis **11** (1980), no. 1, 117–126.

195. J. A. Larson, *The number of one-generated diagonal-free cylindric set algebras of finite dimension greater than two*, Algebra Universalis **16** (1983), no. 1, 1–16.

196. ____, *The number of finitely generated infinite cylindric set algebras of dimension two*, Algebra Universalis **19** (1984), no. 3, 377–396.

197. ____, *The number of one-generated cylindric set algebras*, J. Symbolic Logic **50** (1985), no. 1, 59–71.

198. E. G. K. Lópes-Escobar, *Equivalence between semantics for intuitionism*, I, J. Symbolic Logic **46** (1981), no. 4, 773–780.

199. I. Loureiro, *Axiomatisation et propriétés des algèbres modales tétravalentes*, C. R. Acad. Sci. Paris Sér. I Math. **295** (1982), no. 10, 555–557.

200. ____, *Prime spectrum of a tetravalent modal algebra*, Notre Dame J. Formal Logic **24** (1983), no. 3, 389–394.

201. ____, *Finitely generated free tetravalent modal algebras*, Discrete Math. **46** (1983), no. 1, 41–48.

202. F. Luci, *A theory of potential infinity*. II, Boll. Un. Mat. Ital. D (6) **1** (1982), no. 1, 21–39.

203. C. Luppi, *Sul problema della rappresentabilita par AHT*, Riv. Mat. Univ. Parma **8** (1982), 91–106.

204. ____, *Su alcune varietà notevoli di algebra di Heyting topologiche*, Riv. Mat. Univ. Parma **9** (1983), 59–65.

205. P. F. Lock and G. M. Hardegree, *Connections among quantum logics. Part I, Quantum propositional logics*, Internat. J. Theoret. Phys. **24** (1985), no. 1, 43–53.

206. ____, *Connections among quantum logics. Part 2. Quantum event logics*, Internat. J. Theoret. Phys. **24** (1985), no. 1, 55–61.

207. M. I. Mączyński, *Commutativity and generalized transition properties in quantum logic*, Current Issues in Quantum Logic (Proc. Workshop on Quantum Logic; Erice, 1979), Plenum Press, New York, 1981, pp. 355–364.

208. R. Maddux, *Some varieties containing relation algebras*, Trans. Amer. Math. Soc. **272** (1982), no. 2, 501–526.

209. ____, *A sequent calculus for relation algebras*, Ann. Pure Appl. Logic **25** (1983), no. 1, 73–101.

210. R. Magari, *Algebraic logic and diagonal phenomena*, Logic Colloquium 82 (Florence, 1982), North-Holland, Amsterdam, 1984, pp. 135–144.

211. A. R. Marlow, *Space-time structure from quantum logic*, Current Issues in Quantum Logic (Proc. Workshop on Quantum Logic; Erice, 1979), Plenum Press, New York, 1981, pp. 413–418.

212. R. M. Martin, *On servants, lovers, and benefactors; Peirce's algebra of relations*, J. Philos. Logic **7** (1978), no. 1, 27–48.

213. R. K. Meyer, *More implicative extensions for Belnap lattices*, Rep. Math. Logic (1981), no. 13, 43–51.

214. P. Mittelstaedt, *The dialogic approach to modalities in the language of quantum physics*, Current Issues in Quantum Logic (Proc. Workshop on Quantum Logic; Erice, 1979), Plenum Press, New York, 1981, pp. 259–281.

215. B. Mielnik, *Motion and form*, Current Issues in Quantum Logic (Proc. Workshop on Quantum Logic; Erice, 1979), Plenum Press, New York, 1981, pp. 465–477.

216. B. Molzan, *On the theory of Boolean algebras in the logic with Ramsey quantifiers*, Proceedings of the third Easter conference on model theory (Gross Köris, 1985), Seminarberichte **70**, Humboldt Univ., Berlin, 1985, pp. 186–192.

217. F. Montagna, *A completeness result for fixed-point algebras*, Z. Math. Logik Grundlag. Math. **30** (1984), no. 6, 525–532.

218. L. Monteiro, *M-algèbres de Kleene*, An. Acad. Brasil. Ciênc. **53** (1981), no. 4, 665–672.

219. Ch. G. Morgan, *Probabilistic semantic for orthologic and quantum logic*, Logique et Anal. **26** (1983), no. 1, 103–104, 323–339.

220. Ch. Mortersen, *Model structures and set algebras for Sugihara matrices*, Notre Dame J. Formal Logic **23** (1982), no. 1, 85–90.

221. D. Mundici, *Abstract model theory of many-valued logics and K-theory of certain C^*-algebras*, Proceedings of the second Easter conference on model theory (Wittenberg, 1984), Seminarberichte **60**, Humboldt Univ., Berlin, 1984, pp. 157–204.

222. C. V. Negoita, *Fuzzy sets in topoi*, Fuzzy Sets and Systems **8** (1982), no. 1, 93–99.

223. I. Németi, *The class on neat-reducts of cylindric algebras is not a variety but is closed with respect to HP*, Notre Dame J. Formal Logic **24** (1983), no. 3, 399–409.

224. M. Nowakowska, *A new theory of time: Generation of time from fuzzy temporal relation*, Polish Acad. Sci. Inst. Philos. Sociol. Bull. Sect. Logic **10** (1984), no. 2, 56–62.

225. M. Palasiński, *Representation theorem for commutative BCK-algebras*, Math. Sem. Notes Kobe Univ. **10** (1982), no. 2, 473–478.

226. V. Patryshev, *On the Grothendieck topologies in the toposes of presheaves*, Cahiers Topologie Géom. Différentielle **25** (1984), no. 2, 207–215.

227. Ch. C. Pinter, *Cylindric algebras with a property of Rasiowa and Sikorski*, Mathematical Logic (Proc. First Brazilian Conf., State Univ., Campinas, 1977), Lect. Notes Pure Appl. Math. **39**, Marcel Dekker, New York, 1978, pp. 225–231.

228. A. M. Pitts, *Amalgamation and interpolation in the category of Heyting algebras*, J. Pure Appl. Algebra **29** (1983), no. 2, 155–165.

229. ___, *Fuzzy sets do not form a topos*, Fuzzy Sets and Systems **8** (1982), no. 1, 101–104.

230. C. Platon, *Geometrical models for quantum logics with conditioning*, J. Math. Phys. **25** (1984), no. 10, 2939–2946.

231. D. Ponasse, *Some remarks on the category* Fuz(H) *of M. Eytan*, Fuzzy Sets and Systems **9** (1983), no. 2, 199–265.

232. H. Porta, *Sur quelques algèbres de la logique*, Portugal. Math. **40** (1981), no. 1, 41–77.

233. T. Prucnal and W. Buszkowski, *Topological representation of co-diagonalizable algebras*, Frege conference (Schwerin, 1984), Math. Res. **20**, Akademie-Verlag, Berlin, 1984, pp. 63–65.

234. P. Pták, *Logics with given centers and state spaces*, Proc. Amer. Math. Soc. **88** (1983), no. 1, 106–109.

235. ___, *Weak dispersion-free states and the hidden variables hypothesis*, J. Math. Phys. **24** (1983), no. 4, 839–840.

236. ___, *Spaces of observables*, Czechoslovak Math. J. **34** (1984), no. 4, 552–561.

237. ___, *Categories of orthomodular posets*, Math. Slovaca **35** (1985), no. 1, 59–65.

238. P. Pták and V. Rogalewicz, *Regular full logics and the uniqueness problem for observables*, Ann. Inst. H. Poincaré, **A38** (1983), no. 1, 69–74.

239. ___, *Measures on orthomodular partially ordered sets*, J. Pure Appl. Algebras **28** (1983), no. 1, 75–80.

240. D. Pigozzi, *Varieties of pseudo-Post algebras*, Proc. 10th Internat. Sympos. Multiple-Valued Logic (Evanston, Ill, 1980), IEEE, Long Beach, Calif., 1980, pp. 205–208.

241. S. Pulmanová, *Tensor product of quantum logics*, J. Math. Phys. **26** (1985), no. 1, 1–5.

242. S. Pulmanová and A. Dvurečenskij, *Stochastic processes on quantum logics*, Rep Math. Phys. **18** (1980), no. 3, 303–315.

243. I. Pykacz, *Affine Mączyński logics on compact convex sets of states*, Internat. J. Theoret. Phys. **22** (1983), no. 2, 97–106.

244. R. Quellet, *A categorical approach to polyadic algebras*, Studia Logica **41** (1982), no. 4, 317–327.

245. S. Ribeyre, *Structures flues et algèbres de Post*, Publ. Dep. Math. (Lyon) **15** (1978), no. 1, 1–37.

246. A. Romanowska, *Subdirectly irreducible pseudocomplemented De Morgan algebras*, Algebra Universalis **12** (1981), no. 1, 70–75.

247. A. Rose, *Completeness of sets of three-valued Sheffer functions*, Z. Math. Logik Grundlag Math. **29** (1983), no. 6, 481–483.

248. H. P. Sankappanavar, *Heyting algebras with dual pseudo-complementation*, Pacific J. Math. **117** (1985), no. 2, 405–415.

249. ___, *A characterization of principal congruences of De Morgan algebras and its applications*, Mathematical Logic in Latin America (Proc. IV Latin Amer. Sympos. Math. Logic, Santiago, 1978), North-Holland, Amsterdam, 1980, pp. 341–349.

250. G. Schmidt and T. Ströhlein, *Relation algebras: Concept of points and representability*, Discrete Math. **54** (1985), no. 1, 83–92.

251. W. Schönfeld, *An indecidability result for relation algebras*, J. Symbolic Logic **44** (1979), no. 1, 111–115.

252. D. Schwarts, *Cylindric algebras with filter quantifiers*, Z. Math. Logik Grundlag. Math. **26** (1980), no. 3, 251–254.

253. D. Schweigert and M. Szymańska, *Polynomial functions of correlation lattices*, Algebra Universalis **16** (1983), no. 3, 355–359.

254. ___, *On functionally complete modal algebras related to* Y *and* S4, Demonstration Math. **16** (1983), no. 2, 323–328.

255. A. Sendlewski, *Topological duality for Nelson algebras and its applications*, Polish Acad. Sci. Inst. Philos. Sociol. Bull. Sect. Logic **13** (1984), no. 4, 215–221.

256. ___, *Some investigations of varieties of $\mathcal{N}$-lattices*, Studia Logica **43** (1984), no. 3, 257–280.

257. A. M. Sett, *Partial isomorphism extension for Post-language*, Z. Math. Logik Grundlag. Math. **30** (1984), no. 4, 289–293.

258. C. Smoryński, *Fixed point algebras*, Bull. Amer. Math. Soc. (N.S.) **6** (1982), no. 3, 317–356.

259. ___, *The finite inseparability of the first-order theory of diagonalizable algebras*, Studia Logica **41** (1982), no. 4, 347–349.

260. G. Sereny, *Compact cylindric set algebras*, Polish Acad. Sci. Inst. Philos. Sociol. Bull. Sect. Logic **14** (1985), no. 2, 57–64.

261. M. Servi, *An axiomatization of existential lattices*, Bol. Un. Mat. Ital. A(5) **16** (1979), no. 2, 298–301.

262. R. M. Solovay, *Infinite fixed-point algebras*, Recursion Theory, Proc. Sympos. Pure Math. **42**, Amer. Math. Soc., Providence, RI, 1985, pp. 473–486.

263. E. W. Stachow, *Sequential quantum logic*, Current Issues in Quantum Logic (Proc. Workshop on Quantum Logic; Erice, 1979), Plenum Press, New York, 1981, pp. 173–191.

264. T. Stepień, *A sufficient and necessary condition for Tarski's property in Lindenbaum's extensions*, Z. Math. Logik Grundlag. Math. **30** (1984), no. 5, 447–453.

265. L. Stout, *Topoi and categories of fuzzy sets*, Fuzzy Sets and Systems **12** (1984), no. 2, 169–184.

266. T. A. Sudkamp, *On full cylindric set algebras*, Notre Dame J. Formal Logic **20** (1979), no. 4, 785–793.

267. G. Takeuti, *Quantum set theory*, Current Issues in Quantum Logic (Proc. Workshop on Quantum Logic; Erice, 1979), Plenum Press, New York, 1981, pp. 303–322.

268. G. Takeutiard and S. Titani, *Heyting-valued universes of intuitionistic set theory*, Lecture Notes in Math. no. 891, Springer-Verlag, Berlin and New York, 1981, pp. 189–306.

269. ___, *Intuitionistic fuzzy logic and intuitionistic fuzzy set theory*, J. Symbolic Logic **49** (1984), no. 3, 851–866.

270. B. Tembrowski, *The theory of Boolean algebras with an additional binary operation*, Studia Logica **42** (1983), no. 4, 389–705.

271. ___, *QB and normal algebras*, Polish Acad. Sci. Inst. Philos. Sociol. Bull. Sect. Logic **14** (1985), no. 1, 41–47.

272. F. Thieffine, *Compatible complement in Piron's systems and ordinary modal logic*, Lett. Nuovo Cimento **26** (1983), no. 12, 377–381.

273. E. Trillas and L. Valverde, *A few remarks on some lattice-type properties of fuzzy connectives* (Proc. 12th Internat. Sympos. Multiple-Valued Logic, Paris, May 25–27, 1982), New York, 1982, 228–231.

274. A. Ursini, *Intuitionistic diagonalisable algebras*, Algebra Universalis **9** (1979), no. 2, 229–237.

275. ___, *Decision problems for classes of diagonalizable algebras*, Studia Logica **44** (1985), no. 1, 87–89.

276. *Usporiadané množiny a zväzy*, Bratislava, Vydale Univerzita Komenského, 1985.

277. J. C. Varlet, *Congruences on De Morgan algebras*, Bull. Soc. Roy. Sci. Liège **50** (1981), no. 910, 332–343.

278. ___, *Fixed points in finite De Morgan algebras*, Discrete Math. **53** (1985), no. 1, 265–280.

279. ___, *Relative De Morgan lattices*, Discrete Math. **46** (1983), no. 2, 207–209.

280. Vas de Corval ho Julia, *The subvariety $K_{2,0}$ of Ockham algebras*, Bull. Soc. Roy. Sci. Liège **53** (1984), no. 6, 393–400.

281. L. Wrancken-Mawet, *Le lattis des sous-algèbres d'une algèbre de Heyting finie*, Bull. Soc. Roy. Sci. Liège **51** (1982), no. 1–2, 82–94.

282. A. Wroński, *Quasivarieties of Heyting algebras* (abstract), Polish Acad. Sci. Inst. Philos. Sociol. Bull. Sect. Logic **10** (1981), no. 3, 130–134.

283. F. M. Yaqub, *Post L-algebras and rings*, Math. Nachr. **101** (1981), no. 1, 91–99.

284. ___, *Equationally definable Post L-algebras*, Algebra Universalis **13** (1981), no. 2, 225–232.

285. ___, *Representation of Post L-algebras by rings of sets*, Rend. Istit. Mat. Univ. Trieste **14** (1982), no. 1–2, 32–40.

286. W. Zarebski, *A dual space characterization of P_1- and P_2-lattices of order ω^+*, Colloq. Math. **47** (1982), no. 2, 185–171.

287. A. Zecca, *Product of logics*, Current Issues in Quantum Logic (Proc. Workshop on Quantum Logic; Erice, 1979), Plenum Press, New York, 1981, pp. 405–412.

288. P. Zlatoš, *Two notes on locally finite cylindric algebras*, Comment. Math. Univ. Carolina **25** (1984), no. 1, 181–199.

289. ___, *Two-levelled logic and model theory*, Finite algebra and multiple-valued logic (Szeged, 1979), Colloq. Math. Soc. János Bolyai, vol. 28, North-Holland, Amsterdam and New York, 1981, pp. 825–872.

Translated by BORIS M. SCHEIN

Chapter III
General Theory of Lattices

T. S. FOFANOVA

As in the previous survey [OSL, II], results of general character are included in this paper: lattice constructions, properties of special elements of lattices, congruences and other relations on lattices, properties of modular and similar lattices, geometric problems, and derived structures (lattices of sublattices, of ideals, of congruences, endomorphism monoids, etc.).

In comparison with [OSL, II], a novelty is that results concerning distributive lattices and ortholattices are transferred to Chapter I, while the congruence lattices of lattices are reflected in this paper, in contradistinction to [OSL, IV]. Besides that, the structure of the previous survey remains virtually intact.

§III.1. General topics and certain classes of lattices

1.1. Monographs. The Russian translation [8] of Grätzer's book *General Lattice Theory* contains an appendix written by the author; it includes solutions to problems from the book that had become known by the end of 1979. An appendix of Salii to the Russian translation [7] of the third edition of Birkhoff's *Lattice Theory* tells what happened to the problems from the second edition of that book. The footnotes contain comments on the problems mentioned in the third edition. The second edition of Skornyakov's *Elements of Lattice Theory* [32] contains additional theorems on the cardinality of the set of subsets, problems connected with representation of lattices as lattices of closed subsets, and information on pseudocomplemented lattices. The books by Pracht [299], and Lidl and Pilz [251] are addressed at students, graduate students, and university faculty and oriented at applications.

1991 *Mathematics Subject Classification.* Primary 06B05, 06B10, 06B15, 06B20, 06B25, 06B30, 06B35, 06B99; Secondary 08B15.

The monographs by Kalmbach [232] and Beran [75] are devoted to orthomodular lattices. While studying logical aspects of the theory of ortholattices, it may be useful to consult the book [31] by Sarymsakov et al. on algebras appearing in the modern theory of dynamic quantum systems, as well as the proceedings [379] of a conference on quantum logics.

See Schmidt's book [323] for representation of lattices by congruences and Poguntke's book [294] for finitely generated lattices.

Salii's book [30] considers results connected with the Huntington problem: is every complete uniquely complemented lattice distributive? Auxiliary information necessary to make the book self-contained takes about half of the volume. In particular, axiomatics of Boolean algebras is analyzed in detail, and cut-section completions and representations of complete lattices are considered. All this, as well as interesting historic information and a list of problems, make the book interesting for a wide audience.

A popular, and at the same time very modern, introduction to the theory of ordered sets was written by Erné [151]. The same can be said about the book *Boolean Algebra* [184] by Gumm and Poguntke. The monograph [315] by Rosenstein is devoted to linearly ordered sets. The monographs [85] by Błaszczyk (topological questions of the theory of Boolean algebras), [193] by Heindorf (various languages for the theory of Boolean algebras), and [371] by Weese and Goltz (decision problems for theories of various types connected with Boolean algebras) are of special character. Sections concerning lattices can be found in the books [257], [16], [126], [136], [293], and [314].

In conclusion we note collections of papers based on talks given at international conferences: [377], [378], [380], [382], and [381]. In the end of the last one there are 72 problems with comments, and a bibliography (not numbered, 103 pages long) of the editor, Rival, in which the most important, from his point of view, papers on ordered sets and lattices were included.

1.2. General topics. We note that some general properties of lattices (in particular, those connected with free lattices and free products) are reflected in Chapter IV. McNulty [265] surveyed 15 papers by various authors about the theory of varieties. The survey [127] by Dilworth traces connections between properties of lattices considered, on the one hand, as partially ordered sets and, on the other hand, as universal algebras. The following topics are included: (1) Hasse diagrams as means for picturing and a most important tool of research; (2) the role of partial order in the study of various embedding problems in the theory of lattices; (3) a connection between lattice-theoretical properties of distributive lattices and order-theoretical properties of partially ordered sets of their ∨-irreducible elements; (4) the role of properties of congruence lattices in the theory of lattices. In each of these directions the author mentioned the most significant results, from his point of view, many of them with proofs. In another survey [128] Dilworth considered studies of distributive lattices from the viewpoint of their influence on the entire theory of lattices. He explained the role of distributivity in the general theory of lattices in many respects by the facts that such important lattices as the lattice of varieties of lattices and the lattice Con L of an arbitrary lattice L are distributive. In the final part of the paper he observed that semidistributivity is, after modularity, the most significant generalization of distributivity. A few recent theorems of Nation, Jónsson, and Rival on semidistributive lattices were mentioned.

Grätzer's survey report [176] was devoted to three problems of universal algebra, one of which—the description of transferrable lattices—belongs to the general lattice theory. In the note [229] by Johnson and Moss an almost obvious fact was proved: a bijection of lattices is an isomorphism if and only if it preserves at least one of the operations.

A combinatorial direction of the theory of lattices connected with consideration of simplicial complexes associated with lattices is reflected in §2 of this chapter and in Chapter VI (the fixed point property of partially ordered sets).

Gould et al. [174] proved that an isomorphism of the lattices of all nonempty subsets of lattices L and L', endowed with operations that extend those from L and L', respectively, implies an isomorphism of L and L'. Hansoul and Varlet [191] studied two generalizations of essential extension: strong extension and perfect extension; they proved that every nontrivial lattice possesses arbitrarily large strong extensions. There are also results on extensions of distributive and modular lattices.

Duda [138] used early methods of Birkhoff and Stone to describe an epireflection from the category $\mathscr{L}$ of all lattices into the complete subcategory of distributive lattices, and considered properties of the congruence δ_L induced by this epireflection on lattices $L \in \mathscr{L}$. The monoarity of an algebra $(A|T)$ is defined as the least nonnegative integer n (if it exists) such that $(A|T) = (A|p)$, where p is an n-ary operation from the clone generated by T. Dudek and Płonka [143] showed that the monoarity of nontrivial lattices is three. Bergman and Hrushovski [78] considered conditions for the existence of cofinal distributive sublattices (considered as a possible generalization of cofinal chains) of lattices. They proved that in every variety V of lattices every cofinal sublattice of a free lattice with an uncountable number of generators generates V. It follows, in particular, that cofinal distributive sublattices do not always exist. They proved the existence of the latter in modular lattices with a weakened maximality condition. Bergman [76] considered problems connected with cofinal sublattices.

Smith [343] described certain flat lattices with a stable theory as well as certain flat lattices of height four, whose theory is categorical in the countable cardinality, or in uncountable cardinalities, or totally transcendental. In conclusion he stated open problems connected with stability and categoricity of lattices of finite height.

Höft [205] generalized the concept of the sum of a double system of lattices [OSL, II, p. 59] to partially ordered sets and, applying her results to lattices, obtained the main results of Graczyńska and Grätzer [OSL, II, 234]. Another generalization of this sum was considered by Graczyńska and Pastijn [175].

A triple (G, M, I), where G and M are sets of objects and properties, respectively, and $I \subseteq G \times M$ is a binary relation "to have the property", is called a context. One can define a concept in a given context and introduce hierarchy of concepts in such a way that the set of all concepts becomes a complete lattice. Such lattices were introduced by Wille [373], whose goal was to align lattice-theoretic constructions, results, and methods in accordance with needs and views of possible users of lattice theory. He produced examples of concept lattices in astronomy, health care, medicine, pedagogy, and discussed connections between natural questions arising in these contexts with problems of lattice theory posed by the logic of its internal development. It was proved in [374] that, under certain finiteness conditions imposed upon the context, the concept lattice is representable as a subdirect product of subdirectly irreducible concept lattices. See [333], [334], [335], and [336] for applications of lattice theory to chemistry.

An algebra A is said to be tolerance Hamiltonian if each of its subalgebras is a block of a tolerance (in what follows, tolerances are assumed to be compatible with the operations of the algebra). A variety is said to be tolerance Hamiltonian if each of its algebras has this property. One of the corollaries to a general theorem by Chajda [94] is the following fact: there are no nontrivial tolerance Hamiltonian varieties of lattices (however, the variety of all semilattices is tolerance Hamiltonian). Czédli [115] defined a factor lattice L/ρ of a lattice L modulo a tolerance ρ, thus generalizing the concept of a factor lattice modulo a congruence. The main theorem states that ISTSP(2) is the class of all lattices, and ISTP$_f$(2) is the class of all finite lattices. Here 2 is a two-element lattice, P, P$_f$, S, T, and I are, respectively, the operators of direct products, finite direct products, sublattices, factor lattices modulo tolerances, and isomorphic images. A corollary for finite lattices follows (see §1.4 below).

Bandelt [63] suggested an interpretation of tolerances on lattices as Galois correspondences between the lattices of ideals and of filters. Chajda [97] studied properties of tolerances (regularity, weak regularity, transferability) that generalize corresponding properties of congruences. He and Nieminen [98] generalized certain theorems from [OSL, II, 139] on direct decompositions of tolerances on monoids and distributive lattices to universal algebras with certain additional conditions and, as a corollary, proved that a tolerance on a direct product of lattices is always representable as a direct product of tolerances on the factors. Duda [142] strengthened some results of this paper; he considered conditions for direct decomposability of compatible relations. The same problem for the so-called conditionally ∨-m-complete tolerances, where m is an infinite cardinal, was studied by Chajda and Zelinka [100]. Chajda [95] and Duda ([139], [140]) formulated conditions for the existence of direct decompositions of tolerances and homomorphisms of universal algebras using the diagonal operation on a direct product; there are corollaries for lattices. Pócs [292] showed that if every congruence class of the lattice $L = \prod L_\alpha$ is a $\vee$-closed sublattice of L, then this congruence is decomposable as a direct product of congruences on the lattices L_α. A general theorem of Zlatoš [376] (connected with congruences) produces a characterization of lattices isomorphic to direct sums of finite subdirectly irreducible lattices.

Draškovičová [135] considered interconnection between various properties (commutativity, homogeneity, regularity, etc.) enjoyed by all congruences of a fixed algebra (a study of varieties in which all algebras have this property is more traditional). Her results have applications to modular median algebras connected with modular lattices (see §2) and to certain other lattices. Chajda [96] found a description of weakly regular lattices. Chinthayamma and Parameshwara [109] studied lattices in which all congruences are neutral.

Nöbauer [280] surveyed main results of Austrian algebraists in the theory of interpolation and locally polynomial functions (up to 1978) and suggested some problems. Bandelt [61] obtained a characterization of locally polynomial functions and, as a corollary, conditions under which a lattice is locally isotone polynomially complete (lipc) or locally isotone affine complete (liac). Chajda [93] found four conditions each of which is equivalent to lipc for a lattice (two of these conditions had been known earlier). Dorninger and Eigenthaler [133] studied a connection between the set $C_k(V)$ of all stable k-place functions on a lattice V and the set $OF_k(V)$ of all isotone k-place functions. It is known [OSL, II, 488] that $OF_k(V) \subseteq C_k(V)$ if and only if V is a simple lattice. Here the lattices for which $C_k(V) \not\subseteq OF_k(V)$ are characterized. In the distributive case these are the lattices that contain a nontrivial interval that is a Boolean algebra. These results were applied to liac lattices. Dorninger [132] investigated connections between the lattice $OF_k(V)$ and the lattice $P_k(V)$ of k-place polynomial functions. He considered subsets $D \subseteq OF_k(V)$ such that the set $P_k(V) \cup D$ generates the lattice $OF_k(V)$. The main results relate to finite lattices. Eigenthaler [149] corrected a theorem of Schuff on lattice polynomials (more details are given in Chapter IV, §1.3). A clone of term functions on an algebra A is defined as the subclone of the clone of all functions on A generated by the operations of A. Among the properties of clones of term functions obtained by Schweigert ([329] and [330]) there are results for the case when A is a lattice. Fleischer and Traynor [158] considered mappings of lattices into abelian groups satisfying some natural conditions and called modular. They investigated connections with congruences, obtained a structural description of all modular functions on lattices of finite length, and established that the distances, introduced on lattices by means of these functions, satisfy the triangle inequality. Jakubík [217] investigated metrics on lattices with values in a partially ordered abelian group and found interrelations between isometries and direct decompositions. A certain representation of real-valued functions defined on a direct product of a set and a finite lattice was considered in [150]. Poroshkin and Samorodnitskiĭ ([26], [27], and [28]) studied functions defined on lattices and taking values in topological spaces. Their results are connected with limit values of these functions on special sequences. There are applications to Boolean algebras. A cycle of papers by Bachman and Stratigos, [51], [52], [53], [54], and [55], was devoted to a generalization of known results of measure theory and topology to the case when in the underlying set X a sublattice of the lattice of all subsets of X is distinguished.

Poguntke and Sands [OSL, II, 384] proved that every infinite finitely generated lattice of width 3 contains either a "herringbone" H or a dual lattice H^* as a sublattice. On the other hand, Rival, Ruckelshausen, and Sands [OSL, II, 402] showed that an infinite finitely generated lattice of any finite width contains H or H^* as a subset, but not necessarily as a sublattice (this result was stated in OSL, II with an error). Ruckelshausen and Sands [317] gave an example of an infinite finitely generated lattice of width 4 that does not contain either H or H^* as a sublattice. They considered a certain family $\mathscr{H}_4$ of infinite lattices of width 4. This is their main result: every infinite finitely generated lattice of width 4 contains (up to duality) either H or a lattice from $\mathscr{H}_4$. A corollary for lattice varieties was obtained (see Chapter IV, §1.1). Schmidt [327] proved that a finitely generated lattice is embeddable in any algebraically complete lattice if and only if it was finite, and obtained other results on finitely generated lattices. Rival and Sands [313] established that every countable lattice is embeddable in a simple lattice generated by a four-element antichain (see Chapter IV, §1.3, for other results of this paper). Pouzet and Rival [298] proved that every finite or countable lattice is an order retract of a direct product of chains.

Let $C(L)$ be the covering graph of a lattice L. A graph $(V; H)$ is called an S-graph if the following conditions hold: (1) $|V| \geq 2$; (2) $x, y \in V$ and $d(x, y) = 1$ imply $d(x, z) \neq d(y, z)$ for all $z \in V$; (3) for every $x \in V$ there exists exactly one $y \in V$ such that $d(y, z) = 1$ implies $d(x, y) > d(x, z)$ for all $z \in V$. Gedeonová [166] studied lattices whose covering graph is an S-graph. We note one of the results: if L is semimodular and $C(L)$ an S-graph, then L is a Boolean algebra. She found conditions for an S-graph to be a covering graph of a lattice. A finite graph G is said to be central symmetric, or a CS-graph, if for every $a \in G$ there exists a lattice P with the least element, in which all maximal chains between any two elements have the same length, and a graph isomorphism $f: G \to C(P)$ such that $f(a) = O_P$. A lattice is called a CS-lattice if its covering graph is a CS-graph. Such lattices were considered by Duffus and Rival [145] and Gedeonová [166]. She proved in [167] that every CS-graph is regular and that the sublattice of a CS-lattice generated by the set of atoms is Boolean. Jakubík [218] considered the problem of when lattices are determined by their covering graphs. In [219] and [220] he studied isomorphisms of graphs of locally finite lattices. He singled out properties of cycles, intervals, and subdirect decompositions. Let $T(L)$ be all nonisomorphic types of lattices with graphs isomorphic to that of L. Using weak direct decompositions, he proved that for every cardinal number α there exists a lattice L such that $|T(L)| \geq \alpha$.

Kučera and Trnková [247] considered some problems of universal algebra from the viewpoint of their computational complexity. In particular, this relates to an algorithm checking whether a given bigroupoid is a lattice. Muir and Warner [269] studied applications of the theory of lattices to the theory of automata. Bauer [67] considered applications of algebra in theoretical computer science; in particular, he pointed out a possibility of using lattice theory in programming. Applications of algebra to computational mathematics were considered in the survey [90] by Burmeister. Lomecky [255] investigated programs of algorithms connected with free lattices (see Chapter IV). See also [302].

A theorem of Adams and Sichler on Frattini sublattices is given in Chapter IV, §1.3.

1.3. Complete lattices. Closures and completions. A mapping of a complete lattice into a complete lattice is called relational if it preserves all joins and maps principal ideals onto principal ideals. Fried and Wiegandt [163] considered the category $\mathfrak{A}$ of complete lattices with relational mappings as morphisms. Using a covariant functor from a certain bicategory to $\mathfrak{A}$ they introduced so-called abstract models and investigated their products. Hušek [210] studied a certain complete lattice of subfunctors of the identity functor of a topological category.

Let L be a complete lattice not containing either complete subsemilattices isomorphic to 2^ω or uncountable well-ordered subsets, and let $\{X_i, i \in \mathscr{J}\}$ be a family of nonempty subsets of L. Bergman and Galvin [77] proved that there exist elements $p_i \in X_i$ such that $\sup\{p_i, i \in \mathscr{J}\}$ majorates the set-theoretic union of almost all subsets X_i. Ryabtsev [29] repeated results of Zaretskiĭ on representation of complete lattices by sets (RZhMat 1961, 10A266).

An element a of a lattice L is called regular if $a \wedge x = a \wedge y = 0 \Rightarrow a \wedge (x \vee y) = 0$ for all $x, y \in L$. Saliĭ [319] proved that regular elements of complete uniquely complemented lattices form a complete Boolean sublattice. Moreover, every such lattice can be represented as a direct product of complete sublattices $[0, a]$ and $[0, a']$, where a is any regular element and a' its complement.

Pouzet and Rival [297] established that if some homomorphic image of a complete lattice is not complete, then the lattice contains a subset isomorphic to $P(N)$, where N is the natural series. For complete distributive lattices this condition is necessary and sufficient. If a complete lattice is countable, or of finite dimension, or contains no infinite antichains, then each of its homomorphic images is complete. A homomorphic image of a complete lattice of finite dimension is its retract. Considering conditions for the existence of complete linear extensions of chain complete partially ordered sets, Pouzet and Rival [296] proved, in particular, that complete countable lattices and complete lattices without infinite antichains possess such extensions. There exists, however, a complete distributive lattice of cardinality ω_2 that has no complete linear extension. Complete relations on partially ordered sets that generalize naturally the order relation on complete lattices were studied by Höft [204]. Considering compactifications of partially ordered sets, Abian and Lihová [37] proved, in particular, that a compactification of a complete lattice is a complete lattice.

Let L be an α-complete lattice for some cardinal number α, and X a set of its generators. Goralčík and Koubek [172] defined the complexity $\#(a, X)$ of an element $a \in L$ as, figuratively speaking, the number of operations necessary to generate this element, where β-joins and β-meets could be used for $\beta < \alpha$. They showed that every variety V of α-complete lattices has complexity $\#V = \sup\{\#(a, X) : a \in L, X \text{ generates } L, L \in V\} = \alpha$, if α is a successor cardinal.

Bandelt [60] suggested a version of a sum of a double system (see OSL, II, [234]) for complete lattices, which uses complete congruences and residual mappings. The main theorem of his other paper [62] established implications between eight properties of a complete lattice, one of which is $\mathfrak{M}$-distributivity with respect to a system $\mathfrak{M}$ of its subsets: $\bigwedge_{i \in I}(\bigvee M_i) = \bigvee_{\varphi \in \prod M}(\bigwedge \varphi(i))$ for any set I of indices and $M_i \in \mathfrak{M}$. Under some restrictions these eight properties are equivalent. If $\mathfrak{M} = P(L)$, one obtains known characterizations of completely distributive lattices. There are applications to continuous lattices. In particular, a complete lattice is α-continuous for a regular cardinal number α if and only if it is a complete homomorphic image of an α-algebraic lattice.

Tunnicliffe [357] found a self-dual identity that characterizes distributive lattices in the class of complete lattices. A self-mapping p of a complete lattice is said to $\vee$-determine the lattice if $a = \bigvee\{b | a \not\leq p(b)\}$ for every element a. Lambrou [250] showed that a complete lattice is completely distributive if and only if it possesses a $\vee$-determining mapping. Other results of this paper on completely distributive lattices and Boolean algebras were reflected in Chapter I. The author gave numerous examples, historical information, open questions, and discussed connections with functional analysis.

Note that the majority of results on completely distributive lattices are collected in Chapter I.

Bennett and Birkhoff [74] defined biatomic lattices, examples of which are lattices of convex sublattices of distributive lattices of finite length. Convexity lattices are complete biatomic algebraic lattices in which sublattices generated by three and four atoms satisfy certain conditions. For convexity lattices they proved analogs of theorems of Radon, Helly, and Carathéodory on convex sets in an n-dimensional space. Bennett [73] studied Peano lattices, that is, convexity lattices L in which for any distinct atoms a, b, c, d, and f the conditions $f \leq a \vee b$ and $c \leq a \vee d$ imply $(b \vee c) \wedge (d \vee f) \neq 0$. An example of a Peano lattice is provided by the lattice of all convex sets of a vector space over an ordered division ring. She gave various equivalent forms of the Peano condition in convexity lattices, considered a certain subset of a Peano lattice called its modular core, and described Peano lattices with a modular core of length > 3.

Erné [153] obtained conditions sufficient for the existence of a $\wedge$-decomposition for an element of a complete lattice; in the modular case this condition is necessary as well. This result has interesting applications to Brouwer and continuous lattices. Gierz [170] described colimits and coproducts of continuous lattices. Here is one of the corollaries: a coproduct of algebraic lattices is an algebraic lattice. Isbell [215] proved that every variety of $\wedge$-continuous lattices is closed under direct limits. Let complete lattices be objects of a category Sl and mappings preserving arbitrary sums be its maps, and let Loc be the category whose objects are infinitely distributive complete lattices and morphisms are lattice homomorphism-preserving arbitrary sums. Niefield [277] obtained a number of characterizations of projective lattices $M \in \mathrm{Ob}(\mathrm{Sl})$; we note three of them here: (a) M is completely distributive; (b) M is completely continuous; and (c) M is flat. He established that a lattice $A \in \mathrm{Ob}(\mathrm{Loc})$ is continuous if and only if the functor $A \otimes - : \mathrm{Loc} \to \mathrm{Loc}$ has a left adjoint, and gave another two conditions each of which is equivalent to this property. Bandelt and Erné [65] obtained theorems on continuous partially ordered sets, which could be applied to continuous lattices. In more detail this article, as well as other results on continuous partially ordered sets, are reflected in Chapter VI. In the survey [267] Mislove presented the history of the discovery of continuous lattices by Scott, whose starting point was the logical foundations of the theory of computing, and by Lawson from the structural theory of topological semilattices. He discussed connections between locally compact quasi-balanced topological spaces and continuous distributive lattices. Hofmann's article [203] belongs to the same direction. Kamimura and Tang ([235] and [236]) obtained certain topological properties of continuous lattices. Their other paper [234] is devoted to representations of continuous lattices connected with algebraic congruences on them (see §3.2).

Johnstone [OSL, V, 231] constructed a countable ordered set that admits a balanced topology, but is not balanced in its Scott topology; he raised the question of the existence of a lattice with this property. Isbell [216] answered this question in the affirmative. Breazu and Stănăşilă [88] considered topological spaces in which the intersection of any family of open sets is open, and studied their connections with complete lattices.

Let L and L_1 be complete lattices, and let L_1 be the image of L under a complete homomorphism. Dobbertin, Erné, and Kent [130] obtained results on the connection between order convergence in L and L_1. For example, they proved that the order convergence in L_1 is pretopological if it has this property in L, while topological convergence is not preserved under complete homomorphic images. Greco [179] defined a concept of uniform order convergence for a family of functions defined on a set and taking values in a lattice. He proved that if L is a completely distributive complete lattice, then a family of functions $f_i: X \to L$ uniformly order converges to f if and only if it uniformly converges to f with respect to the interval topology on L. The main goal of another paper by Greco [177] was to show that numerous properties of limits of real functions remain true when real numbers completed by the symbols $+_\infty$ and $-_\infty$ are replaced by an arbitrary completely distributive complete lattice.

Operators of type G [178] are certain operators on a set of mappings of arbitrary sets into complete lattices compatible with the order relation on the mappings and complete homomorphisms of the lattices. Connections between properties of lattices (distributivity, complementability, etc.) and some natural properties of G-operators were clarified. See also [180].

Let $\mathfrak{L}$ be the class of all lattices. A mapping $t: \mathfrak{L} \to \mathfrak{L}$ is called a c-mapping if $t(L)$ is a conditionally complete lattice that contains an isomorphic copy of L for all $L \in \mathfrak{L}$ and $L \cong t(L)$ if and only if L is conditionally complete. The best known example of a c-mapping is the cut completion d. Arnow [OSL, II, 77] obtained $d(L)$ by introducing a lattice structure on the set $L_{<<}$ of a specific set of intervals of the lattice, called "ends". Černák [92] gave another proof of the isomorphism $d(L) \cong L_{<<}$, showed that L might have different systems of "ends" (although the corresponding lattices $L_{<<}$ are isomorphic), and established the existence of pairwise nonequivalent c-mappings. Erné [152] deduced various general properties of sets endowed with a closure operator and obtained from them criteria for distributivity and infinite distributivity of a cut completion of a quasi-ordered set. In another paper [154] he studied a general construction of completion of partially ordered sets, for which the completions of Alexandroff, Dedekind–MacNeille, and Frink are special cases. Papers by Erné and Wilke [155] and Banaschewski and Nelson [59] were devoted to completions of quasi-ordered and partially ordered sets. The latter three papers are related principally to the theory of ordered sets and are considered in more detail in Chapter VI.

For an algebraic closure operator T on a set A, Thron [355] introduced concepts of a T-isolated element and a T-generating subset. Let $H(A)$ be the set of all algebraic closure operators T on a set A for which all finite subsets are T-generated by subsets of their T-isolated elements, and let $K(A)$ be the set of all single-valued mappings $P(A) \to P(A)$. Main results establish a one-to-one correspondence between $H(A)$ and $K(A)$ and describe properties of this correspondence.

Ryšlinková and Sturm [318] studied preservation of n-compactness by m-algebraic closure operators for infinite cardinal numbers n and m and continued their investigation of connections between classes of operators introduced earlier [OSL, II, 407]. Pondelíček [295] obtained a generalization of Ryšlinková's result on this subject. Richter [310] proved a theorem on compact elements of a complete algebraic lattice in which each element is a join of completely ∨-irreducible elements. There are applications to lattices of subgroups. In another paper Richter [311] considered a set of precompact elements of a complete lattice and generalized certain results of Stern. Wood [375] considered a closure operator connected with the so-called Suslin operation. P. Cousot and R. Cousot [112] applied the techniques they worked out earlier to prove Tarski's fixed point theorem to the lattice of all operators on a complete lattice, thus obtaining a constructive characterization of complete lattices of all retractions, closures, and quasiclosures. A subset C of a complete lattice L is called a closure system if $1 \in C$ and $\bigwedge A \in C$ for every nonempty $A \subseteq C$. Dalík [116] studied properties of the principal filter $C(L, N)$ generated by a closure system N in the lattice $C(L)$ formed by all closure systems ordered by inclusion. The results were applied to the so-called lattices of generating systems of partially ordered sets. Dzik ([147] and [148]) considered a certain closure operator on subsets of the set of abstract formulas. Closed subsets are called abstract logics. The set of logics forms a complete lattice $\mathscr{C}$. A representation theorem for $\mathscr{C}$ was stated; it was used for studying the so-called content of this lattice, defined by means of a certain set of propositional formulas of an intuitionistic calculus. An algebraic model of logic that is a complete lattice was suggested in [360]. A property of complete lattices connected with certain aspects of designing the so-called specificator languages for databases and algorithms was considered in [71].

A fuzzy subset of a set X is defined as a mapping $A: X \to L$, where L is a complete lattice. This is a natural generalization of the ordinary definition in which L is the chain of real numbers. Without claims for completeness, we note certain works connected with this concept. Hamburg and Florescu [189] defined a fuzzy lattice as a complete sublattice of the lattice of all fuzzy subsets of a given set and defined topology, uniformity, and proximity on fuzzy lattices as well as other concepts, which coincide with the classical ones when $L = \{0, 1\}$. Hamburg [188] studied connections between fuzzy topological spaces $(\mathscr{L}, \varphi)$ and $(\mathscr{L}', \varphi')$, where $\mathscr{L}$ and $\mathscr{L}'$ are fuzzy lattices. A paper [353] by Sugeno and Sasaki is about a definition of L-fuzzy categories and L-fuzzy functors. Achache [38] considered Galois connections between fuzzy sets. Novák and Nekola [283] suggested constructing a theory of fuzzy sets by introducing a fuzzy logic on residuated lattices. Ying-Ming Liu ([253] and [254]) considered operations over transformations of complete completely distributive lattices to investigate fuzzy dimension. In another paper [252] Wang-Jin Liu studied fuzzy sets in the case when X is endowed with a (group or ring) algebraic structure. Sessa [337], Drewniak [137], and

Di Nola [129] developed a theory of equations with fuzzy relations. See also [208] and the monographs [16] and [136].

1.4. Finite lattices. Crapo [113] introduced the notion of an identity $I(\rho)$ of a correspondence ρ between sets J and M. The negative $I^°(\rho)$ is the identity of the correspondence $\bar{\rho} = (J \times M)/\rho$. Main results relate to the case when J and M are sets of all $\vee$- and $\wedge$-irreducible elements of a finite lattice L, respectively. In this case $I(\rho)$ is a disjunct union $J \cup M$ with a preorder induced by L. He proved that $L \cong d(I(L))$ (here d is cut completion). A set $\mathrm{Sp}(I)$ of certain isotone mappings $I \to 2$, called splitting of identity, was considered. He found a Galois connection between subsets of $d(I)$ and $\mathrm{Sp}(I)$ and proved that $\mathrm{Sp}(I(L)) \cong d(I^0(L))$, hence $L \cong d(I(L)) \cong \mathrm{Sp}(I^0(L))$. Many examples and open problems (with detailed comments) were given.

To every partially ordered set L there corresponds a simplicial complex Symp L, the simplexes of which are finite chains. Kratzer and Thévenaz [246] considered homotopic properties of Symp L in the case when $L = \widetilde{L}\setminus\{0, 1\}$, where $\widetilde{L}$ is a finite lattice. If $\widetilde{L}$ contains an element without a complement, then the complex Symp L is contractible. To describe the homotopic type of Symp L in $\widetilde{L}$, special elements, called axial, were considered. The main theorem states that the homotopic type of Symp L is completely determined by subcomplexes defined by axial elements and their complements. Every maximal element of a modular lattice is axial, and hence, for such a lattice $\widetilde{L}$, the complex Symp L has a homotopical type of a union of $(n-2)$-dimensional spheres, where n is the length of the lattice. These results were applied to the lattices of subgroups. Other papers in this direction are considered in §2 and Chapter VI.

Ashman and Ficker [50] considered all lattices defined on a fixed finite set together with their mappings onto a certain set of $(0, 1)$-matrices.

McKenzie [263] proved that a finite simple lattice L, where $|L| \geq 3$ and the intersection of co-atoms is 0, is not embeddable in the lattice of subvarieties of a locally finite variety. Other important results of this paper, connected with representations of finite lattices as congruence lattices of algebras, are reflected in Chapter V. Sivák [338] constructed an extension of a finite lattice with an isomorphic congruence lattice. He proved two theorems on finite lattices with all elements representable as sums of atoms. They are connected with the notion of a distributive element.

A mapping f of lattices is said to reverse the order of a pair (x, y) if $x < y$ and $f(x) > f(y)$. Let $f: L_1 \to L_2$ be a bijection of finite lattices. Tůma [356] used Pudlák's theorem on symmetric embeddings of finite lattices in finite partition lattices to establish the following: for the existence of mappings φ_i of lattices L_i ($i = 1, 2$) into the partition lattice $\pi(A)$ of a finite set A, for which $\varphi_1(x)$ and $\varphi_2(f(x))$ are isomorphic partitions for all $x \in L$, it is necessary and sufficient that f does not revert the order of any pair in L_1.

Let $w(L)$ denote the Whitman number of a finite lattice L, that is, the least number of elements of a set such that L is representable by its partitions. Peele [285] used properties of multigraphs of a special form to find an upper bound for the Whitman number of certain geometric lattices. The complexity of finding such bounds is clear, for example, from the fact that, for the lattice L_4^* dual to the partition lattice of a four-element set, it is not known whether $w(L_4^*) \leq 2^{1000}$ or the opposite inequality holds.

As observed by Smith [342], Köhler gave a survey of results on representations of finite lattices by congruences of finite algebras. It follows from a general theorem by Czédli [115] that, in particular, every finite lattice is isomorphic to a factor lattice of a distributive lattice modulo a tolerance relation. Perminov's paper [23] on finite diagram lattices is described in §3.3. Grzymala-Busse [181] studied the representation of a finite lattice of special type as the lattice of all admissible partitions of a finite automaton. Nordahl [281] produced an example of a diagram of a finite lattice connected with medicine. Novak [282] established a duality between the concepts "finite" and "directed" of lattice theory in the language of the theory of categories (see also Chapters V and VI about this paper).

1.5. Lattices with additional operations. Observe that a greater part of such lattices is considered in Chapter I and Chapter II.

The survey [OSL, II] has a special section devoted to multiplicative lattices; the main definitions can be found there. Anderson, Johnson, and Johnson [49] showed that numerous important results of the theory of commutative rings with 1 could be obtained in the theory of multiplicative lattices generated by their ∨-principal elements. Becerra and Johnson [70] gave various generalizations of principal ideals in terms of the theory of multiplicative lattices and investigated their interrelations for various classes of rings. Johnson [224] proved that if a locally Noetherian lattice L is a Macaulay lattice, then its completion L^* is also a Macaulay lattice. He gave a condition on joins of prime elements under which the converse holds and showed that the class of Macaulay local lattices is strictly wider than the class of ideal lattices of Cohen-Macaulay local rings. A multiplicative lattice L is called totally quasilocal (Johnson [225]) if it possesses a single maximal element M (that is, it is quasilocal) and $x \leq M$ for every $x \in L \setminus \{1\}$. The author considered various possible definitions of principal elements in such lattices and showed that many of them are equivalent, exactly as in the case of locally Noetherian lattices. He produced an example of a quasilocal but not totally quasilocal multiplicative lattice. An example of a quasilocal r-lattice with a ∨-irreducible element that is neither principal nor completely ∨-irreducible was constructed by Johnson in [226]. In another paper [227] he studied various modifications of the concept of principal element in totally quasilocal multiplicative lattices and their interrelations in various classes of lattices. In [228] he described the structure of r-lattices in which every element is primary. Anderson, Johnson, and Johnson [48] studied properties of small r-lattices (that is, r-lattices with a finite number of prime elements of rank ≤ 1) and strong π-lattices (that is, quasilocal r-lattices in which every principal element possesses a unique representation as a product of prime principals). They studied conditions for embeddability of such lattices into the ideal lattice of a homomorphic image of a polynomial ring over a field.

A frame is a complete lattice in which finite meets distribute over arbitrary joins. (Main results on frames are discussed in Chapter I.) Every frame can be considered as a multiplicative lattice with $\wedge$ as multiplication. Multiplicative lattices appeared as a generalization of ideal lattices of rings, and frames as a generalization of lattices of open sets of topological spaces. Rosický [316] discussed connections between multiplicative lattices and frames and generalized the classical theorem of Krull to continuous multiplicative lattices.

A radical of an element a of a multiplicative lattice L is the element $\sqrt{a} = \bigvee\{x \in L \mid x^s \leq a \text{ for some } s\}$. If a is primary, then $\sqrt{a}$ is a minimal prime element that contains a. Thakare and Manjarekar [354] studied primary decompositions, that is, decompositions

$$a = q_1 \wedge q_2 \wedge \cdots \wedge q_m, \qquad (*)$$

where $q_1, \ldots, q_m$ are primary elements. In Noetherian lattices each element possesses an irreducible primary decomposition. They proved that if a decomposition $(*)$ exists then the prime elements $p_i = \sqrt{q_i}$ are uniquely determined by a. The minimal among these p_i are called isolated primes of a. The corresponding primary components in $(*)$ are called in the same way. One of the results is that the radical of any element equals the intersection of the isolated primes of this element.

A correlation lattice is a bounded lattice with a unary operation σ that is a dual lattice automorphism such that $\sigma^{2n}(x) = x$ for all x and certain fixed odd n. Schweigert and Szymańska [331] studied the variety A_n of all distributive correlation lattices in which $\sigma^n(x)$ is a complement of x. They showed that: (1) all subdirectly irreducible algebras in A_n are prime and finite, and all prime algebras are polynomially complete; (2) for every divisor k of n there exists a prime algebra $\mathfrak{A}_k \in A_n$ with k atoms, and it is isomorphic to the Boolean algebra 2^k; (3) A_n is generated by the algebras $\mathfrak{A}_k$. There are also other results about this variety. In their next paper [332] the same authors consider polynomial functions on correlation lattices and problems of polynomial completeness.

Herrmann [198] characterized distributive polarity lattices in the class of modular polarity lattices and obtained certain results concerning projective covers of modular polarity lattices. Varieties of such lattices are considered in Chapter IV.

Vonkomerová [365] considered the so-called C-lattices that generalize various types of lattices with additional structures introduced in functional analysis, whereas Płonka [291] considered algebras of the form $\mathfrak{A} = (X; +, \cdot, f_1, f_2, \ldots)$, where $(X; +, \cdot)$ is a lattice and $f_1, f_2, \ldots$ unary operations. A congruence of $\mathfrak{A}$ is called bounding if each of its classes is a subalgebra with the greatest and least elements. He described algebras $\mathfrak{A}$ with bounding congruences and established connections between bounding congruences and two functions of a special form. Reis [304] continued consideration of groupoid-lattices [OSL, II]. See also [292].

1.6. Miscellany. A topological lattice is called locally convex if its topology possesses a base of open and order convex sets. Miller [266] obtained certain characterizations of local convexity of a locally connected topological lattice. Wang ([369] and [370]) investigated molecular (that is, in which each element is a join of a finite number of atoms) topological lattices. In the latter paper he considered separation axioms that generalize the T_i axioms in topological spaces and certain fuzzy topological spaces. Kurilic [248] considered the subset $K(L) = \{x \in L | x$ is nontrivially $\wedge$- or $\vee$-irreducible$\}$, the nucleus of the lattice L. He observed that the class $\mathfrak{K} = \{H | K(H) = H\}$ is not a variety and that each lattice is embeddable in a lattice from $\mathfrak{K}$. Van de Vel [359] observed that by choosing the set of convex sublattices of an arbitrary lattice L as the system of convex sets, we turn L into a convex structure. (By definition, this is a set A with a distinguished system of subsets that contains both $\varnothing$ and A and is closed under intersections and unions of ascending chains.) He found properties of a convex structure that are equivalent to the distributivity and completeness of L.

If f is an n-ary operation of an algebra A, then the mapping $x \to f(a_1, \ldots, a_{i-1}, x, a_{i+1}, \ldots, a_n)$ is called an elementary translation. A translation is defined as a product of elementary translations. A subset L of an algebra A is called distinguishing if, for any distinct $x, y \in A$, there exists a translation f such that exactly one of the elements $f(x)$ and $f(y)$ belongs to L. Kopeček [243] showed that if a lattice is relatively complemented then each of its singletons is a distinguishing subset. A number of corollaries was given. A subset R of a chain A is distinguishing if and only if $[a, b] \cap \mathscr{R} \neq \varnothing$ and $[a, b] \cap (A \backslash \mathscr{R}) \neq \varnothing$ for every interval $[a, b]$, where $a < b$. Saliĭ [319] showed that every atom in a uniquely complemented lattice is regular and established other properties of regular elements in such lattices. Results of this paper related to complete lattices are considered in §1.3.

A standard element of order two in a lattice is a standard element of its standard ideal. Grätzer and Schmidt asked whether standard elements of order two form a lattice. Iqbalunnisa and Akilandam [214] answered this question in the negative, namely, they showed that the meet of any two standard elements of order two is standard of order two, but their join does not necessarily possess this property. A subset L_1 of a lattice L is called a

∧-sublattice if it is a sublattice in which the operation ∧ coincides with this operation in L. Ramana Murty and Engelbert [271] obtained necessary and sufficient conditions for an ideal of a ∧-sublattice $L_1 \subseteq L$ that is standard in L to be standard in L_1. They proved some corollaries and theorems on connections between standard ideals of a direct product of lattices and standard ideals of the factors. They proved that if v is an isotone and 0-preserving valuation on a partially complemented lattice, then $v^{-1}(0)$ is a standard ideal. The results of Fourneau [159] on ideals of molecular lattices are reflected in §3.1. Stern [351] obtained a technical result on standard ideals. A preprint of Gorbunov and Tumanov [OSL, II, 15] on existence of prime ideals in semidistributive lattices has been published [173].

Adams and Sichler [45] proved the following theorem: there exist chains A and B_α ($\alpha < 2^{2^{\aleph_0}}$) that satisfy the conditions: (a) $|A| = 2^{\aleph_0}$; (b) $B_\alpha \cong B_\beta$ if and only if $\alpha = \beta$; (c) for every α and every n the chain A contains n disjoint copies of B_α, but not an infinite number of them. For distributive lattices this result was established by Koubek [OSL, II, 315], and for universal algebras with certain conditions by Korec [244]. Gacsályi [164] remarked on two chains in a lattice that satisfies the maximality condition.

Cornish and Noor [111] investigated standard and neutral elements in near-lattices and obtained new properties of such elements also for lattices. An element a in a ∧-semilattice is called prime if $a \geq b \wedge c$ implies $a \geq b$ or $a \geq c$. Hoo [206] proved, in particular, that a complete atom generated lattice is implicative if and only if each of its elements is representable as a meet of a set of prime elements. Reilly [303] defined exceptional elements of a lattice (these are neutral elements that satisfy an additional condition) and considered representations of a bounded lattice as a sublattice of a direct product of three lattices connected with the chain $0 = a_0 < a_1 < a_2 < a_3 = 1$ of exceptional elements.

The so-called m-lattices with the ascending chain condition are considered in [131]. Aissen and Shoy [46] considered some properties of the lattice of all finite sequences of natural numbers connected with the algorithm that, for every such sequence, constructs a mapping $F^n \to F^k$, where F is a free groupoid with a single generator, $k \leq n$, and k and n are uniquely determined by the sequence.

§III.2. Modular, semimodular and similar lattices. Geometric topics of lattice theory

Freese [OSL, III, 81] showed that the word problem is algorithmically unsolvable in a free modular lattice (ml) $FM(n)$ with $n \geq 5$, and Herrmann [197] proved this result for $n = 4$. This solved a problem of the theory of ml that had remained open for nearly half a century. The geometric methods used in solving this problem are discussed further. See also the survey [161]. The proof of Freese and Herrmann uses gluing ml, and hence it cannot be generalized straightforwardly to varieties of ml. There are known examples of subvarieties of ml with a decidable word problem, but the decidability of the word problem for a free Arguesian lattice remains open. Kut'in [18] studied identities similar to the modular one. Chandran and Lakshmanan [102] showed that supermodular lattices could be defined by a balanced identity. Schmidt [322] obtained conditions necessary for finite projectivity of a finite ml in certain varieties of ml. Configurations containing frames M_3 and M_4 (see in the sequel) play an important role in these conditions. See [352] for representation of lattices M_n as congruence lattices of finite algebras. The paper [110] by Chuong on certain properties of ml of breadth two is reflected in Chapter IV. Freese [162] considered modular and distributive varieties of algebras and, in particular, produced a new example of an ml that is not embeddable in a complemented ml. For strong extensions (§1) of ml see [191].

Theory of varieties and of certain other classes of ml are reflected principally in Chapter IV.

Herrmann [196] showed that a variety $\mathscr{K}$ of algebras is polynomially equivalent to the variety of modules over a ring if $\Theta(A)$ is a complemented ml for every algebra $A \in \mathscr{K}$. His preprints [OSL, II, 256 and 259] were published in [194] and [199]. The latter of these papers is connected with results of Gel'fand-Ponomarev ([168] and [169]) on representations of free ml. In particular, he described neutral and perfect elements in $FM(4)$. The paper [35] by Tsyl'ke [Zilke] is connected with the Gel'fand-Ponomarev conjecture on the sublattice of perfect elements. Stekol'shchik [34] suggested a construction of invariant elements connected with invariant spaces and perfect elements.

A primitive length of a lattice is defined by means of a sequence of its projective intervals. For an ml L with all intervals having the same primitive length, Schmidt and Wille [325] proved the equivalence of the following: (1) every operation on L compatible with congruences preserves the order, and (2) L does not contain nontrivial Boolean intervals. Brunner [89] obtained an upper bound for projective distances between two projective intervals of any lattice of the variety generated by all ml of length $\leq n$ and, as a corollary, he obtained a known result on this variety being finitely based.

Schmidt [321] found an ml L with the following properties: (1) L is a subdirect product of finite lattices; (2) there exists a function $f\colon L \to L$ compatible with all congruences but not isotone; (3) no proper interval of L is a complemented lattice. The following problems remain open: (a) for which ml does the compatibility of a k-place function with the congruences imply that the function preserves the order? (b) for which ml of finite primitive length is every operation compatible with the congruences a polynomial function?

A compatible tolerance that satisfies an additional condition is called a central relation on a lattice. Schweigert [328] proved that a simple ml of finite length is a projective geometry if and only if it has no central relations. It is known that every finite distributive lattice is representable as the congruence lattice of an ml M. Schmidt [324] showed that one can choose a complemented ml for M, namely, a sublattice of the lattice of subspaces of a vector space over a two-element field of an infinite countable dimension. Certain investigations connected with the problem of representation of affine geometries from Wille's book [372] can be found in Schmidt's work [326]. Gumm's book [182] is devoted to a geometric approach to studying congruences of a universal algebra. The papers by Ihringer [213] and Duda [141] are connected with Wille's book.

Albu [47] identified a class of ∨-continuous ml, for which the properties of being Artinian and Noetherian are equivalent.

A mapping f of the set of ∨-irreducible elements of a finite lattice into its set of ∧-irreducible elements is called a matching if it is one-to-one and $x \le f(x)$ for every ∨-irreducible x. Duffus [144] investigated the following conjecture: every finite ml has a matching. For certain classes of ml the conjecture holds. Another class of ml that have a matching was pointed out by Reuter [305].

Nicholson, Grubb, and Sharma [273] studied properties of so-called regular ∨-endomorphisms of complemented ml. Here is one of their results: a linear transformation of a finite-dimensional vector space induces a regular ∨-endomorphism of the lattice of its subspaces. Let L be a bounded ml, z' a complement of $z \in L$, and $T(a, b)$ the least compatible tolerance under which the elements a and b of L are related. Chajda and Zelinka [101] proved that an element $z \in L$ is central if and only if $T(0, z)$ and $T(0, z')$ are mutual complements in the lattice of tolerances. Călugăreanu [91] defined torsion lattices L as lattices in which every interval $[x, 1]$, where $x \ne 1$, contains atoms. He proved, in particular, that a complete pseudocomplemented torsion ml is atomic. Weight functions on finite ml were considered by Pezzoli [289].

Slatinský [339] found an ml G and a family M_i, $i \in G$, of ml such that the lexicographic sum $\sum_{i \in G} M_i$ is not an ml. He found some conditions under which a lexicographic sum of an ordered system of ordered sets is an ml.

Fourneau [160] called a lattice with 0 weakly modular if the conditions $a \le c$ and $b \wedge c \ne 0$ imply the equality $a \vee (b \wedge c) = (a \vee b) \wedge c$ for its elements. He deduced some properties of these lattices analogous to properties of ml. Jayaram [223] transferred some properties of ideals and complements of various types in ml to 1-modular lattices.

Hedlíková [192], Bandelt and Hedlíková [66], and Draškovičová [134] studied ternary algebras connected with the operation of median in ml. The latter paper is a publication of a preprint reflected in the previous survey (OSL, II, [178]). In another paper Draškovičová [135] considered varieties of modular median algebras found earlier as examples of applications of general theorems about properties of congruences. Tolerances on median algebras were considered in [64].

Let Ω be the set of atoms of an atomic lattice and let the sets Ω^n be defined inductively as follows: $\Omega^1 = \Omega$, $\Omega^{n+1} = \Omega^n \vee \Omega$. Now, let $F = \bigcup_{n=1}^{\infty} \Omega^n \cup \{0\}$ and, for any subsets $A, B \subseteq L$, let $(A, B)M$ mean that $(a, b)M$ for all $a \in A$ and $b \in B$. Here are typical results from the papers [259] and [260] by Maeda: (a) if $\Omega \subseteq A \subseteq L$ then $(A, \Omega^n)M \Leftrightarrow (A, \Omega^{n-1})M^*$ for all $n \geq 2$; (b) $(A, F)M \Leftrightarrow (A, F)M^*$; (c) $(\Omega^n, L)M^* \Leftrightarrow (F, L)M^*$ for every $n \geq 2$. In the subsequent paper [261] Maeda defined semidistributive and distributive pairs of elements and established connections between these notions and modularity and dual modularity of the pairs. Jones [230] considered various types of elements of a lattice connected with modularity. In particular, he proved the following theorem: in an M-symmetric lattice each distributive and dually distributive element is neutral.

Pezzoli [290] investigated systems of axioms that define a generalization of matroids on a partially ordered set and showed that these systems of axioms coincide for ml with locally finite chains. In particular, for this class of lattices he defined the notions of independence, generators, and the rank function. Stern ([346] and [348]) studied generalized matroid lattices close to Maeda's AC-lattices and obtained for them results analogous to those he had proved earlier for Baer lattices [OSL, II, 444–446] and to results of Soltan [33]. Stern's results on Baer lattices were published without proof in [347]. Kohl and Stern [239] proved that, in a generalized matroid lattice, the rank function preserves the order if and only if each of its $\vee$-irreducible elements is a cycle. A greater part of results of Richter's articles [306], [307], and [309] is contained in the preprints [OSL, II, 397–399]. In his subsequent paper [308], which is close to these articles, he proved that the Kurosh-Ore theorem and certain other properties hold in a class of lattices wider than ml. See also [345].

Let L be a lattice of finite length, $J(L)$ the set of its nonzero $\vee$-irreducible elements, and u' the only element covered by an element $u \in J(L)$. A lattice L is called strong if, for any $x \in L$ and $u \in J(L)$, the condition $u \leq x \vee u'$ implies $u \leq x$. This notion was introduced by Faigle [OSL, II, 203]. Richter and Stern [312] proved that a lattice of finite length is strong if and only if it does not contain a sublattice that is a pentagon such that, in the middle of its four-element chain, there are u' and u for some $u \in J(L)$. An analogous characterization was obtained for strong lattices in the class of all semimodular lattices of finite length. Stern ([349] and [350]) and Faigle, Richter, and Stern [156] investigated connections between the Steinitz-Mac Lane exchange property and its various modifications and the semimodularity of a lattice. The latter paper contains, in particular, the following theorems: (1) a lattice L is strong semimodular if and only if it satisfies the exchange property (EP_2): if $a \in L$, $u, v \in J(L)$, $v \leq a \vee u$ and $v \not\leq a \vee u'$, then

$u \leq a \vee v$; (2) L is modular if and only if it and its dual satisfy (EP$_2$); (2) L is ∨-symmetric in the sense of Gaskill and Rival [OSL, II, 223] if and only if it is strong semimodular. Extending the notion of a minimal pair of Gaskill–Rival, the authors defined a base of an element and introduced a natural condition of exchange of a base. They proved that a semimodular lattice of finite length possesses the base exchange property if and only if it is strong.

Jakubík [221] found a condition necessary and sufficient for semimodular lattices of locally finite length with isomorphic graphs to differ only by duality of a direct factor. Quackenbush [301] showed that the tensor product $M_3 \otimes B$, where M_3 is a three-element lattice with three atoms and B is a bounded ml, both considered as ∨-semilattices with 0, is a semimodular lattice that is modular if and only if B is distributive. A large article by Crapo [114] studied applications of the theory of formal languages in programming and in representation of semimodular lattices that reflects the process logic. A series of papers by Percsy [286], [287], and [288], Herzer [201], and Kahn [231] are dedicated to problems of embedding certain semimodular lattices in projective spaces.

Lately, geometric ideas have produced many significant results in the theory of modular lattices. Day ([118] and [119]) surveyed this direction; a great merit of his surveys is a description of connections between classical concepts of Birkhoff and von Neumann and modern techniques of Huhn, Herrmann, Freese, and Day applied to study of varieties of ml. A substantial bibliography is given.

A spanning n-frame of a bounded ml is a system $(d_1, \ldots, d_{n+1})$ of its elements satisfying the conditions (1) $\bigvee_{j \neq i} d_j = 1$ for all i; and (2) $d_i \wedge (\bigvee_{k \neq i, j} d_k) = 0$ for all $i \neq j$. Together with homogeneous frames of von Neumann and n-diamonds of Huhn, this concept is a formalization of $(n-1) + 2$ points in general position for an $(n-1)$-dimensional projective geometry. Projective geometries of dimension $(n-1)$ are precisely those ml that possess a spanning n-frame. The survey [119] begins with a popular introduction to lattice interpretation of projective geometry and explanation of the principal ideas of coordinatization theory. Then the author discusses Jónsson's Arguesian identity and equivalent conditions, properties of frames, and gives a recent result of the author and Pickering [124] on a construction of a ring coordinatizing hyperplanes of Arguesian lattices. Finally, there is an exposition of applications of n-frame theory and coordinatization to equational theory of ml. The central concepts here are those of projective configuration (it is defined for an arbitrary variety of algebras) and "glueing" of ml. The author traced development of these concepts from Hall, Dilworth (examples of ml nonembeddable in complemented ml), and Jónsson (a characterization of Arguesian lattices of length ≤ 4; Arguesian lattices not representable as lattices of subgroups of abelian groups) to Huhn (a connection between projective geometries and n-distributivity [OSL, II, 272], Freese and Herrmann (unsolvability of the word problem in a free ml). Four parts of survey [118] are devoted mainly to the following topics: (1) frames; (2) problems of the theory of the variety of Arguesian lattices (connected with work of Freese and Herrmann); (3) lattice-theoretic characterization of commutativity of the coordinatizing ring and Pappian lattices; and (4) variety of quasiplanes (the author defined a quasiplane as an ml that does not contain two-dimensional "glueing" of three-diamonds). Among the new results of the final part we note the following: coordinatizable quasiplanes are subdirect products of projective planes. Geometric aspects of lattice theory are reflected in the survey [79] by Birkhoff. See also [183].

Day [120] proved an important technical lemma on ml with a spanning n-frame, connected with a set of lattice elements, that is the underlying set of the Day-Pickering [124] coordinatizing ring. This lemma was applied to describe some freely generated ml and Arguesian lattices. In another paper [121] Day gave a definition of Arguesian lattices by means of four natural geometric conditions that connect any two triangles of an ml. A triangle is understood as an arbitrary triple of the lattice elements. A corollary: every 2-distributive ml is Arguesian. Herrmann [199] described a class of subdirectly irreducible Arguesian lattices with four generators, which contains, in particular, all finite-dimensional projective geometries over prime fields. Results of his paper [200] are reflected in Chapter IV. See also [195].

Day and Jónsson [123] characterized non-Arguesian ml: an ml is non-Arguesian if and only if its interval lattice contains 10 intervals P_α and 10 intervals Q_β, which are nondegenerate projective planes, and their points $p_\alpha \in P_\alpha$ and lines $q_\beta \in Q_\beta$, which, in a sense, form a certain "classic" non-Arguesian configuration. Earlier, Day [OSL, III, 67] introduced two identities, each of which could be considered as a lattice equivalent of Pappus' projective theorem. Continuing studying these identities he obtained [122], in particular, a new solution of Skornyakov's problem: a regular ring R is commutative if and only if the lattice $L(_R R^3)$ satisfies the LP-Pappian identity.

Halperin [186] analyzed the proof of von Neumann's coordinatization theorem and surveyed [185] books of von Neumann on continuous geometry. He prepared for publication von Neumann's book [187] on the axiomatics of a logico-probabilistic scheme that can be identified with a certain set of projections of a Hilbert space.

It is known that a complemented ml L is coordinatized by a strongly regular Baer semigroup $P(L)$. Nambooripad and Pastijn [272] considered a semigroup with division $B(L)$ of transformations that contains $P(L)$ and proved that a partially ordered set L with 0 and 1 is a complemented ml if and only if $B(L)$ is regular and coordinatizes L. A semigroup S is a strongly regular Baer semigroup that coordinatizes a complemented ml L if and only if it is isomorphic to a full regular subsemigroup of $B(L)$. The papers of Faulkner [157], Hannah [190], and Müller [270] are devoted to problems of coordinatization of projective planes, continuous geometries, and related regular rings.

The papers by Björner [81]–[84] and Baclawski [56] and [57] belong to a combinatorial-geometric direction connected with consideration of a simplicial complex whose vertices are elements of an arbitrary partially ordered set and simplexes are its chains. This approach produced interesting results on fixed points of isotone maps (see Chapter VI and OSL, II). Also, geometric lattices with 0 and 1 removed are examples of Cohen-Macauley partially ordered sets, whose theory has been actively developed lately. We note the following of the lattice-theoretic results in this direction: if a lattice of finite length does not contain a three-element interval, it is relatively complemented [83]. Using the concept of blowing up cycles, which is one of the fundamental notions in the theory of Cohen-Macaulay sets, Björner [84] generalized Folkman's theorem on cohomology groups of geometric lattices to the infinite case, gave a method for construction of a Z-base for the group $H_{r-2}(L)$, where r is the rank of a geometric lattice L, and found a homological proof of Greene's inequality for the Möbius function. Voigt [364] obtained certain combinatorial properties of regressing (that is, such that $f(x) \leq x$ for all x) mappings of lattices of subspaces of affine and projective spaces; Stanley [344] found a combinatorial property of geometric lattices. A paper [78] of Bergman and a Hrushovski on cofinal distributive sublattices (see §1.2) contains a series of interesting examples of ml, mainly of combinatorial-geometric character. Delandtsheer [125] described geometric lattices of dimension $n \geq 3$ whose automorphism groups act transitively on unordered pairs of disjoint hyperplanes. These are either Boolean lattices, or a projective space, or an affine space, or w_{22} (the lattice obtained from a Steiner system). Bennett [72] characterized lattices of subspaces of an affine space as Hilbert lattices satisfying certain conditions. See Nikitin [22] for a construction of a free projective plane. Urquhart [358] studied geometric lattices connected with logic. He used the undecidability of the word problem for free ml in this article. A partition of type n is defined as a partition of a set such that every n-element subset is contained in precisely one block and each block has no less than n elements. Let $\text{LP}_n(S)$ denote the lattice of all partitions of type n on S. Earlier Hartmanis characterized $\text{LP}_2(S)$ by five geometric axioms. Modifying these axioms, Chên Jung Hsü [207] obtained an analogous result for $\text{LP}_n(S)$ for any $n \geq 2$. Służalec [341] found some properties of independent elements of projective and affine spaces considered as complemented complete lattices. Nachev [21] considered a connection between the global dimension of the incidence ring $I(X, R)$, where X is a finite ml and R is an associative ring, and the dimension of X.

Jambu and Terao [222] studied properties of a bunch of hyperplanes determined by the properties of a geometric lattice associated with that bunch.

See [284] for cohomologies of this geometric lattice. Löwen [256] considered certain geometries embedded in an n-dimensional projective space over one of the classical fields and, as the main result, proved that every automorphism of such a geometry is induced by an automorphism of the entire space. See [262] for so-called A-systems of projections in a finite-dimensional projective space. See also Chapter V for lattices of subspaces.

§III.3. Related structures

3.1. Sublattice and ideal lattices of lattices. Chen, Koh, and Lee [105] studied properties of the lattice Sub L of sublattices of a finite distributive lattice L connected with its Frattini sublattice $\Phi(L)$. They called the lattice Sub L pure if the interval $[\Phi(L), L]$ is a Boolean sublattice of Sub L; it was called double pure if the interval $[\varnothing, \Phi(L)] \cong \text{Sub}\,\Phi(L)$ also possesses the same property. A delta-sum of lattices was defined as their ordinal sum in which the identity of each summand, except the last one, is identified with the zero of the preceding summand. The goal of [105] is obtaining conditions sufficient for the purity of Sub L and a characterization of finite ordinal and delta-ordinal irreducible distributive lattices L with double pure Sub L. Continuing this research Chen and Koh [104] defined strongly pure Sub L and, for finite distributive lattices, obtained two conditions each of which is equivalent to strong purity. Chen, Koh, and Teo [108] investigated a connection between delta-decompositions of lattices with isomorphic sublattice lattices and the existence of pseudocomplementations in such lattices. They proved that if every interval $[0, x]$, $x \in L$, is a complemented lattice or if L is a semimodular lattice of finite length, then every isomorphism Sub $L \cong$ Sub K induces an isomorphism or a dual isomorphism between L and K. Balbes [58] characterized maximal sublattices and completely $\wedge$-irreducible sublattices of Sub L for a distributive lattice L. Düntsch [146] found a condition sufficient for the simplicity of the lattice Sub B for a Boolean algebra B and gave a number of examples when this lattice is not simple.

Let $M(L)$ and $J(L)$ be the sets of $\wedge$-irreducible and $\vee$-irreducible elements of L, and let $l(\text{Sub}\,L)$ be the length of Sub L. Studying linear extension of the sets $M(L)$ and $J(L)$ for a finite distributive lattice L, Chen and Koh [103] deduced the formula $l(\text{Sub}\,L) = \sum_{i=1}^{n} |J((x_i] - \bigcup_{j<i}(x_j])|$, where $x_1 < \cdots < x_n = 1$ is any linear extension of a partially ordered set $M(L) = \{x_1, \ldots, x_n\}$. An analogous formula can be written applying a linear extension of $J(L)$. Let $u_k(L)$ $[v_k(L)]$ be the number of elements of a finite distributive lattice L with exactly k covering [covered] elements. Koh [237] introduced an order on the set $A(P)$ of all antichains of a partially ordered set P, which turns this set into a lattice such that every finite distributive lattice is isomorphic to $A(P)$ for suitable P. Using the above formula, he obtained two more formulas: $l(\text{Sub}\,L) = 1 + u_1(L) + u_2(L) = 1 + v_1(L) + v_2(L)$

and $l(\operatorname{Sub} A(P)) = 1 + |P| + \mathscr{L}_2(P)$, where $\mathscr{L}_2(P)$ denotes the number of two-element chains in P. He found corollaries for direct products of finite distributive lattices and for the free distributive lattice $FD(n)$. The lattice $A(P)$ was considered by Koh in [238].

If $l^*(\operatorname{Sub} L)$ [$l_*(\operatorname{Sub} L)$] is the length of the longest [shortest] maximal chain in $\operatorname{Sub} L$ for a finite lattice L, then let $g(L) = l^*(\operatorname{Sub} L) - l_*(\operatorname{Sub} L)$ and $\mathscr{L}(L) = L - l_*(\operatorname{Sub} L)$. Earlier Lakser [249] characterized finite lattices with $g(L) = 0$. Chen, Koh, and Lee [106] used delta-sums to describe finite distributive lattices with $\mathscr{L}(L) = 0$ or 1 or with $g(L) = 0$ or 1.

Igoshin continued (see [OSL, II, 22–25]) to study the lattice $\operatorname{Si}(L)$ of closed intervals of a lattice L. In the papers [12], [13], [14], [211], and [212] results deposited earlier were included. Also, in the last of these papers the equivalence of the following conditions was proved: (1) $\operatorname{Si}(L)$ is a geometric lattice; (2) $\operatorname{Si}(L)$ is an algebraic lattice; (3) all chains in L are finite. He proved that properties of weak atomicity are equivalent for L and $\operatorname{Si}(L)$. The lattice $\operatorname{Si}(L)$ is strongly atomic if and only if L is both strongly atomic and strongly co-atomic. In another paper [211] a lattice identity was found that holds in L if and only if it holds in $\operatorname{Si}(L)$. He proved [15] that a lattice L with 0 is isomorphic to $\operatorname{Si}(L')$ for a lattice L' with 0 if and only if it contains an atom p that satisfies the conditions: (a) $p \vee s = p \vee t \Rightarrow s = t$ for any atoms $s, t \in L$; (b) if $x \neq 0$ and x is not an atom, then there exists a unique representation $x = s \vee s'$, where s and s' are atoms, $s \neq s'$, $p \vee s < p \vee x$, $p \vee s' = p \vee x$ and, if $p < x$, then $s = p$; (c) if $x = s \vee s'$ and $y = t \vee t'$ are representations from (b), then $x \leq y \Leftrightarrow p \vee t \leq p \vee s \,\&\, p \vee x \leq p \vee y$; and if s is an atom, then $s \leq y \Leftrightarrow p \vee t \leq p \vee s \leq p \vee y$. Kolibiar [241] and Slavík [340] studied when lattices are determined by their interval lattices. Here are three theorems:

(1) The following conditions are equivalent for lattices A and B: (a) $\operatorname{Si}(A) \cong \operatorname{Si}(B)$; (b) $\operatorname{CS}(A) \cong \operatorname{CS}(B)$ (here $\operatorname{CS}(L)$ is the lattice of convex sublattices of a lattice L); (c) there exist lattices A_1, A_2, B_1, B_2 such that $A = A_1 \times A_2$, $B = B_1 \times B_2$ and $A_1 \cong B_1$, $A_2 \stackrel{d}{\cong} B_2$ (here $\stackrel{d}{\cong}$ is a dual isomorphism; observe that the direct products here can be replaced by subdirect ones).

(2) If $f \colon \operatorname{Si}(A) \to \operatorname{Si}(B)$ is an isomorphism and $f' \colon A \to B$ is the bijection it induces (that is, $f'(a) = l \Leftrightarrow f([a, a]) = [b, b]$), then the following conditions are equivalent: (a) f' is either an isomorphism or a dual isomorphism; (b) A is indecomposable into a nontrivial direct product.

(3) The following conditions are equivalent for the lattices A and B: (a) $\operatorname{Si}(A) \cong \operatorname{Si}(B)$ implies $A \cong B$ or $A \stackrel{d}{\cong} B$; (b) either A is directly indecomposable or $A = A_1 \times A_2$, and Goth lattices A_1 and A_2 are self-dual.

They observed that one of Igoshin's results follows from the last theorem and gave some corollaries.

A lattice is called molecular if each of its elements is a join of a finite number of atoms. Fourneau [159] showed that the ideal lattice Id(L) is atomic if and only if L is molecular and obtained conditions equivalent to an ideal of a molecular lattice being standard. An ideal A of L is called complete if it contains all least upper bounds of its subsets that exist in L. Bishop and Schreiner [80] studied conditions under which joins of ideals in the lattices of all ideals and of all complete ideals coincide. They found a number of properties of a lattice L inherited by the second of these two lattices. Nauryzbaev and Omarov [20] found an example of the ideal lattice of a distributive lattice that is not weakly atomic. Beazer [69] gave another proof of Katriňak's theorem: the ideal lattice Id(L) of a bounded distributive lattice L is Stone if and only if L is a Stone algebra with a complete center. From this result he deduced conditions equivalent to Id(L) being relatively Stone.

Bandelt [63] considered Galois connections between lattices of ideals and filters of lattices. Nieminen [279] introduced a notion of a 2-ideal of a finite lattice, which generalizes ordinary ideals. A relative annihilator in a lattice L is a set $\langle a, b \rangle = \{x \in L | x \wedge a \leq b\}$. He proved that the lattice Id$_2(L)$ of all 2-ideals of a finite lattice L is distributive if and only if every relative annihilator in this lattice is a 2-ideal in it. An example showing that Id$_2(L)$ does not determine L up to isomorphism was constructed. In another paper [278] Nieminen considered partially ordered sets called $\aleph_{\text{mub}}$-lattices, which are a generalization of lattices. The set CS(L) of all convex sublattices of a lattice L can serve as an example. He proved that a lattice L is distributive or modular if and only if the $\aleph_{\text{mub}}$-lattice CS(L) has the same property. Considering so-called k-distributive partially ordered sets, Hickman and Monro [202] showed that a set is k-distributive if and only if the lattice of its k-ideals is a complete Heyting algebra. See [165] for the lattice of order ideals of some partially ordered set. Bauer's talk (mentioned in [342]) on semilattices with distributive ideal lattices is reflected in the next subsection §3.2.

Vrancken-Mawet ([366] and [367]) investigated properties of the congruence lattices of partially ordered sets and used them for subalgebra lattices of finite Heyting algebras and finite Brouwerian semilattices. In particular, it was proved that the subalgebra lattice of a finite Heyting algebra A is always lower semimodular. It is a Boolean algebra if and only if $A \cong \underset{\sim}{n} \times 2$ for some n, where $\underset{\sim}{n}$ denotes the n-element chain. For a finite Brouwerian semilattice A the following are equivalent: (a) A is a chain; (b) Sub A is a Boolean algebra; (c) Sub A is distributive; (d) Sub A is uniquely complemented. Vrancken-Mawet [368] considered the lattice $S_R(L)$ of the so-called R-subalgebras of a bounded distributive lattice L and described those L for which $S_R(L)$ is modular, or distributive, or belongs to some other natural classes. He also characterized lattices of the form $S_R(L)$ in the classes of distributive lattices and modular lattices. Köhler's paper [240] on the lattice of the so-called total subalgebras of a Brouwerian semilattice was reflected in Chapter I. Shiryaev [36] proved that the lattice Sub S of subsemilattices of a semilattice S is $\wedge$-semidistributive if and only if S is a chain. He introduced a notion of a special chain and proved that Sub S is $\vee$-semidistributive if and only if S contains no special chains. Chen, Koh, and Teo [107] obtained a formula for the length of the subalgebra lattice of finite Stone algebras.

3.2. Lattices of congruences and tolerances. Congruence lattices of algebras were considered from the viewpoint of the theory of representations of lattices in Schmidt's monograph [323]. The third chapter is devoted specifically to congruences of lattices. Schmidt's paper [324] was devoted to representations of finite distributive lattices by congruences of modular lattices. The result was stated in §2. It was repeated in Schmidt's talk mentioned in [342]. In Bauer's talk at the same conference, as observed by Smith [342], it was shown that every semilattice with the distributive ideal lattice is isomorphic to the lattice of compact congruences of a lattice. Some results strengthening the conjecture that every distributive semilattice could be represented as the semilattice of all compact congruences of a suitable lattice can be found in Huhn's paper [209].

The main result of Bosbach [87] is the theorem: every completely $\vee$-distributive algebraic lattice can be represented as the congruence lattice of a Brouwerian semilattice. Lystad and Stralka [258] proved that the lattice of closed congruences of algebraic lattices is bialgebraic. Pudlák [300] considered an old problem: is every distributive algebraic lattice isomorphic to the congruence lattice of a lattice? The author wanted to reduce this problem to investigating congruence lattices of finite lattices. To this end he introduced a general notion of representation and proved a number of theorems on the lattice Con L for a finite lattice L. Kamimura and Tang [234] considered continuous congruences of continuous (in the sense of Scott) lattices. If L is an algebraic lattice and θ a continuous congruence on it, then the continuous lattice L/θ need not be algebraic. If it is algebraic, θ is called an algebraic congruence on L. Let ACon L be the set of such congruences. They proved that the class of all lattices of the form ACon L coincides with the class of all lattices of the form Sub C, where C is a $\vee$-semilattice with zero. Kurinnoĭ [17] described semilattices defined on a given set A, whose congruence lattices coincide as subsets of the set $A \times A$.

In his survey [128] Dilworth considered, in particular, the role of Con L in studying lattice varieties. A theory of irreducible $\wedge$-decompositions for compactly generated distributive lattices was constructed, which, when applied to Con L, produced some theorems on the structure of L. Sivák [338] showed that, for every finite lattice, it is possible to construct a finite extension with an isomorphic congruence lattice and such that all of its elements are joins of atoms. This paper is considered also in §1.4.

As a generalization of distributive lattices, Davey and McCarthy [117] considered the variety w_3 generated by a three-element set with some pseudo-order; they described, in particular, congruence lattices of arbitrary algebras from w_3. Baranskiĭ's theorem on the independence of congruence lattices and automorphism groups of lattices is stated in §3.3, below. A survey [268] by Mitsch on the congruence lattices of semigroups has a section devoted to semilattices S with various types of lattices $\operatorname{Con} S$ (modular, strictly semimodular, pseudocomplemented, etc.). Zhitomirskiĭ ([10] and [11]) obtained some characterizations of the class K of all lattices isomorphic to the congruence lattices of semilattices. These characterizations follow from the basic theorem on embedding lattices with 0 and 1, and satisfying certain additional conditions, in lattices from K. It is known that K is not elementarily axiomatizable. It turned out that the Boolean part of this class, which consists of all lattices isomorphic to the congruence lattices of locally finite trees, is elementarily axiomatizable in the class of all complete lattices. The structure of those congruences of a semilattice S that are distributive or Boolean elements of $\operatorname{Con} S$ was described in [9].

We mention briefly papers on the congruence lattices of algebras close to lattices (see also Chapters I and II). Varlet [362] characterized regular congruences in the congruence lattices of p-algebras and p-semilattices and described irregular p-algebras (that is, those whose only regular congruence is the universal one). Congruence lattices of p-algebras and double p-algebras were considered by Beazer [68]. Sankappanavar [320] described those p-semilattices whose congruence lattices are semimodular or distributive, while Goldberg [171] described distributive double p-algebras whose congruence lattices are chains.

Varlet [361] observed that there exist precisely two minimal De Morgan algebras whose congruence lattice is isomorphic to 2^n $(n \geq 1)$, and that such an algebra has $2n$ elements. Vas de Carvalho [363] proved, in particular, that the congruence lattice of every finite algebra from a certain variety $K_{2;0}$ of Ockham algebras is Boolean. Blyth and Varlet [86] established certain properties of the congruence lattices of MS-algebras.

Let $\operatorname{LT}(A)$ be the lattice of stable tolerances of an algebra A and $\operatorname{LD}(A)$ the lattice of all stable reflexive binary relations contained in a fixed order $\leq$ of A. It is known that these lattices are isomorphic if A is a lattice and $\leq$ is the natural order in it. It turned out that this result does not hold for semilattices. Chajda, Nieminen, and Zelinka [99] showed that the isomorphism $\operatorname{LT}(S) \cong \operatorname{LD}(S)$ holds for a finite semilattice S if and only if S is a chain. In the general case there exists an isomorphism of $\operatorname{LD}(S)$ onto a sublattice of $\operatorname{LT}(S)$, and the case when $\operatorname{LD}(S)$ is isomorphic both to $\operatorname{LT}(S)$ and a proper sublattice of it is not excluded. The result of [101] is stated in §2.

Niederle [276] showed that the lattice $\operatorname{LT}(L)$ of a distributive lattice L is Boolean if and only if L is pseudocomplemented and of locally finite length.

He [274] studied properties of certain special elements of LT(L) for a finite distributive lattice L, and, in [275], proved that the tolerance lattice of a direct product of a finite number of lattices is isomorphic to the product of the tolerance lattices of the factors.

3.3. Endomorphism monoids, automorphism groups. Independence and compatibility.
McKenzie and Tsinakis [264] showed that a bounded distributive lattice is determined by its endomorphism monoid up to an isomorphism or a dual isomorphism.

It is known that the varieties of pseudocomplemented distributive lattices form a chain $B_{-1} \subset B_0 \subset B_1 \subset \cdots \subset B_\omega$, where B_{-1} is the trivial variety, and B_0 and B_1 are the varieties of Boolean and Stone algebras, respectively. Adams, Koubek, and Sichler [41] proved that every algebra $A \in B_1$ is uniquely determined by its monoid End A. In B_2 there exist nonisomorphic algebras with isomorphic endomorphism monoids, but at most two nonisomorphic algebras have isomorphic endomorphism monoids. In this and papers subsequent to [40], properties of End A followed from general theorems on universality of varieties. The goal of [40] was to investigate the problem when an infinite distributive pseudocomplemented lattice A has a finite monoid End A. The authors proved, in particular, that in B_3 there exist algebras of arbitrarily large cardinality with only a finite number of endomorphisms.

Perminov [24] published proofs of results on rigid lattices announced earlier [OSL, II, 48]. He [23] also proved that every finite lattice is embeddable in a diagram lattice (that is, in a lattice whose automorphism group is isomorphic to the automorphism group of its covering graph; the description of such lattices is Birkhoff's Problem 6). Let L be a lattice with 0 and 1. Koubek and Sichler [245] considered the category $Q(L)$ of all bounded lattices admitting a homomorphism onto L, with $(0, 1)$-homomorphisms. For a lattice L with 0 and 1 that does not generate the variety of all lattices the following conditions were proved to be equivalent: (1) L has no homomorphisms onto a free $(0, 1)$-lattice; (2) L has no prime ideals; (3) $Q(L)$ contains rigid (that is, without nontrivial $(0, 1)$ endomorphisms) lattices; (4) $Q(L)$ contains rigid lattices of any cardinality; (5) for every monoid M the class $Q(L)$ contains a proper subclass C consisting of pairwise nonisomorphic lattices for each of which the monoid of its $(0, 1)$-endomorphisms is isomorphic to M; and (6) $Q(L)$ is a universal category.

Kalmbach [233] proved that every group is isomorphic to the automorphism group of a suitable orthomodular lattice. This result was known for finite groups. A series of interesting papers on automorphism groups of partially ordered sets appeared recently. All of them are reflected in Chapter VI.

Baranskiĭ's survey [1] contained, without proofs, results of the author on independence of derived structures in the classes of semigroups and lattices obtained by him before 1982. In [2] he constructed a class of complete distributive lattices indistinguishable in the first-order language in which every group is representable as the automorphism group. In other words, he proved that there exist complete distributive lattices whose elementary theory is compatible with any group. In [3] Problem 19 from Grätzer's book *General Lattice Theory* was solved, namely, Baranskiĭ proved that Aut and Con are independent in the class of all finite lattices. The same paper contains a partial answer to Problem 18 of Grätzer: he proved that for every group G and every bi-algebraic lattice L there exists a lattice T such that $G \cong \operatorname{Aut} T$ and $L \cong \operatorname{Con} T$. Barnanskiĭ [4] proved the independence of automorphism groups and equational theories in the class of all lattices, that is, he showed that for every group G and every nontrivial lattice S there exists a lattice T that is equationally equivalent to S and such that $G \cong \operatorname{Aut} T$. In [5] he proved, in particular, the independence of automorphism groups and retracts in any quasivariety of lattices and in the class of all finite lattices of any quasivariety of lattices. In other words, he proved that, for every nontrivial quasivariety $\mathfrak{M}$ of lattices, every group G, and every lattice $L \in \mathfrak{M}$, there exists a lattice $T \in \mathfrak{M}$ (which is finite, if G and L are finite), such that $G \cong \operatorname{Aut} T$ and L is a retract in T. The independence of automorphism groups and quasi-equational theories in the class of all lattices, that is, in particular, the result of [4], follows from this result. In [6] a natural formalization was given for the notion of independence of properties of derived structures of one type from derived structures of another type in a class of algebraic systems. We note that the independence of a pair (π_1, π_2) of derived structures in a given class of algebraic structures always implies the independence of properties of derived structures of type π_1 from properties of derived structures of type π_2 in this class. He proved that, in the class of all lattices, properties of automorphism groups do not depend on those of congruence lattices and those of elementary theories.

Perminov [25] showed that, for every finite lattice S and every [finite] monoid M, there exists a [finite] bounded lattice L such that $\operatorname{End}_{0,1} L \cong \operatorname{End}_{nc} L \cong M$ and S is a sublattice of L (here $\operatorname{End}_{0,1} L$ and $\operatorname{End}_{nc} L$ are monoids of the endomorphisms that preserve 0 and 1 and, respectively, of nonconstant endomorphisms). Lender [19] proved the following theorems:

(1) for any [finite] lattices L_1 and L_2 and any [finite] group G there exists a [finite] lattice L such that $\operatorname{Aut} L \cong G$ and L_1 is a sublattice of L while L_2 is a homomorphic image of L;

(2) for any three groups G_1, G_2, and G_3 there exist lattices L_1, L_2, and L_3 such that $\operatorname{Aut} L_i \cong G_i$ for $i = 1, 2, 3$, and L_1 is a sublattice of L_3, and L_2 a homomorphic image of L_3.

Let L be a lattice, G a group, and either $|L|, |G| > 1$, or $|L| = 1$ and $|G| \geq \aleph_0$, and let α be a cardinal number such that $\alpha \geq \max\{|L|, |G|\}$. Adams and Sichler [44] proved that there exist 2^α nonisomorphic lattices C

such that $|C| = \alpha$, L is embeddable in C, $\operatorname{Aut} C \cong G$, and $L = \{x | x \in C, \varphi(x) = x \text{ for all } \varphi \in \operatorname{Aut} C\}$. They showed that an analogous theorem holds for partially ordered sets. Let G be a finite group different from Z_2^2, Z_2^3, Z_2^4, Z_8^2 and the quaternion group, and let L be a finite distributive lattice without nontrivial automorphisms in which each chain can have more than two $\vee$-irreducible elements. Adams, Babai, and Sichler [39] proved that there exists a finite distributive lattice L' such that $\operatorname{Aut} L' \cong G$ and $S(L') \cong L$, where $S(L') = \{x | x \in L', \varphi(x) = x \text{ for all } \varphi \in \operatorname{Aut} L'\}$. The main result of Adams, Pigozzi, and Sichler [42] is a theorem on the endomorphism monoids, which, when applied to lattices, produces the following result: for every ordinal number α and a set $\{M_\gamma; \gamma \leq \alpha\}$ of monoids there exist bounded lattices G_γ $(\gamma \leq \alpha)$ satisfying the conditions: (1) G_{γ_1} is a $(0, 1)$-sublattice of G_{γ_2} for all $\gamma_1 \leq \gamma_2$; (2) $G_\lambda = \bigcup \{G_\gamma; \gamma < \lambda\}$ for limit λ; and (3) $\operatorname{End} G_\gamma \cong M_\gamma$ for all $\gamma \leq \alpha$.

Bibliography

1. V. A. Baranskiĭ, *On independence of related structures of algebraic systems*, Izv. Vyssh. Uchebn. Zaved. Mat. **1982**, no. 11, 75–77; English transl. in Soviet Math. (Iz. VUZ) **26** (1982), no. 11.

2. _____, *Algebraic systems whose elementary theory is compatible with an arbitrary group*, Algebra i Logika **22** (1983), no. 6, 599–607; English transl. in Algebra and Logic **22** (1983), no. 6.

3. _____, *On independence of lattices of congruence and groups of automorphisms of lattices*, Izv. Vyssh. Uchebn. Zaved. Mat. **1984**, no. 12, 12–17; English transl. in Soviet Math. (Iz. VUZ) **28** (1984), no. 12.

4. _____, *Independence of equational theories and of groups of lattice automorphisms*, Sibirsk Mat. Zh. **26** (1985), no. 4, 3–10; English transl. in Siberian Math. J. **26** (1985), no. 4.

5. _____, *Independence of groups of automorphisms and retracts for semigroups and lattices*, Izv. Vyssh. Uchebn. Zaved. Mat. **1986**, no. 2, 52–54; English transl. in Soviet Math. (Iz. VUZ) **30** (1986), no. 2.

6. _____, *Independence of properties of groups of automorphisms from properties of other derivative structures*, Izv. Vyssh. Uchebn. Zaved. Mat. **1986**, no. 3, 17–22; English transl. in Soviet Math. (Iz. VUZ) **30** (1986), no. 3.

7. G. Birkhoff, *Lattice Theory*, Colloquium Publications, vol. 25, Amer. Math. Soc., Providence, RI, 1979.

8. G. Grätzer, *General Lattice Theory*, Academic Press, New York, 1978.

9. G. I. Zhitomirskiĭ, *Boolean elements of the congruence lattice of a semilattice*, Teoriya Polugrupp i ee Prilozheniya, No. 7, Saratov Gos. Univ., Saratov, 1984, pp. 9–16. (Russian)

10. _____, *A characterization of the congruence lattices on semilattices*, Preprint No. 8142-84, deposited at VINITI by the editors of Izv. Vyssh. Uchebn. Zaved. Mat. **1984**. (Russian)

11. _____, *A characterization of congruence relations on semilattices*, Izv. Vyssh. Uchebn. Zaved. Mat. (1985), no. 5, 85.

12. V. I. Igoshin, *Lattices of intervals of chains*, Ordered Sets and Lattices, No. 8, Saratov Gos. Univ., Saratov, 1982, pp. 50–55. (Russian)

13. _____, *Interval projections of lattices*, Algebraic Actions and Orderings, Leningrad. Gos. Ped. Inst., Leningrad, 1983, pp. 42–49. (Russian)

14. _____, *An abstract characterization of two classes of lattices of intervals of lattices*, Teoriya Polugrupp i ee Prilozheniya, No. 7, Saratov Gos. Univ., Saratov, 1982, pp. 24–33. (Russian)

15. _____, *An algebraic characterization of interval lattices*, Uspekhi Mat. Nauk **40** (1985), no. 3, 205–206; English transl. in Russian Math. Surveys **40** (1985), no. 3.

16. A. Kaufmann, *Introduction to the theory of fuzzy sets*, Masson, Paris, 1977. (French)

17. G. Ch. Kurinnoĭ, *Semilattices with shared congruences*, Preprint No. 284k-85, deposited at the Ukrainian NIINTI, 1985. (Russian)

18. A. M. Kut'in, *Classes of lattices with identities generalizing the modular law*, Preprint No. 1953-85, deposited at VINITI, 1985. (Russian)

19. V. B. Lender, *On automorphism groups of lattices, semigroups and rings*, Abstract of Report from the IX Symposium on the Theory of Groups, Moscow, 1984, p. 217. (Russian)

20. K. A. Nauryzbaev and A. I. Omarov, *On equational compactness of distributive lattices*, Investigations in the Theory of Models, Alma-Ata, 1982, pp. 35–41. (Russian)

21. N. A. Nachev, *The global dimension of incidence rings*. I, Plovdiv. Univ. Nauchn. Trud. **18** (1980), no. 1, 19–41; II, Plovdiv. Univ. Nauchn. Trud. **18** (1981), no. 1, 43–63. (Bulgarian)

22. A. A. Nikitin, *On freely generated projective planes*, Algebra i Logika **22** (1983), no. 1, 61–78; English transl. in Algebra and Logic **22** (1983), no. 1.

23. E. A. Perminov, *On diagram lattices*, Preprint No. 340-85, deposited at VINITI, 1985. (Russian)

24. _____, *On rigid lattices*, Preprint No. 847-84, deposited at VINITI, 1984. (Russian)

25. _____, *On the independence of the endomorphism monoid of a lattice from the property "to be a sublattice"*, Talk at the Sverdlovsk seminar on April 11, 1985. (Russian)

26. A. G. Poroshkin, *On certain classes of functions on lattices*, Problems in Functional Analysis, Petrozavodsk Gos. Univ., Petrozavodsk, 1980, pp. 71–85. (Russian)

27. A. G. Poroshkin and A. A. Poroshkin, *Functions on lattices.* II, Ordered Sets and Operator Equations, Perm', 1982, pp. 127–135. (Russian)

28. A. G. Poroshkin and A. Samorodnitskiĭ, *On functions on lattices.* III, Preprint No. 1710-84, deposited at VINITI, 1984. (Russian)

29. A. A. Ryabtsev, *Representations of lattices by sets*, Preprint No. 5366-82, deposited at VINITI, 1982. (Russian)

30. V. N. Saliĭ, *Lattices with unique complements*, "Nauka", Moscow, 1984; English transl., Transl. Math. Monographs, vol. 69, 1988, Amer. Math. Soc., Providence, RI.

31. T. A. Sarymsakov, Sh. A. Ayupov, Dzh. Khadzhiev, and V. I. Chilin, *Ordered algebras*, Fan, Tashkent, 1983. (Russian)

32. L. A. Skornyakov, *Elements of Lattice Theory*, 2nd ed., "Nauka", Moscow, 1982. (Russian)

33. V. P. Soltan, *Jordan elements of a lattice, and subordinate sums*, Mat. Issled. **10** (1975), no. 2 (36), 230–237, 287. (Russian)

34. R. B. Stekol'shchik, *Invariant elements in a modular lattice*, Funktsional.Anal. i Prilozhen. **18** (1984), no. 1, 82–83; English transl. in Functional Anal. Appl. **18** (1984), no. 1.

35. A. A. Tsyl'ke, *On perfect elements of free modular lattices*, Funktsional. Anal. i Prilozhen. **16** (1982), no. 1, 87–88; English transl. in Functional Anal. Appl. **16** (1982), no. 1.

36. V. M. Shiryaev, *Semilattices with semidistributive lattices of subsemilattices*, Vestnik Beloruss. Gos. Univ. Ser. I Fiz. Mat. Mekh. (1985), no. 1, 61–64. (Russian)

37. A. Abian and J. Lihová, *Compact partially ordered sets and compactification of partially ordered sets*, Math. Slovaca **32** (1982), no. 4, 321–325.

38. A. Achache, *Galois connection of a fuzzy subset*, Fuzzy Sets and Systems **8** (1982), no. 2, 215–218.

39. M. E. Adams, L. Babai, and J. Sichler, *Automorphism groups of finite distributive lattices with a given sublattice of fixed points*, Monatsh. Math. **90** (1980), no. 4, 259–266.

40. M. E. Adams, V. Koubek, and J. Sichler, *Pseudocomplemented distributive lattices with small endomorphism monoids*, Bull. Austral. Math. Soc. **28** (1983), no. 3, 305–318.

41. _____, *Homomorphisms and endomorphisms in varieties of pseudocomplemented distributive lattices (with applications to Heyting algebras)*, Trans. Amer. Math. Soc. **285** (1984), no. 1, 57–79.

42. M. E. Adams, D. Pigozzi, and J. Sichler, *Endomorphisms of direct unions of bounded lattices*, Arch. Math. (Basel) **36** (1981), no. 3, 221–229.

43. M. E. Adams and J. Sichler, *Frattini sublattices in varieties of lattices*, Colloq. Math. **44** (1981), no. 2, 181–184.

44. _____, *Automorphism groups of posets and lattices with a given subset of fixed points*, Monatsh. Math. **93** (1982), no. 3, 173–190.

45. _____, *Disjoint sublattices of lattices*, Acta Sci. Math. (Szeged) **46** (1983), no. 1-4, 77-83.
46. M. Aissen and B. Shoy, *Nonassociative multiplication*, Ann. New York Acad. Sci. **231** (1979), 3-8.
47. T. Albu, *Certain Artinian lattices are Noetherian. Applications to the relative Hopkins-Levitzki theorem*, Math. Ring. Theory, Reidel, Dordrecht, 1984, pp. 37-52.
48. D. D. Anderson, E. W. Johnson, and J. A. Johnson, *Structure and embedding theorems for small strong π lattices*, Algebra Universalis **16** (1983), no. 2, 147-152.
49. _____, *Join-principally generated multiplicative lattices*, Algebra Universalis **19** (1984), no. 1, 74-82.
50. C. J. Ashman and V. Ficker, *Families of sets on a finite set*, J. Austral. Math. Soc. Ser. A **37** (1984), no. 3, 405-412.
51. C. Bachman and P. D. Stratigos, *A general measure decomposition theorem by means of the generalized Wallman remainder*, J. Austral. Math. Soc. Ser. A **37** (1984), no. 1, 87-105.
52. _____, *Lattice repleteness and some of its applications to topology*, J. Math. Anal. Appl. **99** (1984), no. 2, 472-493.
53. _____, *On general lattice repleteness and completeness*, Illinois J. Math. **27** (1983), no. 4, 535-561.
54. _____, *Criteria for σ-smoothness, τ-smoothness, and tightness of lattice regular measures, with applications*, Canad. J. Math. **33** (1981), no. 6, 1498-1525.
55. G. Bachman and Mabel Szeto, *On strongly replete lattices, support of a measure, and the Wallman remainder*, Period. Math. Hungar. **15** (1984), no. 2, 127-155.
56. K. Baclawski, *Combinatorics: Trends and examples*, New Directions in Applied Mathematics (Cleveland, OH, 1980), Springer-Verlag, New York and Berlin, 1982, pp. 1-10.
57. _____, *Cohen-Macaulay connectivity and geometric lattices*, European J. Combin. **3** (1982), no. 4, 293-305.
58. R. Balbes, *Maximal and completely meet irreducible sublattices of a distributive lattice*, Algebra Universalis **17** (1983), no. 3, 317-328.
59. B. Banaschewski and E. Nelson, *Completions of partially ordered sets*, SIAM J. Comput. **11** (1982), no. 3, 521-528.
60. H.-J. Bandelt, *Płonka sum of complete lattices*, Simon Stevin **55** (1981), no. 3, 169-171.
61. _____, *Local polynomial functions on lattices*, Houston J. Math. **7** (1981), no. 3, 317-325.
62. _____, *$\mathfrak{M}$-distributive lattices*, Arch. Math. (Basel) **39** (1982), no. 5, 436-442.
63. _____, *Toleranzrelationen als Galoisverbindungen*, Acta Sci. Math. (Szeged) **46** (1983), no. 1-4, 55-58.
64. _____, *Tolerances on median algebras*, Czechoslovak Math. J. **33(108)** (1983), no. 3, 344-347.
65. H.-J. Bandelt and Marcel Erné, *The category of Z-continuous posets*, J. Pure Appl. Algebra **30** (1983), no. 3, 219-226.
66. H.-J. Bandelt and Jarmila Hedlíková, *Median algebras*, Discrete Math. **45** (1983), no. 1, 1-30.
67. F. L. Bauer, *Algorithms and algebra*, Lecture Notes in Comput. Sci., vol. 122, Springer-Verlag, New York and Berlin, 1981, pp. 421-429.
68. R. Beazer, *On congruence lattices of some p-algebras and double p-algebras*, Algebra Universalis **13** (1981), no. 3, 379-388.
69. _____, *Lattices whose ideal lattice is Stone*, Proc. Edinburgh Math. Soc. (2) **26** (1983), no. 1, 107-112.
70. L. Becerra and J. A. Johnson, *A note on quasi-principal ideals*, Tamkang J. Math. **15** (1984), no. 1, 77-82.
71. K. Benecke, *Spezifikation parametrisierter Datentypen*, Z. Math. Logik Grundlag. Math. **29** (1983), no. 1, 83-96.
72. M. K. Bennett, *Affine geometry: A lattice characterization*, Proc. Amer. Math. Soc. **88** (1983), no. 1, 21-26.
73. _____, *Separation conditions on convexity lattices*, Universal algebra and lattice theory (Charleston, S.C., 1984), Lecture Notes in Math., vol. 1149, Springer-Verlag, New York and Berlin, 1985, pp. 22-36.

74. M. K. Bennett and G. Birkhoff, *Convexity lattices*, Algebra Universalis **20** (1985), no. 1, 1–26.

75. L. Beran, *Orthomodular Lattices. Algebraic Approach*, Reidel, Dordrecht, 1985.

76. G. M. Bergman, *A question on lattices*, Preprint, Univ. California, Berkeley, 1984.

77. G. M. Bergman and F. Galvin, *Transversals of families in complete lattices, and torsion in product modules*, Order **3** (1987), no. 4, 391–403.

78. G. M. Bergman and E. Hrushovski, *Identities of cofinal sublattices*, Order **2** (1985), no. 2, 173–191.

79. G. Birkhoff, *Ordered sets in geometry*, Ordered Sets (Banff, Alta., 1981), NATO Adv. Study Inst. Ser. C: Math. Phys. Sci., vol. 83, Reidel, Dordrecht, Boston, 1982, pp. 407–443.

80. A. A. Bishop and E. A. Schreiner, *The lattice of complete ideals of an atomic lattice*, Algebra Universalis **14** (1982), no. 1, 55–63.

81. A. Björner, *On Whitney numbers and matchings in infinite geometric lattices*, Preprint, Mat. Inst. Stockholm. Univ. (1976), no. 7.

82. _____, *Some combinatorial properties of infinite geometric lattices*, Preprint, Mat. Inst. Stockholm Univ. (1977), no. 3.

83. _____, *On complements in lattices of finite length*, Discrete Math. **26** (1981), no. 3, 325–326.

84. _____, *On the homology of geometric lattices*, Algebra Universalis **14** (1982), no. 1, 107–128.

85. A. Błaszczyk, *Aspekty topologiczne algebr Boole'a* (1982), Uniwersytet Śląski, Katowice.

86. T. S. Blyth and J. C. Varlet, *Congruences on MS-algebras*, Bull. Soc. Roy. Sci. Liège **53** (1984), no. 6, 341–362.

87. B. A. Bosbach, *A representation theorem for completely join distributive algebraic lattices*, Period Math. Hungar. **13** (1982), no. 2, 113–118.

88. V. Breazu and C. Stănăşilă, *Overtopologies and their algebraic significance*, Bul. Inst. Politehn. Bucureşti Ser. Electrotehn. **43** (1981), no. 3, 3–10.

89. J. Brunner, *A finite base for M^n and maximal projective distance in M^n*, Algebra Universalis **14** (1982), no. 1, 99–106.

90. P. Burmeister, *Partial algebras—survey of a unifying approach toward a two-valued model theory for partial algebras*, Algebra Universalis **15** (1982), no. 3, 306–358.

91. Grigore G. Călugăreanu, *Torsion in lattices*, Mathematica (Cluj) **25(48)** (1983), no. 2, 127–129.

92. S. Černák, *On the completion of a lattice by ends*, Math. Slovaca **33** (1983), no. 4, 341–346.

93. I. Chajda, *Two characterizations of locally order-polynomially complete lattices*, Algebra Universalis **13** (1981), no. 3, 395–396.

94. _____, *Tolerance Hamiltonian varieties of algebras*, Acta Sci. Math. (Szeged) **44** (1982), no. 1–2, 13–16.

95. _____, *Varieties with directly decomposable diagonal subalgebras*, Ann. Univ. Sci. Budapest. Eötvös Sec. Math. **25** (1982), 193–201.

96. _____, *Weakly regular lattices*, Math. Slovaca **35** (1985), no. 4, 387–391.

97. _____, *Transferable tolerances and weakly tolerance regular lattices*, Lectures in Universal Algebra (Szeged, 1983), Colloq. Math. Soc. János Bolyai, vol. 43, North-Holland, Amsterdam and New York, 1986, pp. 27–40.

98. I. Chajda and J. Nieminen, *Direct decomposability of tolerances on lattices, semilattices and quasilattices*, Czechoslovak Math. J. **32** (1982), no. 1, 110–115.

99. I. Chajda, J. Nieminen, and Bohdan Zelinka, *Tolerances and orderings on semilattices*, Arch. Math. (Basel) **19** (1983), no. 3, 125–131.

100. I. Chajda and B. Zelinka, *Directly decomposable tolerances on direct products of lattices and semilattices*, Czechoslovak Math. J. **33** (1983), no. 4, 519–521.

101. _____, *Complemented tolerances on lattices*, Časopis Pěst. Mat. **109** (1984), no. 1, 54–59.

102. V. R. Chandran and Vimala Lakshmanan, *On supermodular lattices*, Indian J. Pure Appl. Math. **14** (1983), no. 9, 1128–1130.

103. C. C. Chen and K. M. Koh, *On the length of the lattice of sublattices of a finite distributive lattice*, Algebra Universalis **15** (1982), no. 2, 233–241.

104. _____, *On the strong purity of the sublattice-lattice of a finite distributive lattice*, Tokyo J. Math. **6** (1983), no. 2, 381–388.

105. C. C. Chen, K. M. Koh, and S. C. Lee, *On the purity of the lattice of sublattices of a finite distributive lattice*, Algebra Universalis **15** (1982), no. 2, 258–271.

106. _____, *On finite distributive lattices of grade one*, Bull. Inst. Math. Acad. Sinica **10** (1982), no. 3, 289–298.

107. C. C. Chen, K. M. Koh, and K. L. Teo, *On the length of the lattice of subalgebras of a finite Stone algebra*, Bull. Malaysian Math. Soc. (2) **5** (1982), no. 2, 101–104.

108. _____, *On the sublattice-lattice of a lattice*, Algebra Universalis **19** (1984), no. 1, 61–73.

109. Malliah Chinthayamma and Bhatta S. Parameshwara, *Lattices all of whose congruences are neutral*, Proc. Amer. Math. Soc. **94** (1985), no. 1, 49–51.

110. H. M. Chuong, *Remark on finitely projected modular lattices of breadth two*, Wiss. Z. Martin-Luther-Univ. Halle-Wittenberg Math.-Natur. Reihe (1981), no. 21, 127–132.

111. W. H. Cornish and A. S. A. Noor, *Standard elements in a near-lattice*, Bull. Austral. Math. Soc. **26** (1982), no. 2, 185–213.

112. P. Cousot and Radhia A. Cousot, *A constructive characterization of the lattice of all retractions, preclosure, quasiclosure, and closure operations on a complete lattice*, Portugal. Math. **38** (1979), no. 1–2, 185–198.

113. H. Crapo, *Unities and negation: On the representation of finite lattices*, J. Pure Appl. Algebra **23** (1982), no. 2, 109–135.

114. _____, *Selectors: A theory of formal languages, semimodular lattices, and branching and shelling processes*, Adv. Math. **54** (1984), no. 3, 233–277.

115. G. Czédli, *Factor lattices by tolerances*, Acta Sci. Math. (Szeged) **44** (1982), no. 1–2, 35–42.

116. J. Dalík, *On semimodular lattices of generating systems*, Arch. Math. (Basel) **18** (1982), no. 1, 1–8.

117. B. A. Davey and M. J. McCarthy, *A representation theory for the variety, generated by the triangle*, Acta Math. Hungar. **38** (1981), no. 1–4, 241–255.

118. A. Day, *Equational theories of projective geometries*, Contributions to Lattice Theory (Szeged, 1980), Colloq. Math. Soc. János Bolyai, vol. 33, North-Holland, Amsterdam and New York, 1983, pp. 277–316.

119. _____, *Geometrical application in modular lattices*, Lecture Notes in Math., vol. 1004, Springer-Verlag, New York, 1983, pp. 111–141.

120. _____, *A lemma on projective geometries as modular and/or Arguesian lattices*, Canad. Math. Bull. **26** (1983), no. 3, 283–290.

121. _____, *A note on Arguesian lattices*, Arch. Math. (Basel) **19** (1983), no. 3, 117–123.

122. _____, *On some geometrical properties of defining classes of rings and varieties of modular lattices*, Algebra Universalis **17** (1983), no. 1, 21–33.

123. A. Day and B. Jónsson, *A structural characterization of non-Arguesian lattices*, Order **2** (1986), no. 4, 335–350.

124. A. Day and D. Pickering, *The coordinatization of Arguesian lattices*, Trans. Amer. Math. Soc. **278** (1983), no. 2, 507–522.

125. A. Delandtsheer, *Finite geometric lattices with highly transitive automorphism groups*, Arch. Math. (Basel) **42** (1984), no. 4, 376–383.

126. Keith J. Devlin, *Constructibility*, Perspectives in Mathematical Logic, vol. XII, Springer-Verlag, Berlin and New York, 1984.

127. R. P. Dilworth, *The role of order in lattice theory*, Ordered Sets (Banff, Alta., 1981), NATO Adv. Sci. Study Inst. Ser. C: Math. Phys. Sci., vol. 83, Reidel, Dordrecht, 1982, pp. 333–353.

128. _____, *Aspects of distributivity*, Algebra Universalis **18** (1984), no. 1, 4–17.

129. A. Di Nola, *On functionals measuring the fuzziness of solutions in relational equations*, Fuzzy Sets and Systems **14** (1984), no. 3, 249–258.

130. H. Dobbertin, M. Erné, and D. C. Kent, *A note on order convergence in complete lattices*, Rocky Mountain J. Math. **14** (1984), no. 3, 647–654.

131. H. H. Domingues and Maria Cecília Costa e Silva, *m-lattices with the ascending chain condition*, Metrica **1** (1979), no. 2.

132. D. Dorninger, *On generating sets of order-preserving functions over finite lattices*, Contributions to Lattice Theory (Szeged, 1980), Colloq. Math. Soc. János Bolyai, vol. 33, North-Holland, Amsterdam and New York, 1983, pp. 317–324.

133. D. Dorninger and G. Eigenthaler, *On compatible and order-preserving functions on lattices*, Universal Algebra and Applications (Warsaw, 1978), Banach Center Publ., 9, PWN, Warsaw, 1982, pp. 97–104.

134. H. Draškovičová, *Modular median algebras*, Math. Slovaca **32** (1982), no. 3, 269–281. (Russian summary)

135. _____, *Connections between some congruence properties in a single algebra*, Contributions to General Algebra, 3 (Vienna, 1984), Hölder-Pichler-Tempsky, Vienna, 1985, pp. 103–114.

136. J. Drewniak, *Podstawy teorii zbiorów rozmytych*, Skrypty Uniwersytetu Slaskiego, 347, Uniwersytet Slaski, Katowice, 1984.

137. _____, *Fuzzy relation equations and inequalities*, Fuzzy Sets and Systems **14** (1984), no. 3, 237–247.

138. J. Duda, *Categorical meaning of the Birkhoff-Stone construction*, Bull. Acad. Polon. Sci. Sér. Sci. Math. **29** (1981), no. 7-8, 331–336. (Russian summary)

139. _____, *An application of diagonal operations: Direct decomposability of homomorphisms*, Demonstratio Math. **14** (1981), no. 4, 989–996.

140. _____, *Solution of the problem of directly decomposable homomorphisms*, Časopis Pěst. Mat. **107** (1982), no. 3, 289–293, 308.

141. _____, *Regularity of algebras with applications to congruence class geometry*, Arch. Math. (Basel) **19** (1983), no. 4, 199–208.

142. _____, *Directly decomposable compatible relations*, Glas. Mat. Ser. III **19** (1984), no. 2, 225–229.

143. J. Dudek and J. Płonka, *On the monoarity of algebras*, Contributions to General Algebra, 2 (Klagenfurt, 1982), Hölder-Pichler-Tempsky, Vienna, 1983, pp. 71–75.

144. D. Duffus, *Matching in modular lattices*, J. Combin. Theory Ser. A **32** (1982), no. 3, 303–314.

145. D. Duffus and I. Rival, *Path length in the covering graph of a lattice*, Discrete Math. **19** (1977), no. 2, 139–158.

146. I. Düntsch, *Congruences in the lattice of subalgebras of a Boolean algebra*, Colloq. on Ordered Sets, vol. 7, Szeged, 1985.

147. W. Dzik, *On the content of lattices of logics. I*, Rep. Math. Logic (1981), no. 13, 17–27.

148. _____, *On the content of lattices of logics. II*, Rep. Math. Logic (1982), no. 14, 29–47.

149. G. Eigenthaler, *Eine Bemerkung zur Darstellung von Polynomen über Verbänden*, Math. Nachr. **103** (1981), 299–300.

150. W. J. R. Eplett, *An additive representation for real functions on the product of a set and a lattice*, Proc. Math. Soc. **81** (1981), no. 1, 23–26.

151. M. Erné, *Einführung in die Ordnungstheorie*, Bibliographisches Institut, Mannheim, 1982.

152. _____, *Distributivegesetze und Dedekindsche Schnitte*, Abh. Braunschweig. Wiss. Ges. **33** (1982), no. 117–145.

153. _____, *On the existence of decompositions in lattices*, Algebra Universalis **16** (1983), no. 3, 338–343.

154. _____, *Adjunctions and standard constructions for partially ordered sets*, Contributions to General Algebra, 2 (Klagenfurt, 1982), Hölder-Pichler-Tempsky, Vienna, 1983, pp. 77–106.

155. M. Erné and G. Wilke, *Standard completions for quasiordered sets*, Semigroup Forum **27** (1983), no. 1-4, 351–356.

156. U. Faigle, G. Richter, and Manfred Stern, *Geometric exchange properties in lattices of finite length*, Algebra Universalis **19** (1984), no. 3, 355–365.

157. John R. Faulkner, *Coordinatization of Moufang-Veldkamp planes*, Geom. Dedicata **14** (1983), no. 2, 189–202.

158. I. Fleischer and T. Traynor, *Group-valued modular functions*, Algebra Universalis **14** (1982), no. 3, 287–291.

159. R. Fourneau, *Lattis moléculaires*, Bull. Soc. Roy. Sci. Liège **50** (1981), no. 9-10, 328-331.

160. _____, *Modularité et distributivité affaiblies dans les lattis*, Comment. Math. Univ. Carolin. **23** (1982), no. 3, 607-612.

161. R. Freese, *Some order-theoretic questions about free modular lattices*, Ordered Sets (Banff, Alta., 1981), NATO Adv. Study Inst. Ser. C: Math. Phys. Sci., vol. 83, Reidel, Dordrecht, Boston, 1982, pp. 355-377.

162. _____, *On Jónsson's theorem*, Algebra Universalis **18** (1984), no. 1, 70-76.

163. E. Fried and R. Wiegandt, *Abstract rational structures. I General theory*, Algebra Universalis **15** (1982), no. 1, 1-21.

164. S. Gacsályi, *A remark on lattices satisfying the maximum condition*, Publ. Math. Debrecen **31** (1984), no. 1-2, 127-128.

165. E. R. Gansner, *On the lattice of order ideals of an up-down poset*, Discrete Math. **39** (1982), no. 2, 113-122.

166. E. Gedeonová, *Lattices whose covering graphs are S-graphs*, Contributions to Lattice Theory (Szeged, 1980), Colloq. Math. Soc. János Bolyai, vol. 33, North-Holland, Amsterdam and New York, 1983, pp. 407-435.

167. _____, *Lattices with centrally symmetric covering graphs*, Contributions to General Algebra 2 (Klagenfurt, 1982), Hölder-Pichler-Tempsky, Vienna, 1983, pp. 107-113.

168. I. M. Gel′fand and V. A. Ponomarev, *Free modular lattices and their representations*, London Math. Soc. Lecture Note Ser., vol. 69, Cambridge University Press, Cambridge, 1982, pp. 173-228.

169. _____, *Lattices, representations, and algebras connected with them*. I, London Math. Soc. Lecture Note Ser., vol. 69, Cambridge University Press, Cambridge, 1982, pp. 229-247; II, London Math. Soc. Lecture Note Ser., vol. 69, Cambridge University Press, Cambridge, 1982, pp. 249-272.

170. G. Gierz, *Colimits of continuous lattices*, J. Pure Appl. Algebra **23** (1982), no. 2, 137-144.

171. M. S. Goldberg, *Distributive double p-algebras whose congruence lattices are chains*, Algebra Universalis **17** (1983), no. 2, 208-215.

172. P. Goralčík and V. Koubek, *On generative complexity of α-complete lattices*, Contributions to Lattice Theory (Szeged, 1980), Colloq. Math. Soc. János Bolyai, vol. 33, North-Holland, Amsterdam and New York, 1983, pp. 437-448.

173. V. A. Gorbunov and V. I. Tumanov, *On the existence of prime ideals in semidistributive lattices*, Algebra Universalis **16** (1983), no. 2, 250-252.

174. M. Gould, J. A. Iskra, and C. Tsinakis, *Globally determined lattices and semilattices*, Algebra Universalis **19** (1984), no. 2, 137-141.

175. E. Graczyńska and F. Pastijn, *A generalization of Płonka sums*, Fund. Math. **120** (1984), no. 1, 53-62.

176. G. Grätzer, *Universal algebra and lattice theory: A story and three research problems*, Universal Algebra and its Links with Logic, Algebra, Combinatorics and Computer Science (Darmstadt, 1983), R & E Res. Exp. Math., 4, Heldermann Verlag, Berlin, 1984, pp. 1-13.

177. Gabriele H. Greco, *Limitoids and complete lattices*, Ann. Univ. Ferrara Sez. VII (N.S.) **29** (1983), 153-164. (Italian)

178. _____, *Operators of type G on complete lattices*, Rend. Sem. Mat. Univ. Padova **72** (1984), 277-288. (Italian)

179. _____, *Uniform order-convergence for complete lattices*, Proc. Amer. Math. Soc. **90** (1984), no. 4, 657-658.

180. _____, *Semifilter decompositions and sequential Γ-limits in completely distributive lattices*, Ann. Mat. Pura Appl. (4) **137** (1984), 61-81. (Italian)

181. J. W. Grzymala-Busse, *On the representation of finite lattices in the class of finite automata*, MTA Számitástechn. És Autom. Kut. Intéz. Tanul. (1982), no. 137, 199-204.

182. H.-P. Gumm, *Geometrical methods in congruence modular algebras*, Mem. Amer. Math. Soc. **45** (1983), no. 286.

183. _____, *Geometrical reasoning and analogy in universal algebra*, Universal Algebra and its Links with Logic, Algebra, Combinatorics and Computer Science (Darmstadt, 1983), R & E Res. Exp. Math., 4, Heldermann, Berlin, 1984, pp. 14-28.

184. H.-P. Gumm and W. Poguntke, *Boolesche Algebra*, Bibliographisches Institut, Mannheim, 1981.

185. I. Halperin, *A survey of John von Neumann's book on continuous geometry*, Order **1** (1985), 301–305.

186. _____, *Von Neumann's coordinatization theorem*, Acta Sci. Math. (Szeged) **45** (1983), no. 1–4, 213–218.

187. I. Halperin and J. von Neumann, *Continuous geometries with a transition probability*, Mem. Amer. Math. Soc. **34** (1981), no. 252.

188. P. Hamburg, *Subspatii fuzzy*, Lucr. Semin. Itiner. Ecuatii Funct., Approxim. si Convexit. (Timisoara, 1980) **1** (1980), 231–242. (Romanian)

189. P. Hamburg and L. Florescu, *Topological structures in fuzzy lattices*, J. Math. Anal. Appl. **101** (1984), no. 2, 475–490.

190. J. Hannah, *Putting coordinates on lattices*, Irish Math. Soc. Bull. **8** (1983), 21–28.

191. G. Hansoul and J. C. Varlet, *Essential, strong and perfect existensions of lattices*, Bull. Soc. Roy. Sci. Liège **52** (1983), no. 1, 22–34.

192. J. Hedlíková, *Ternary spaces, media and Chebyshev sets*, Czechoslovak Math. J. **33** (1983), no. 3, 373–389.

193. L. Heindorf, *Beiträge zur Modelltheorie der Booleschen Algebren*, Seminarberichte, vol. 53, Humboldt Universität, Berlin, 1984.

194. Ch. Herrmann, *Rahmen und erzeugende Quadrupel in modularen Verbänden*, Algebra Universalis **14** (1982), no. 3, 357–387.

195. _____, *Über die von vier Moduln erzeugte Dualgruppe*, Abh. Braunschweig. Wiss. Ges. **33** (1982), 157–159.

196. _____, *On varieties of algebras having complemented modular lattices of congruences*, Algebra Universalis **16** (1983), no. 1, 129–130.

197. _____, *On the word problem for the modular lattice with four free generators*, Math. Ann. **265** (1983), no. 4, 513–527.

198. _____, *A characterization of distributivity for modular polarity lattices*, Contributions to Lattice Theory (Szeged, 1980), Colloq. Math. Soc. János Bolyai, vol. 33, North-Holland, Amsterdam and New York, 1983, pp. 473–490.

199. _____, *On elementary Arguesian lattices with four generators*, Algebra Universalis **18** (1984), no. 2, 225–259.

200. _____, *On the arithmetic of projective coordinate systems*, Trans. Amer. Math. Soc. **284** (1984), no. 2, 759–785.

201. A. Herzer, *Semimodular locally projective lattices of rank 4 from v. Stoudt's point of view*, NATO Adv. Study Inst. Ser. C: Math. Phys. Sci., vol. 70, Reidel, Dordrecht, Boston, 1981, pp. 373–400.

202. R. C. Hickman and G. P. Monro, *Distributive partially ordered sets*, Fund. Math. **120** (1984), no. 2, 151–166.

203. K. H. Hofmann, *Stably continuous frames and their topological manifestations*, Categorical Topology (Toledo, Ohio, 1983), Sigma Ser. Pure Math., vol. 5, Heldermann, Berlin, 1984, pp. 282–307.

204. Hartmut F. W. Höft, *Crossed and complete binary relations*, Rev. Roumaine Math. Pures Appl. **28** (1983), no. 8, 703–708.

205. M. H. Höft, *Sums of double systems of partially ordered sets*, Demonstratio Math. **16** (1983), no. 1, 229–238.

206. C. S. Hoo, *Atoms, primes and implicative lattices*, Canad. Math. Bull. **27** (1984), no. 3, 279–285.

207. Chên Jung Hsü, *On the characterization of the partition lattice $LP_n(S)$ as a geometric lattice*, Chinese J. Math. **9** (1981), no. 1, 37–46.

208. Shi Geng Hu, *Proximities, contiguities and nearness on lattices*, J. Huazhong Univ. Sci. Tech. **10** (1982), no. 5, 1–6.

209. A. P. Huhn, *On the representation of distributive algebraic lattices*, Acta Sci. Math. (Szeged) **45** (1983), no. 1–4, 239–246.

210. M. Hušek, *Applications of category theory to uniform structures*, Lecture Notes in Math., vol. 962, Springer, New York, 1982, 138–144.

211. V. I. Igoshin, *Identities in interval lattices of lattices*, Contributions to Lattice Theory (Szeged, 1980), Colloq. Math. Soc. János Bolyai, vol. 33, North-Holland, Amsterdam and New York, 1983, pp. 491–501.

212. _____, *On lattices with restrictions on their interval lattices*, Lectures in Universal Algebra (Szeged, 1983), Colloq. Math. Soc. János Bolyai, vol. 43, North-Holland, Amsterdam and New York, 1986, pp. 209–216.

213. T. Ihringer, *On groupoids having a linear congruence class*, Math. Z. **180** (1982), no. 3, 395–411.

214. Iqbalunnisa and S. Akilandam, *Standard elements of order two of a lattice*, Studia Sci. Math. Hungar. **14** (1979), no. 4, 453–454.

215. J. Isbell, *Direct limits of meet-continuous lattices*, J. Pure Appl. Algebra **23** (1982), no. 1, 33–35.

216. _____, *Completion of a construction of Johnstone*, Proc. Amer. Math. Soc. **85** (1982), no. 3, 333–334.

217. J. Jakubík, *On isometries of lattices*, Math. Slovaka **34** (1984), no. 2, 177–184.

218. _____, *On lattices determined up to isomorphisms by their graphs*, Czechoslovak Math. J. **34(109)** (1984), no. 2, 305–314.

219. _____, *On isomorphisms of graphs of lattices*, Czechoslovak Math. J. **35(110)** (1985), no. 2, 188–200.

220. _____, *On weak direct product decompositions of lattices and graphs*, Czechoslovak Math. J. **35(110)** (1985), no. 2, 269–277.

221. _____, *Graph isomorphisms of semimodular lattices*, Math. Slovaka **35** (1985), no. 3, 229–232.

222. M. Jambu and H. Terao, *Arrangements libres d'hyperplans et treillis hyper-résolubles*, C. R. Acad. Sci. Paris Sér. I Math. **296** (1983), no. 15, 623–624.

223. C. Jayaram, *1-modular lattices*, Rev. Roumaine Math. Pures Appl. **29** (1984), no. 2, 163–169.

224. J. A. Johnson, *Completions of Macaulay local lattices*, Algebra Universalis **14** (1982), no. 1, 44–54.

225. _____, *Totally quasilocal multiplicative lattices*, Tamkang J. Math. **13** (1982), no. 1, 91–104.

226. _____, *Nonprincipal join-irreducible elements in r-lattices*, Algebra Universalis **14** (1982), no. 2, 265–266.

227. _____, *Principal elements in multiplicative lattices*, Boll. Un. Mat. Ital. B **1** (1982), no. 2, 673–681.

228. _____, *The structure of a class of r-lattices*, Comment. Math. Univ. St. Paul **32** (1983), no. 2, 189–194.

229. J. A. Johnson and K. V. Moss, *Bijections and homomorphisms*, Semigroup Forum **28** (1984), no. 1–3, 373–374.

230. P. R. Jones, *Distributive modular and separating elements in lattices*, Rocky Mountain J. Math. **13** (1983), no. 3, 429–436.

231. J. Kahn, *Locally projective-planar lattices which satisfy the bundle theorem*, Math. Z. **175** (1980), no. 3, 219–247.

232. G. Kalmbach, *Orthomodular Lattices*, London Mathematical Society Monographs, vol. 18, Academic Press, London and New York, 1983.

233. _____, *Automorphism groups of orthomodular lattices*, Bull. Austral. Math. Soc. **29** (1984), no. 3, 309–313.

234. T. Kamimura and A. Tang, *Algebraic relations and presentations*, Theoret. Comput. Sci. (1983), no. 1–2, 39–60.

235. _____, *Effectively given spaces*, Lecture Notes in Comput. Sci., vol. 154, Springer-Verlag, New York and Berlin, 1983, pp. 385–396.

236. _____, *Total objects of domains*, Theoret. Comput. Sci. **34** (1984), no. 3, 275–288.

237. K. M. Koh, *On the length of the sublattice-lattice of a finite distributive lattice*, Algebra Universalis **16** (1983), no. 3, 282–286.

238. _____, *On the lattice of maximum-sized antichains of a finite poset*, Algebra Universalis **17** (1983), no. 1, 73–86.

239. L. Kohl and M. Stern, *A characterization of certain finite lattices in which every element is a join of cycles*, Contributions to Lattice Theory (Szeged, 1980), Colloq. Math. Soc. János Bolyai, vol. 33, North-Holland, Amsterdam and New York, 1983, pp. 575–589.

240. P. Köhler, *Brouwerian semilattices: The lattice of total subalgebras*, Universal Algebra and Applications (Warsaw, 1978), Banach Center Publ., 9, PWN, Warsaw, 1982, pp. 47–56.

241. M. Kolibiar, *Intervals, convex sublattices and subdirect representations of lattices*, Universal Algebra and Applications (Warsaw, 1978), Banach Center Publ., 9, PWN, Warsaw, 1982, 335–339.

242. M. Konstantinidou, *On the hyperlattice-ordered groupoids*, Boll. Un. Mat. Ital. A **2** (1983), no. 3, 343–350.

243. I. Kopeček, *Distinguishing subsets in lattices*, Arch. Math. (Basel) **18** (1982), no. 3, 145–149.

244. I. Korec, *On systems of isomorphic copies of an algebra in another algebra*, Acta Univ. Palack. Olomuc. Fac. Rerum Natur. Math. (1979), no. 34, 213–222.

245. V. Koubek and J. Sichler, *Quotients of rigid* $(0, 1)$-*lattices*, Arch. Math. (Basel) **44** (1985), no. 5, 403–412.

246. Ch. Kratzer and J. Thévenaz, *Type d'homotopie des treillis et treillis des sous-groupes d'un groupe fini*, Comment Math. Helv. **60** (1985), no. 1, 85–106.

247. L. Kučera and V. Trnková, *The computational complexity of some problem of universal algebra*, Universal Algebra and its Links with Logic, Algebra, Combinatorics and Computer Science (Darmstadt, 1983), R & E Res. Exp. Math., 4, Heldermann Verlag, Berlin, 1984, pp. 216–229.

248. M. Kurilic, *Nucleus of a lattice*, Colloq. on Ordered Sets, vol. 18, Szeged, 1985.

249. H. Lakser, *A note on the lattice of sublattices of a finite lattice*, Nanta Math. **6** (1973), no. 1, 55–57.

250. M. S. Lambrou, *Completely distributive lattices*, Fund. Math. **119** (1983), no. 3, 227–240.

251. R. Lidl and G. Pilz, *Applied Abstract Algebra*, Undergraduate Texts in Mathematics, Springer-Verlag, New York and Berlin, 1984.

252. Wang-Jin Liu, *Operations on fuzzy ideals*, Fuzzy Sets and Systems **11** (1983), no. 1, 31–41.

253. Ying-Ming Liu, *Intersection operation on union-preserving mappings in completely distributive lattices*, J. Math. Anal. Appl. **84** (1981), no. 1, 249–255.

254. _____, *The inverse operation on union-preserving mappings in lattices and its application to fuzzy uniform spaces*, Fuzzy Math. **1** (1981), no. 2, 21–28.

255. Z. Lomecky, *Algorithms for the computation of free lattices*, Computer Algebra (Marseille, 1982), Lecture Notes in Comput. Sci., vol. 144, Springer, Berlin-New York, 1982, pp. 223–230.

256. R. Löwen, *A local "Fundamental Theorem" for classical topological projective spaces*, Arch. Math. (Basel) **38** (1982), no. 3, 286–288.

257. H. Lugowski, *Fundamentals of universal algebra*, Teubner Texts in Mathematics, 3, BSB B. G. Teubner Verlagsgesellschaft, Leipzig, 1982. (German)

258. G. S. Lystad and A. R. Stralka, *Lawson semilattices with bialgebraic congruence lattices*, General Topology and Modern Analysis (Proc. Conf., Univ. California, Riverside, CA, 1980), Academic Press, New York, 1981, pp. 247–254.

259. Sh. A. Maeda, *A note on modularity in atomistic lattices*, Proc. Japan Acad. Ser. A Math. Sci. **58** (1982), no. 7, 287–289.

260. _____, *On modularity in atomistic lattices*, Contributions to Lattice Theory (Szeged, 1980), Colloq. Math. Soc. János Bolyai, vol. 33, North-Holland, Amsterdam and New York, 1983, pp. 627–636.

261. _____, *On distributive pairs in lattices*, Acta Math. Hungar. **45** (1985), no. 1–2, 133–140.

262. P. Mayrhofer, *Der Verband der A-Primärsysteme eines A-Systems*, Sitzungsber. Österr. Akad. Wiss. Math.-Natur. Kl. **191** (1982), no. 1–3, 23–33.

263. R. McKenzie, *Finite Forbidden Lattices*, Lecture Notes in Math., vol. 1004, Springer-Verlag, New York and Berlin, 1983, pp. 176–205.

264. R. McKenzie and C. Tsinakis, *On recovering a bounded distributive lattice from its endomorphism monoid*, Houston J. Math. **7** (1981), no. 4, 525–529.

265. G. F. McNulty, *Fifteen possible previews in equational logic*, Lectures in Universal Algebra (Szeged, 1983), Colloq. Math. Soc. János Bolyai, vol. 43, North-Holland, Amsterdam and New York, 1986, pp. 307–331.

266. J. B. Miller, *Local convexity in topological lattices*, Portugal. Math. **38** (1979), no. 3–4, 19–31.

267. M. W. Mislove, *An introduction to the theory of continuous lattices*, Ordered Sets (Banff, Alta., 1981), NATO Adv. Study Inst. Ser. C: Math. Phys. Sci., vol. 83, Reidel, Dordrecht, Boston, 1982, pp. 379–406.

268. H. Mitsch, *Semigroups and their lattice of congruences*, Semigroup Forum **26** (1983), no. 1–2, 1–63.

269. A. Muir and M. W. Warner, *Lattice valued relations and automata*, Discrete Appl. Math. **7** (1984), no. 1, 65–78.

270. B. J. Müller, *Continuous geometries, continuous regular rings, and continuous modules*, Proceedings of the Conference on Algebra and Geometry (Kuwait, 1981), Kuwait Univ., Kuwait, 1982, pp. 49–52.

271. P. V. Ramana Murty and T. Engelbert, *Standard ideals of sublattices and product lattices*, Math. Sem. Notes Kobe Univ. **10** (1982), no. 2/2, 697–505.

272. K. S. S. Nambooripad and F. J. Pastijn, *The fundamental representation of a strongly regular Baer semigroup*, J. Algebra **92** (1985), no. 2, 283–302.

273. G. E. Nicholson, A. Grubb, and C. S. Sharma, *Regular join endomorphisms on a complemented modular lattice of finite rank*, Discrete Math. **52** (1984), no. 2–3, 235–242.

274. J. Niederle, *On skeletal and irreducible elements in tolerance lattices of finite distributive lattices*, Časopis Pěst. Mat. **107** (1982), no. 1, 23–29.

275. _____, *A note on tolerance lattices of products of lattices*, Časopis Pěst. Mat. **107** (1982), no. 2, 114–115.

276. _____, *A note on tolerance lattices*, Časopis Pěst. Mat. **107** (1982), no. 3, 221–224.

277. S. B. Niefield, *Exactness and projectivity*, Lecture Notes in Math., vol. 962, Springer, New York, 1982, pp. 221–227.

278. J. Nieminen, *On χ_{mub}-lattices and convex substructures of lattices and semilattices*, Acta Math. Hungar. **44** (1984), no. 3–4, 229–236.

279. _____, *2-Ideals of finite lattices*, Tamkang J. Math. **16** (1985), no. 2, 23–27.

280. W. Nöbauer, *Local polynomial functions: Results and problems*, Universal Algebra and Applications (Warsaw, 1978), Banach Center Publ., 9, PWN, Warsaw, 1982, pp. 197–202.

281. T. Nordahl, *Lattice representation in exchange transfusion*, Semigroup Forum **23** (1981), no. 3, 275.

282. D. Novak, *On a duality between the concepts "finite" and "directed"*, Houston J. Math. **8** (1982), no. 4, 545–563.

283. V. Novák and J. Nekola, *Basic operations with fuzzy sets from the point of fuzzy logic*, Fuzzy Information, Knowledge Representation and Decision Analysis (Marseille, 1983), IFAC Proc. Ser., 6, IFAC, Laxenburg, 1984, pp. 249–253.

284. P. Orlik and L. Solomon, *Combinatorics and topology of complements of hyperplanes*, Invent. Math. **56** (1980), no. 2, 167–189.

285. R. Peele, *On finite partition representation of lattices*, Discrete Math. **42** (1982), no. 2–3, 267–280.

286. N. Percsy, *Embedding geometric lattices in a projective space*, Finite Geometries and Designs (Proc. Conf., Chelwood Gate, 1980), London Math. Soc. Lecture Note Ser., vol. 49, Cambridge University Press, Cambridge, 1981, pp. 304–315.

287. _____, *Locally embeddable geometries*, Arch. Math. (Basel) **37** (1981), no. 2, 184–192.

288. _____, *Une condition nécessaire et suffisante de plongeabilité pour les treillis semi-modulaires*, European J. Combin. **2** (1981), no. 2, 173–177.

289. L. Pezzoli, *Weight functions on modular lattices*, Boll. Un. Mat. Ital. A (1980), no. 2, 341–353.

290. _____, *Modular independence systems*, Boll. Un. Mat. Ital. B **18** (1981), no. 2, 575–590.

291. J. Płonka, *On bounding congruences in some algebras having the lattice structure*, Universal Algebra and Applications (Warsaw, 1978), Banach Center Publ., 9, PWN, Warsaw, 1982, pp. 203–207.

292. J. Pócs, *Congruence relations on direct products of lattices*, Math. Slovaka **32** (1982), no. 2, 173–175.

293. W. A. Pogorzelski and P. Wojtylak, *Elements of the theory of completeness in propositional logic*, Prace Nauk. Uniw. Śląsk. Katowic. (1982), no. 512.

294. W. Poguntke, *Endlich erzeugte Verbände*, Mitt. Math. Sem. Giessen (1982), no. 151.

295. B. Pondelíček, *Relative compact elements in lattices*, Contributions to Lattice Theory (Szeged, 1980), Colloq. Math. Soc. János Bolyai, vol. 33, North-Holland, Amsterdam and New York, 1983, pp. 667–674.

296. M. Pouzet and I. Rival, *Which ordered sets have a complete linear extension?*, Canad. J. Math. **33** (1981), no. 5, 1245–1254.

297. _____, *Quotients of complete ordered sets*, Algebra Universalis **17** (1983), no. 3, 393–405.

298. _____, *Every countable lattice is a retract of a direct product of chains*, Algebra Universalis **18** (1984), no. 3, 295–307.

299. E. Pracht, *Algebra of lattices*, Uni Paperbacks, 958, Ferdinand Schöningh, Paderborn, 1980. (German)

300. P. Pudlák, *On congruence lattices of lattices*, Algebra Universalis **20** (1985), no. 1, 96–114.

301. R. W. Quackenbush, *Non-modular varieties of semi-modular lattices with a spanning M_3*, Discrete Math. **53** (1985), 193–205.

302. V. J. Rayward-Smith, *On embedding a lattice in its power lattice*, Internat. J. Math. Ed. Sci. Tech. **13** (1982), no. 3, 253–259.

303. N. R. Reilly, *Representations of lattices via neutral elements*, Algebra Universalis **19** (1984), no. 3, 341–354.

304. M. Raquel P. da Costa Reis, *Sur les éléments de Dilworth dans les groupoïdes-treillis*, Mathematics Today (Luxembourg, 1981), Gauthier-Villars, Paris, 1982, pp. 259–262.

305. K. Reuter, *A matching result for modular lattices*, Colloq. on Ordered Sets, vol. 23, Szeged, 1985.

306. G. Richter, *Strong purity in lattices*, Beiträge Algebra Geom. **12** (1982), 7–16.

307. _____, *Generalization of a theorem of Kertész*, Beiträge Algebra Geom. **12** (1982), 117–121.

308. _____, *The Kuros-Ore theorem, finite and infinite decompositions*, Studia Sci. Math. Hungar. **17** (1982), no. 1–4, 243–250.

309. _____, *On the structure of lattices in which every element is a join of join-irreducible elements*, Period. Math. Hungar. **13** (1982), no. 1, 47–69.

310. _____, *Application of some lattice theoretic results in group theory*, Contributions to General Algebra, 2 (Klagenfurt, 1982), Hölder-Pichler-Tempsky, Vienna, 1983, pp. 305–317.

311. _____, *Standard and neutral ideals in prealgebraic lattices*, Studia Sci. Math. Hungar. **18** (1983), no. 2–4, 221–228.

312. G. Richter and M. Stern, *Strongness in (semimodular) lattices of finite length*, Wiss. Z. Martin-Luther-Univ. Halle-Wittenberg Math.-Natur. Reihe **33** (1984), no. 4, 73–77.

313. I. Rival and B. Sands, *How many four-generated simple lattices?*, Universal Algebra and Applications (Warsaw, 1978), Banach Center Publ., 9, PWN, Warsaw, 1982, pp. 67–72.

314. A. B. Romanowska and J. D. H. Smith, *Modal theory: An algebraic approach to order, geometry, and convexity*, R & E Res. Exp. Math., 9, Heldermann Verlag, Berlin, 1985.

315. J. G. Rosenstein, *Linear Orderings*, Academic Press, New York and London, 1982.

316. J. Rosický, *Multiplicative lattices and frame*, Preprint, Purkyne University, Brno, 1985.

317. W. Ruckelshausen and B. Sands, *On finitely generated lattices of width four*, Algebra Universalis **16** (1983), no. 1, 17–37.

318. J. Ryšlinková and T. Sturm, *Two closure operators which preserve m-compacticity*, Universal Algebra and Applications (Warsaw, 1978), Banach Center Publ., 9, PWN, Warsaw, 1982, pp. 113–119.

319. V. N. Salii, *Regular elements in complete uniquely complemented lattices*, Universal Algebra and Applications (Warsaw, 1978), Banach Center Publ., 9, PWN, Warsaw, 1982, pp. 15–19.

320. H. P. Sankappanavar, *Congruence-semimodular and congruence-distributive pseudocomplemented semilattices*, Algebra Universalis **14** (1982), no. 1, 68–81.

321. E. T. Schmidt, *Remark on compatible and order-preserving function on lattices*, Studia Sci. Math. Hungar. **14** (1979), no. 1–3, 139–144.

322. _____, *On finitely projected modular lattices*, Acta Math. Acad. Sci. Hungar. **38** (1981), no. 1–4, 45–51.

323. _____, *A survey on congruence lattice representations*, Teubner Texts in Mathematics, 42, BSB B. G. Teubner Verlagsgesellschaft, Leipzig, 1982.

324. _____, *Congruence lattices of complemented modular lattices*, Algebra Universalis **18** (1984), no. 3, 386–395.

325. E. T. Schmidt and R. Wille, *Note on compatible operations of modular lattices*, Algebra Universalis **16** (1983), no. 3, 395–397.

326. J. Schmidt, *Clones and semiclones of operations*, Universal Algebra (Esztergom, 1977), Colloq. Math. Soc. János Bolyai, vol. 29, North-Holland, Amsterdam and New York, 1982, pp. 705–723.

327. P. H. Schmitt, *Algebraically complete lattices*, Algebra Universalis **17** (1983), no. 2, 135–142.

328. D. Schweigert, *Central relations on lattices*, J. Austral. Math. Soc. Ser. A **35** (1983), no. 3, 369–372.

329. _____, *On varieties of clones*, Semigroup Forum **26** (1983), no. 3–4, 275–285.

330. _____, *Clones of term functions of lattices and abelian groups*, Algebra Universalis **20** (1985), no. 1, 27–33.

331. D. Schweigert and M. Szymańska, *A completeness theorem for correlation lattices*, Z. Math. Logik Grundlag. Math. **29** (1983), no. 5, 427–434.

332. _____, *Polynomial functions on correlation lattices*, Algebra Universalis **16** (1983), no. 3, 355–359.

333. K. Seitz, *On special structure-lattices*, Notes on Algebraic Systems III, vol. 3, Karl Marx Univ. Econom., Budapest, 1981, pp. 23–32.

334. _____, *On certain lattice-theoretical aspects of the theory of systems in chemical engineering*, Notes on Algebraic Systems III, vol. 3, Karl Marx Univ. Econom., Budapest, 1981, pp. 69–82.

335. _____, *Notes on Algebraic Systems* I, vol. 6, Karl Marx Univ. Econom., Budapest, 1978.

336. K. Seitz and J. Bulazs, *Notes on algebraic systems* II, vol. 5, Karl Marx Univ. Econom., Budapest, 1979.

337. S. Sessa, *Some results in the setting of fuzzy relation theory*, Fuzzy Sets and Systems **14** (1984), no. 3, 281–297.

338. B. Sivák, *Congruences on finite lattices*, Math. Slovaka **32** (1982), no. 3, 283–290.

339. E. Slatinský, *Die Abgeschlossenheit der lexikographischen Summe in der Klasse modularer Verbände*, Arch. Math. (Brno) **20** (1984), no. 4, 205–210.

340. V. Slavík, *On lattices with isomorphic interval lattices*, Czechoslovak Math. J. **35** (1985), no. 4, 550–554.

341. A. Służalec, *Projection systems described in lattice theory*, Geom. Dedicata **18** (1985), no. 1, 35–41.

342. J. D. Smith, *Universelle Algebra*, Tagungsber. Math. Forschungs. Inst. Oberwolfach (1982), no. 1–16.

343. K. W. Smith, *Stability and categoricity of lattices*, Canad. J. Math. **33** (1981), no. 6, 1380–1419.

344. R. P. Stanley, *Some aspects of groups acting on finite posets*, J. Combin. Theory Ser. A **32** (1982), no. 2, 132–161.

345. M. Stern, *On atomic algebraic lattices with covering property*, Preprint No. 45 (1981), Math. Dept., Martin Luther University, Halle.

346. _____, *Generalized matroid lattices*, Colloq. Math. Soc. János Bolyai, vol. 25, North-Holland, Amsterdam and New York, 1981, pp. 727–748.

347. _____, *On the theory of Baer lattices*, Universal Algebra and Applications (Warsaw, 1978), Banach Center Publ., 9, PWN, Warsaw, 1982, pp. 73–74.

348. _____, *On derivations in generalized matroid lattices*, Acta Sci. Math. (Szeged) **44** (1982), no. 3–4, 281–286.

349. _____, *Semimodularity in lattices of finite length*, Discrete Math. **41** (1982), no. 3, 287–293.

350. _____, *Exchange properties in lattices of finite length*, Wiss. Z. Martin-Luther-Univ. Halle-Wittenberg Math.-Natur. Reihe **31** (1982), no. 3, 15–26.

351. _____, *An isomorphism theorem for standard ideals in lattices*, Algebra Universalis **19** (1984), no. 1, 133–134.

352. M. G. Stone and R. H. Weedmark, *On representing $M_n S$ by congruence lattices of finite algebras*, Discrete Math. **44** (1983), no. 3, 299–308.

353. M. Sugeno and M. Sasaki, *L-fuzzy category*, Fuzzy Sets and Systems **11** (1983), no. 1, 43–64.

354. N. K. Thakare and C. S. Manjarekar, *Radicals and uniqueness theorems in multiplicative lattices with chain conditions*, Stud. Sci. Math. Hungar. **18** (1983), no. 1, 13–19.

355. R. Thron, *Algebraic closure operators for which every finite set is generated by the set of its isolated elements*, Beiträge Algebra Geom. **14** (1983), 77–85. (German)

356. Jiří Tůma, *On a question of K. Leeb*, Comment. Math. Univ. Carolin. **23** (1982), no. 3, 589–591.

357. W. R. Tunnicliffe, *On defining "Completely distributive"*, Algebra Universalis **19** (1984), no. 3, 397–398.

358. A. Urquhart, *Relevant implication and projective geometry*, Logique et Anal. (N.S.) **26** (1983), no. 103–104, 345–357.

359. M. van de Vel, *Binary convexities and distributive lattices*, Proc. London Math. Soc. **48** (1984), no. 1, 1–33.

360. B. C. van Fraasen, *Quantification as an act of mind*, J. Philos. Logic **11** (1982), no. 3, 343–369.

361. J. C. Varlet, *Congruences on De Morgan algebras*, Bull. Soc. Roy. Sci. Liège **50** (1981), no. 9–10, 332–343.

362. _____, *Regularity in p-algebras and p-semilattices*, Universal Algebra and Applications (Warsaw, 1978), Banach Center Publ., 9, PWN, Warsaw, 1982, pp. 369–378.

363. J. Vaz de Carvalho, *The subvariety $K_{2,0}$ of Ockham algebras*, Bull. Soc. Roy. Sci. Liège **53** (1984), no. 6, 393–400.

364. B. Voigt, *Combinatorial properties of regressive mappings*, Discrete Math. **47** (1983), no. 1, 97–108.

365. M. Vonkomerová, *A note on C-lattices*, Acta Math. Univ. Comenian. **40(41)** (1982), 33–44.

366. L. Vrancken-Mawet, *Le lattis des sous-algèbres d'une algèbre de Heyting finie*, Bull. Soc. Roy. Sci. Liège **51** (1982), no. 1–2, 82–94.

367. _____, *Sur des congruences d'un ensemble ordonné. Application à l'étude du lattis des sous-algèbres d'un demi-lattis de Brouwer fini*, Bull. Soc. Roy. Sci. Liège **51** (1982), no. 5–8, 174–187.

368. _____, *The lattice of R-subalgebras of a bounded distributive lattice*, Comment. Math. Univ. Carolin. **25** (1984), no. 1, 1–17.

369. Guo Jun Wang, *Topological molecular structure. 1*, I Sanshi Sida Xuebao **1** (1979), 1–15. (Chinese)

370. _____, *Separation axioms in topological molecular lattices*, J. Math. Res. Exposition **3** (1983), no. 2, 9–16. (Chinese)

371. M. Weese and H.-J. Goltz, *Boolean algebras*, Seminarberichte, vol. 62, Humboldt Universität, Berlin, 1984.

372. R. Wille, *Kongruenzklassengeometrien*, Lecture Notes in Math., vol. 113, Springer-Verlag, New York and Berlin, 1970.

373. _____, *Restructuring lattice theory: An approach based on hierarchies of concepts*, Ordered Sets (Banff, Alta., 1981), NATO Adv. Study Inst. Ser. C: Math. Phys. Sci., vol. 83, Reidel, Dordrecht, Boston, 1982, pp. 445–470.

374. _____, *Subdirect decomposition of concept lattices*, Algebra Universalis **17** (1983), no. 3, 275–287.

375. R. Y. Wood, *Stability of the Souslin operation*, Algebra Universalis **19** (1984), no. 2, 250–254.

376. P. Zlatoš, *Unitary congruence adjunctions*, Lectures in Universal Algebra (Szeged, 1983), Colloq. Math. Soc. János Bolyai, vol. 43, North-Holland, Amsterdam and New York, 1986, pp. 587–647.

377. A. P. Huhn and E. T. Schmidt (eds.), *Contributions to lattice theory*, Colloq. Math. Soc. János Bolyai, vol. 33, North-Holland, Amsterdam and New York, 1983.

378. G. Eigenthaler, H. K. Kaiser, W. B. Muller, and W. Nöbauer (eds.), *Contributions to general algebra*, 2 (Klagenfurt, 1982), Hölder-Pichler-Tempsky, Vienna; B. G. Teubner, Stuttgart, 1983.

379. Enrico G. Beltrametti and Bastiaan C. van Fraasen (eds.), *Current issues in quantum logic*, (Proc. Workshop on quantum logic; Erice, 1979), Plenum Press, New York and London, 1981.

380. L. Szabo and A. Szendrei (eds.), *Lectures in universal algebra* (Szeged, 1983), Colloq. Math. Soc. János Bolyai, vol. 43, North-Holland, Amsterdam and New York, 1986.

381. Ivan Rival (ed.), *Ordered Sets*, (Banff, Alta, 1981), NATO Adv. Study Inst. Ser. C: Math. Phys. Soc., vol. 83, Reidel, Dordrecht, Boston, 1982.

382. Ralph S. Freese and Octavio C. Garcia (eds.), *Universal algebra and lattice theory*, (Puebla, 1982), Lecture Notes in Math., vol. 1004, Springer-Verlag, New York and Berlin, 1983, pp. 1–308.

Translated by BORIS M. SCHEIN

Chapter IV
Classes of Lattices and Related Algebras

V. I. IGOSHIN

This article contains a survey of results related to various classes of lattices: varieties, quasivarieties, prevarieties, etc., and properties of lattices in them. Further, analogous classes of algebras related to lattices are considered. Finally, there is a discussion of free lattices and free and ordered products of lattices and related algebras.

§IV.1. Varieties and other classes of lattices

From the beginning of the seventies problems connected with basability of varieties of (modular) lattices generated by classes of lattices of limited length or width (see previous surveys [11], p. 43 and [OSL, p. 155]) were being solved. Hsuech [78] proved a theorem that completed a classification of such varieties with respect to their basability: lattices of width three generate a lattice variety without a finite basis. Brunner [42] found an identity which, in the variety of all modular lattices, defined the variety generated by all modular lattices of length $\leq n + 1$.

Day [51] surveyed recent results about varieties of modular lattices connected with projective geometries. In particular, he discussed properties of the variety of Arguesian lattices and asked whether it was generated by its finite members and whether the word problem was decidable in free lattices of that variety. Herrmann [77] showed that a variety of modular lattices that contains rational projective geometries could not be both finitely based and finitely generated by its finite-dimensional members. García and Peña [64] proved that the variety of modular lattices is not generated by its W-lattices (see the end of §3.1 of this chapter for a definition). Kut′in [12] gave an example of an identity that defines a variety of modular lattices, investigated this variety, and established its connection with lattices of projective geometries.

1991 *Mathematics Subject Classification.* Primary 06B20, 18B35; Secondary 06B25, 08B15.

He [13] found identities generalizing the modular law and studied properties of the varieties of lattices so defined. In particular, he transferred certain theorems known for modular lattices to lattices from the variety defined by the identity

$$(x \vee y) \wedge (x \vee z) = x \vee ((x \vee y) \wedge ((x \wedge y) \vee z)).$$

Jónsson and Rival [OSL, III, 129] proved that the least nonmodular variety of lattices (it is generated by the pentagon) is covered by precisely 15 join-irreducible varieties, each of which is generated by a single finite lattice L_i ($i = 1, 2, \ldots, 15$). Rose ([103] and [104]) constructed eight infinite sequences of join-irreducible varieties of lattices, in each of which every successor was the only join-irreducible cover of its predecessor. The first terms of these sequences are the varieties generated by the lattices L_i, where $i = 6, 7, 8, 9, 10, 13, 14, 15$. He mentioned a series of open problems connected with the lattices L_{11} and L_{12}.

Negru [19] called a lattice quasidistributive if it satisfies both the identity

$$(x \vee y) \wedge (z \vee u) = (x \wedge (z \vee u)) \vee (y \wedge (z \vee u)) \vee (z \wedge (x \vee y)) \vee (u \wedge (x \vee y))$$

and its dual. He found examples of nontrivial identities that are consequences of the quasidistributivity identity. He found [21] the least (by the number of its elements) lattice that does not satisfy one of these identities, namely,

$$(x \vee (y \wedge z)) \wedge (y \vee z)$$
$$= (x \wedge (y \vee z)) \vee (y \wedge (x \vee (y \wedge z))) \vee (z \wedge (x \vee (y \wedge z))).$$

It is an eight-element lattice consisting of the elements $a, b, c, b \wedge c, a \wedge b \wedge c, a \vee (b \wedge c), (a \vee (b \wedge c)) \wedge (b \vee c), a \vee b \vee c$. In this paper a variety of lattices defined by the identity

$$x \wedge (y \vee z) = (x \wedge (y \vee (x \wedge z))) \vee (x \wedge (z \vee (x \wedge y))) \qquad (*)$$

was considered. He observed that a lattice satisfying the inequality $x \wedge y \leq x \wedge z$ (or $x \wedge z \leq x \wedge y$) satisfies $(*)$ if and only if it contains neither of the five finite sublattices mentioned by him.

Ruckelshausen and Sands [105] proved that every infinite finitely generated lattice of width four generates a variety of infinite height in the lattice of lattice varieties. Bergman and Hrushovski [39] showed that every variety of lattices is generated by any cofinal sublattice of a free in this variety lattice with a countable number of free generators. Czédli [50] introduced a notion of a factor lattice of a lattice modulo a tolerance relation and proved that ISTSP(2) is the variety of all lattices and ITSP$_f$(2) the variety of all finite lattices. Here 2 is the two-element lattice, and $\check{P}, P_f, S, T, I$ are the operators of direct product, finite direct product, sublattices, factor lattices modulo tolerances, and isomorphic lattices, respectively.

Igoshin [8] proved that the identity dual for (∗) holds in the interval lattice of a lattice L if and only if L is a chain. Identity (∗) holds if and only if L is a singleton or two-element.

1.2. Properties of lattice varieties. Day and Ježek [52] showed that only three lattice varieties possess the amalgamation property: the trivial one, the variety of all distributive lattices, and the variety of all lattices. An earlier result of Slavík [114] follows as a simple corollary: every nontrivial and nondistributive lattice variety with the amalgamation property contains all primitive lattices.

A concrete category is called universal (in other terminology, binding) if every category of algebras and their homomorphisms can be embedded in it as a full subcategory. A concrete category is called almost universal if a maximal subcategory contained in the class of nonconstant morphisms of the given category is universal. Koubek and Sichler [86] proved that there exist infinitely many almost universal lattice varieties each of which is generated by a single finite lattice and in finitely many universal varieties of bounded lattices each of which is generated by a single finite bounded lattice. Koubek [85] constructed two universal varieties of bounded lattices whose intersection is not universal, and asked whether his varieties are minimal universal. Goralčik, Koubek, and Pröhle [69] found a condition necessary for the universality of an arbitrary variety of bounded lattices. Werner [119] studied, as a category, the lattice variety generated by the five-element three-atom lattice M_3.

Baker [34] proved that the projectivity and weak projectivity of intervals in nondistributive lattice varieties could not be expressed by formulas of the first-order language.

Lachlan [87] introduced a notion of separated distributive lattices and proved that the elementary theory of finite separated distributive lattices is decidable. Gurevich [75] showed that the elementary theory of all separated distributive lattices is undecidable. Schmerl [106] established that the theory of ordered sets of breadth n can be interpreted in the theory of distributive lattices of ∧-breadth n, and the theory of distributive lattices of ∧-breadth 2 is undecidable.

Baranskiĭ [2] proved independence of the equational theory and automorphism groups in the class of all lattices, that is, he showed that for every group G and every nontrivial lattice variety V there exists a lattice L for which $G \cong \text{Aut}(L)$ and $V = V(L)$.

Dilworth [54] considered how properties of the lattice $\text{Con}\, L$ could be applied in studying lattice varieties.

1.3. Properties of lattices in varieties. Nauryzbaev and Omarov [18] proved that distributive 1-equationally compact lattices are precisely complete continuous lattices. With every complete distributive lattice Nauryzbaev [17] associated a sublattice of its ultrapower and, in these terms, gave a criterion for its equational compactness. Fleischer [61] gave a new proof of a theorem that gives two characterizations of equationally compact with respect to single variable distributive lattices and stems from the work of Beazer [36], and Bulman-Fleming and Fleischer [44]. Schmerl [106] proved that every $\aleph_0$-categorical distributive lattice of finite $\wedge$-breadth has a finitely axiomatizable theory.

Chuong [46] gave certain characterizations of lattices of breadth two finitely projected in the variety of all modular lattices [OSL, pp. 103–104]. Also, he characterized lattices of breadth two splittable in that variety. Rival and Sands [102] proved that there exists a continuum of pairwise nonisomorphic simple lattices generated by a four-element antichain, and Freese [62] strengthened that result by showing that a continuum of such lattices exists in the variety of modular lattices. Brunner [42] showed that the number $2n - 1$ is the upper bound for projective distances between two projective intervals in any lattice from the variety of lattices generated by all modular lattices of length $\leq n + 1$. Negru [19] proved that every modular sublattice of the lattice of Post classes is quasidistributive. He [20] studied the lattice SP_k of all classes of functions of k-valued logic closed under suppositions, and proved that SP_2 is neither distributive, nor modular, nor p-modular, nor satisfies certain nontrivial lattice identities. He found a family of identities with a parameter n, where n is the number of variables. These identities are nontrivial for $n \geq 3$ and hold on SP_2 for $n \geq 16$. He gave another example of a simpler identity of three variables valid in SP_2. Finally, he proved that, for every nontrivial lattice identity, there exists k such that this identity does not hold on SP_k. He asked whether there is a nontrivial identity that holds on SP_3.

Adams and Sichler [32] proved that, for any nontrivial lattice variety V, each of its lattices is isomorphic to the Frattini sublattice of a suitable lattice from V, thus generalizing an analogous result for the variety of distributive lattices obtained earlier by the former author [28]. Eigenthaler [59] corrected a theorem from Schuff [108], proving that the set of principal (in Schuff's sense) formulas is not always a system of normal forms for the algebra of polynomials in the variety of all lattices, except in very special cases, which he mentioned.

1.4. Other classes of lattices. Tumanov [27] proved that the 10-element modular lattice $M_{3\text{-}3}$ has no independent basis for quasi-identities, and every finite lattice is embeddable in a finite lattice that has no independent basis for quasi-identities. Igoshin [8] proved that the quasi-identity

$$x \vee y = x \vee z \to x \vee y = x \vee (y \wedge z)$$

of $\vee$-semidistributivity holds in the lattice $\mathrm{Int}(L)$ of all intervals of a lattice L if and only if L is a chain. The dual quasi-identity of $\wedge$-semidistributivity holds if and only if L is a singleton or has two elements. He [9] announced that there exists no proper quasivariety Q of lattices such that the quasivariety $\text{Int}(Q)$ generated by the interval lattices of all lattices from Q coincides with the variety of all lattices. On the other hand, there exists a continuum of quasivarieties Q that are not varieties and such that $\text{Int}(Q) = Q$. Finally, selfdual quasivarieties of lattices and only they are closed under interval projective images, that is, they satisfy the condition: if $\text{Int}(L_1) \cong \text{Int}(L_2)$ and $L_1 \in Q$, then $L_2 \in Q$. Negru [20] showed that, for every nontrivial quasi-identity κ, there exists a natural number k such that κ does not hold on the lattice SP_k of all classes of functions of k-valued logic closed under superposition. He raised the question of the existence of a nontrivial quasi-identity in SP_3. Baranskiĭ [3] announced that the automorphism groups and quasi-equational theories were independent in the class of all lattices; that is, for every group G and every lattice L that is not a singleton there exists a lattice L such that $G \cong \text{Aut}(T)$ and every quasi-identity holds on L if and only if it holds on T.

Lender ([14] and [15]) studied Mal'tsev multiplication of prevarieties of lattices. In the former paper he showed that a product of any finite number of prevarieties of lattices is idempotent if and only if one of the factors is idempotent and it contains all other factors. This statement fails for infinite products; a number of corresponding examples were given. In the latter paper he called a prevariety of lattices bounded if the variety generated by it differs from the class of all lattices, and proved that the product of any finite number of bounded prevarieties of lattices is a bounded prevariety, and the product of any infinite number of nontrivial prevarieties of lattices is not bounded.

Schmitt [107] proved that the class of existentially complete lattices is not elementary, while Lienkamp [92] proved that the class of finite amalgams $P = P(A, B; U)$, for which amalgamated free products $F(P)$ are finite, is not finitely universally axiomatizable in the class of all finite amalgams.

Jambu-Giraudet [66] proved that arithmetic could be interpreted in certain lattices, and hence elementary theories of certain classes of lattices are undecidable. For interpretations of various theories in lattices see her paper [67].

1.5. Hypervarieties of lattices. Let $\mathscr{E}$ be an identity in a variety V. Formally speaking, a hyperidentity is the same thing as an identity. However, a variety V is said to satisfy a hyperidentity $\mathscr{E}$ if, when all symbols of operations in $\mathscr{E}$ are replaced by arbitrary polynomials of suitable arity, one obtains identities valid in V. Padmanabhan and Penner [98], pointed out for every $m \geq 2$, a finite basis of semilattice hyperidentities, whose operation symbols have arities $\leq m$, and proved that the variety of all semilattices, and nontrivial varieties of lattices have no finite bases of hyperidentities. The last two statements were obtained by Penner [99] as a corollary to his theorem: for every lattice variety (and also for the variety of all semilattices) V and arbitrary natural numbers m and n there exists a lattice (respectively, semilattice) hyperidentity that is valid in V but not deducible from $H^m(V) \cup H_n(V)$, where $H^m(V)$ is the set of all hyperidentities in V in which operation symbols of arities $\leq m$ participate, while $H_n(V)$ is the set of all hyperidentities in V with at most n different variables. By the way, there is no finite basis of hyperidentities for the variety of all monoids [38].

Taylor [116] showed that a class of varieties is determined by hyperidentities if and only if it is closed under products of varieties, subvarieties, reducts of varieties, and natural equivalence (such classes of varieties are called hypervarieties). He proved the existence of a continuum of different lattice hypervarieties.

Schweigert [109] pointed out an identity valid in the variety $T(N_5)$ generated by the clone of term operations on the five-element nonmodular lattice N_5, and an identity valid in the variety $T(M_3)$, where M_3 is the five-element modular lattice. Each of these identities is a hyperidentity on the varieties generated by the lattices N_5 and M_3, respectively.

§IV.2. Varieties and other classes of algebras related to lattices

2.1. Pseudocomplemented distributive lattices and related algebras. Adams, Koubek, and Sichler ([29] and [30]) investigated endomorphism monoids of algebras from non-Boolean varieties of pseudocomplemented distributive lattices and problems connected with the notion of universality of a category or a variety. It is known (Lee [91]) that varieties of pseudocomplemented distributive lattices form a chain $B_{-1} \subset B_0 \subset B_1 \subset \cdots \subset B_\omega$, where B_{-1} is the trivial variety, B_0 the variety of Boolean algebras, and B_1 the variety of Stone algebras. These authors showed that all varieties B_n, $n \geq 3$, are almost universal [30] and, if $n \geq 4$, they are (ω, B_0)-universal [29]. The variety B_3 is neither (ω, B_0)-universal nor (ω, B_1)-universal, but it is (ω, B_2)-universal [29]. These properties of varieties were used to obtain properties of endomorphism monoids of algebras from these varieties. For example, it was proved in [29] that in the variety B_3 there exist algebras of arbitrarily large cardinality with only finitely many endomorphisms. It was shown in [30] that every algebra is uniquely determined by its endomorphism monoid in B_1, while there exist nonisomorphic algebras with isomorphic endomorphism monoids in B_2, but no more than two nonisomorphic algebras could have isomorphic endomorphism monoids. In the same paper they proved the universality of the variety of Heyting algebras.

Dziobiak ([57] and [58]) produced examples of finite Heyting algebras (relatively pseudocomplemented distributive lattices) that generate quasivarieties without a finite basis of quasi-identities. Bordalo [41] introduced and studied classes K_n of strongly n-normal lattices, which play the same role for the variety of Heyting algebras as the classes of n-normal lattices for the variety of all pseudocomplemented distributive lattices. (A distributive lattice with 0 is called n-normal if each of its prime filters is contained in no more than n pairwise incomparable filters.) It was shown, in particular, that $H_n = H \cap K'_n$, where H and H_n are the varieties of all and all strongly n-normal Heyting algebras, respectively. In this paper a variety M of type $\langle 2, 2, 2, 0 \rangle$ was studied; it consists of distributive lattices with 1 endowed with another binary operation that generalizes Heyting's implication. A chain $N_1 \subseteq \cdots \subseteq N_n \subseteq \cdots \subseteq N$ of quasivarieties was constructed such that the lattices in the variety N_n were n-normal for every n. Iturrioz [81] studied free symmetric Heyting algebras of order k.

Meskhi [16] introduced and studied a variety HRI of involuted Heyting algebras (i.e., those having a lattice anti-isomorphism of order two), in which the operation of involution coincides with the Boolean complementation on regular elements. He produced an inner characterization of subdirectly irreducible algebras in HRI, which implies that HRI is generated by the set of its finite algebras. He proved that HRI has a continuum of subvarieties and investigated the notion of injectivity in subvarieties of HRI.

Urquhart [118] proved that there exist a continuum of different varieties of distributive double p-algebras (distributive lattices with two pseudocomplementations), pointed out a variety of such algebras not generated by its finite members, and raised a question about the existence of finitely based varieties of such algebras with an undecidable word problem. Zimmermann and Köhler [121] showed that the Mal'tsev product of two finitely based varieties of Brouwerian semilattices is finitely based.

2.2. Algebras related to Boolean algebras. Rybakov [24] proved the decidability of the universal theory of free algebras in the variety of topo-Boolean algebras corresponding to the modal logic λ for $\lambda = S4 + \sigma_k$ for $k < \omega$, and that free algebras of every variety eq($S4 + \sigma_k$) for $k \geq 2$ do not have a basis of quasi-identities in finitely many variables and have a hereditarily undecidable elementary theory. He [23] showed that a subdirectly irreducible finite modal algebra has a finite basis of quasi-identities and that there exists a finite modal algebra without a basis for quasi-identities in finitely many variables. (A Boolean algebra with two unary operations x' and x^* is called a modal algebra if $1^* = 1$, $x^* \to x = 1$, and $(x \to y)^* \to (x^* \to y^*) = 1$, where $x \to y = x' \vee y$.) Rybakov [25] proved the hereditary undecidability of the theory of free algebras of countable rank in varieties of topo-Boolean and pseudo-Boolean algebras and some of their subvarieties.

Galanter [5] announced that among the subvarieties of the variety of pseudo-Boolean algebras defined by the identity $x' \vee x'' = 1$ there exists a continuum of indecomposable non-Gödelian varieties, and the variety of Boolean algebras is the greatest among them. Sendlewski ([111] and [112]) investigated varieties of N-lattices (quasi-pseudo-Boolean algebras). In the former article he proved that there exist precisely three pretabular varieties of N-lattices, and that all of them are locally finite. In the latter article he characterized primitive varieties of N-lattices (a variety is called primitive if each of its subquasivarieties is a variety): these are the varieties that do not contain any of a finite number of finite N-lattices mentioned in the paper. In the same paper all preprimitive varieties of N-lattices were listed; there are six of them. (A variety is called preprimitive if it is not primitive, but each of its proper subvarieties is primitive.)

Abashidze [1] proved the residual finiteness of the intersection of splittable varieties corresponding to splittable algebras in the variety of all Magari algebras. Grigoliya [6] described free Magari algebras of finite nonzero rank in a variety generated by a free Magari algebra of rank 0. Another paper [7] of the same author contains a complete description of free S4.3-algebras with finitely many generators. Schweigert and Szymańska [110], studying varieties of correlation lattices, characterized their subdirectly irreducible algebras, found generating algebras, and established the decidability of the word problem in them.

Urquhart [117] proved that the assembly of varieties of Ockham lattices is not countable, and that every such variety is generated by its finite members. (An Ockham lattice is a bounded distributive lattice endowed with a dual endomorphism.) Blyth and Varlet [40] investigated varieties of MS-algebras; there are 20 of them. (An MS-algebra is a bounded distributive lattice endowed with a unary operation x° such that $x \leq x^{\circ\circ}$, $(x \wedge y)^\circ = x^\circ \vee y^\circ$, and $1^\circ = 0$.) For each of them the defining identities were found.

2.3. Orthomodular lattices. Beran [37] produced the following system of axioms for the variety of orthomodular lattices of type $\langle 2, 2, 1 \rangle$:

(1) $x \vee (y \wedge y') = x$;
(2) $(x \vee y) \vee z = (z' \wedge y')' \vee x$;
(3) $((x \vee y) \wedge (x \vee z)) \vee ((x \vee y) \wedge x') = x \wedge y$.

The derived operations $x + y = (x \vee y) \wedge (x' \vee y')$ and $xy = (x \wedge y) \vee (x' \wedge y')$ in orthomodular lattices were studied by Šimon [113], who, in particular, showed that if one of these two introduced operations is associative then the orthomodular lattice is a Boolean algebra.

Godowski [68] proved that the class K of orthomodular lattices with a full set of two-valued states is a variety and, in the variety generated by the class C of orthomodular lattices with a strongly full set of states, no subvariety containing K has a finite basis of identities. Carrega [45] proved that the variety of Boolean algebras could be defined in the variety of orthomodular lattices by the condition that it does not contain sublattices isomorphic to two fixed finite lattices. Dorninger, Länger, and Mącyński [55] proved that a σ-continuous orthomodular lattice L considered as a quantum logic and the lattice $\mathscr{D}(L)$ of all observables of a quantum mechanics system with logic L generate the same lattice variety.

Bruns announced in an abstract of a conference [115] and proved in [43] that every variety of modular ortholattices that is not contained in [MO2] contains [MO3]. (Here [MOn] denotes the variety of orthomodular lattices with a complementation operator generated by the lattice M_{2n} considered as an ortholattice.)

2.4. Other algebras related to lattices. Quackenbush [101] considered the class V_3 of all lattices with five nullary operations that he introduced, isomorphic to lattices of the form $M_3 \otimes B$, $B \in V$, where V is a variety of bounded modular lattices and $M_3 \otimes B$ is a tensor product of ∨-semilattices with 0. He proved that V_3 is a variety which, as a category, is equivalent to V. Also, V_3 is finitely based if and only if V is finitely based. Isbell [80] showed that every variety of ∨-irreducible lattices is closed under direct limits.

Cornish [48] introduced a notion of S-algebra as an algebra with a single ternary operation satisfying five identities and showed that if a derived operation $s(x, y, z) = x \wedge (y \vee z)$ is introduced on each lattice, the class of all lattices becomes a variety of S-algebras. With respect to that operation, distributive lattices form a subvariety, defined by an identity he mentioned. He gave estimates for axiomatic ranks of these varieties. Bandelt [35] considered varieties of all rings and distributive lattices as subvarieties of the variety of all semirings with an inverse additive semigroup and proved that their Mal'tsev product coincides with their lattice-theoretic join. Lomecky [93] produced algorithms for computing a corresponding initial algebra (if it is finite) in a class of extensions of lattice theory generated by symbols of unary functions and equational axioms (including modularity), and produced data to compare how his program worked.

Kamara ([83] and [84]) studied polarity lattices, that is, lattices with an additional unary operation (polarity) that is idempotent and satisfies the De Morgan identities. In the former article he described all upper neighbors in the lattice of varieties of polarity lattices of the variety of distributive polarity lattices; there are 12 of them. In the latter paper he studied properties of polarity lattices of splitting in varieties, of being projective, or of having projective covers.

Verkhozina [4] announced that there are precisely two pretabular varieties among the varieties of implicatures; one of them was generated by all finite linearly ordered implicatures, while the second was generated by all indecomposable finite implicatures. A class of ordered sets closed under direct products and retracts is called an order variety. Nevermann [97] showed that if V is an order variety generated by a set of ordered sets that coincides neither with the class of all singletons nor with the class of all antichains, then V is not axiomatizable. Cornish, Sturm, and Traczyk [49] showed that the operation * of an arbitrary BCK-algebra A can be extended to a similar operation on the lattice A°, which is a free extension of A in the class of all distributive lattices; they studied interrelations between the BCK algebras A and A° and showed, in particular, that these algebras belong to the same variety of BCK-algebras.

§IV.3. Free lattices and products

3.1. Free lattices. By proving that the word problem for a free modular lattice with four generators is algorithmically unsolvable, Herrmann [76] completed an almost half a century long story of one of the most famous problems of lattice theory.

Nation [95] solved another of the most famous problems of lattice theory: he described the finite sublattices of a free lattice. As conjectured by Jónsson [82], who suggested that problem, a finite lattice is embeddable in a free lattice if and only if it is semidistributive (that is, it satisfies both the ∨-semidistributivity quasi-identity and its dual) and Whitman's condition

$$x \wedge y \leq u \vee v \Leftrightarrow x \leq u \vee v \text{ or } y \leq u \vee v \text{ or }$$
$$x \wedge y \leq u \text{ or } x \wedge y \leq v$$

holds in it.

Finite sublattices of free lattices were studied by Day and Nation [53]. Freese and Nation [63] suggested an algorithm which, for every given element of a free lattice $FL(n)$, finds all elements covering it and covered by it (if they exist). They showed that the element $[(x \wedge (y \vee z)) \vee (y \wedge z)] \wedge [(y \wedge (x \vee z)) \vee (x \wedge z)]$ has no covers and itself covers no element. Finite intervals in $FL(n)$, except a finite number of them, have one, two, or three elements, and the length of a chain in which each element covers its predecessor does not exceed four.

Freese [62] surveyed known results, as well as open problems, and included new facts about ordered sets in free lattices, coverings in free lattices, lattices freely generated by ordered sets, fixed points of lattice polynomials, and free modular lattices.

Nation [96] investigated the problem of order-preserving embeddability of ordered sets in a free lattice. He found a few sufficient and a few necessary conditions for such embeddability. The main results were obtained for ordered sets that are a union of two infinite antichains. Cibulskis [47] introduced a concept of a semipartial lattice and produced algorithms for constructing the free lattice $F(P)$ generated by a semipartial lattice P, and for determining when $F(P)$ is finite in the case of finite P. In addition, he constructed the operation tables in $F(P)$.

Tropin [26] proved that a free lattice is embeddable in the lattice of quasivarieties of pseudocomplemented distributive lattices. Negru [19] proved that a free quasidistributive lattice with three generators is infinite and, in the abstract [21] announced that that result is true for any variety of lattices defined by an identity in three variables, which follows from two identities defining the variety of quasidistributive lattices. Dwinger [56] described the structure of a free dense in itself completely distributive complete lattice with an arbitrary number of generators and showed that if the number of generators is infinite, then the lattice is a subdirect product of copies of a segment of the real line.

Mönting [94] developed methods for solving the word problem in a variety of lattices and certain varieties of algebras related to lattices.

García and Peña [64] called a lattice L a W-lattice if there exists a finite lattice B that satisfies the Whitman condition and whose homomorphic image is L. They found necessary and sufficient conditions for the semidistributivity of a W-lattice.

3.2. Free products. A cycle of papers by Grätzer and Huhn, [70], [71], and [72], was devoted to investigation of amalgamated free products of lattices. A lattice $F(P)$ freely generated by an amalgam $P(A, B; U)$ of two lattices A and B with intersection $A \cap B = U$ is called an amalgamated free product of A and B and denoted by $A *_U B$. In the first paper they introduced the notion of a common refinement for two decompositions of a lattice into amalgamated free products and gave necessary and sufficient conditions for the existence of a common refinement. In the second paper they estimated the number of elements in a generating set of an amalgamated free product with the least number of elements. Finally, in the third paper they showed that a free generating set of an amalgamated product of two lattices does not have to contain free generators of the set of its components. Lienkamp [92] clarified conditions under which the amalgamated free product $A *_U B$ of two finite lattices A and B with intersection $A \cap B = U$ is finite.

Eigenthaler [60] studied a general concept of a free algebra over a partial algebra in a class of algebras and applied his results to obtain some properties of coproducts of lattices.

Powell and Tsinakis [120] studied free products of abelian l-groups as free products in the category of distributive lattices with a fixed element. They [100] surveyed results on free products of l-groups, giving some of the proofs. Adams and Sichler [31] investigated an R-reduced V-free product of bounded lattices they introduced earlier [OSL, III, 37]. They found a sufficient condition for the common extension property for this product in the variety V to hold and showed that there exist a continuum of varieties in which the common extension property fails.

Grätzer and Kelly [73] defined free m-products of m-complete lattices, where m is an infinite regular cardinal number, and showed that a greater part of the properties of ordinary free products carry over to free m-products of m-lattices. Adams and Trnková [33] studied free products in the category of bounded distributive lattices with $(0, 1)$-homomorphisms, and Huhn [79] considered reduced free products of distributive lattices.

3.3. Ordered products. Ladzianska [88] studied ordered products (o-products) of lattices considering the following topics: the word problem for o-products, linear representation for elements of o-products, sublattices of o-products of lattices similar to free lattices, decompositions of lattices in o-products, o-products and direct (or inverse) limits of lattices. In another paper [90] she found a sufficient condition for any two representations of a lattice as an o-product to have a common refinement in an arbitrary nontrivial variety. (An analogous result for decompositions free in a variety was proved by Grätzer and Sichler [74].) Romanovich [22] introduced a definition of a local dimension of an arbitrary lattice which, in the distributive case, coincides with a definition of an analogous dimension due to Gavalec [65] and studied this concept for certain ordered products of lattices.

Ladzianska [89] described the structure of an m-ordered product of a family of m-lattices, thus generalizing to ordered products of lattices a result of Grätzer and Kelly [73], who studied free products of m-lattices.

BIBLIOGRAPHY

1. M. A. Abashidze, *On varieties of Magari algebras*, Logico-Methodolog. Issled., Tbilisi, 1983, pp. 121–134. (Russian)

2. V. A. Baranskiĭ, *Independence of equational theories and of groups of lattice automorphisms*, Sibirsk. Mat. Zh. **26** (1985), no. 4, 3–10; English transl. in Siberian Math. J. **26** (1985), no. 4.

3. _____, *Independence of automorphism groups and retracts for semigroups and lattices*, Abstract of the XVIII All-Union Algebraic Conference (Kishinev, 1985), Part I, 1985, p. 30. (Russian)

4. M. I. Verkhozina, *Pretabular varieties of implicatures*, Abstract of the XVIII All-Union Algebraic Conference (Kishinev, 1985), Part I, 1985, p. 91. (Russian)

5. G. I. Galanter, *On irreducible elements of the lattice of varieties of pseudo-Boolean algebras*, Abstract of the XVIII All-Union Algebraic Conference (Kishinev, 1985), Part I, 1985, p. 106. (Russian)

6. R. Sh. Grigoliya, *Free Magari algebras with a finite number of generators*, Logico-Methodolog. Issled. (1983), Tbilisi, 135–149. (Russian)

7. _____, *Free S4.3-algebras with a finite number of generators*, Studies in Nonclassical Logics and Formal Systems, "Nauka", Moscow, 1983, pp. 281–287. (Russian)

8. V. I. Igoshin, *Lattices of intervals of chains*, Ordered Sets and Lattices, No. 8, Izdat. Saratov Univ., Saratov, 1982, pp. 50–55. (Russian)

9. _____, *Interval properties of quasivarieties of lattices*, Abstract of the XVIII All-Union Algebraic Conference (Kishinev, 1985), Part I, 1985, p. 212. (Russian)

10. _____, *Varieties and quasivarieties of lattices*, Ordered Sets and Lattices, No. 7, Izdat. Saratov Univ., 1983, pp. 57–68. (Russian)

11. V. I. Igoshin and T. S. Fofanova, *Varieties of lattices. Category problems*, Ordered Sets and Lattices, No. 3, Izdat. Saratov Univ., 1975, pp. 41–49. (Russian)

12. A. M. Kut′in, *On submodular varieties of lattices*, Preprint No. 132-85, deposited at VINITI, 1985. (Russian)

13. _____, *Classes of lattices with identities generalizing the modular law*, Preprint No. 1953-85, deposited at VINITI, 1985. (Russian)

14. V. B. Lender, *A theorem on multiplication of prevarieties of lattices*, Ural. Gos. Univ. Mat. Zap. **13** (1982), no. 1, 69–74. (Russian)

15. _____, *Bounded prevarieties of lattices*, Ural. Gos. Univ. Mat. Zap. **13** (1983), no. 3, 87–94. (Russian)

16. V. Yu. Meskhi, *A discriminator variety of Heyting algebras with involution*, Algebra i Logika **21** (1982), no. 5, 537–552; English transl. in Algebra and Logic **21** (1985), no. 5.

17. K. A. Nauryzbaev, *A criterion of equational compactness of distributive lattices*, Issledovaniya po Konstruktivnym Modelyam, Alma-Ata, 1982, pp. 46–53. (Russian)

18. K. A. Nauryzbaev and A. I. Omarov, *On equational compactness of distributive lattices*, Investigations in the Theory of Models, Alma-Ata, 1982, pp. 35–41. (Russian)

19. I. S. Negru, *Quasi distributivity of lattices*, Mat. Issled. (1982), no. 66, 113–127. (Russian)

20. _____, *Algebraic properties of the lattice of Post classes and their many-valued generalizations*, Studies in Nonclassical Logics and Formal Systems, "Nauka", Moscow, 1983, pp. 300–315. (Russian)

21. _____, *On a problem of Grätzer*, Abstract of the XVIII All-Union Algebraic Conference (Kishinev, 1985), Part II, 1985, p. 62. (Russian)

22. V. A. Romanovich, *Dimension of an ordered product of lattices*, Abstract of the 7th Regional Conference Mathematics and Mechanics (Tomsk, 1981), Section of Algebra, Tomsk, 1981, pp. 37–38. (Russian)

23. V. V. Rybakov, *Bases of quasi-identities of finite modal algebras*, Algebra i Logika **21** (1982), no. 2, 219–227; English transl. in Algebra and Logic **21** (1982), no. 2.

24. _____, *Decidability of the problem of admissibility in finite-layered modal logics*, Algebra i Logika **23** (1984), no. 1, 100–116; English transl. in Algebra and Logic **23** (1984), no. 1.

25. _____, *Elementary theories of free topo-Boolean and pseudo Boolean algebras*, Mat. Zametki **37** (1985), no. 6, 797–802; English transl. in Math. Notes **37** (1985).

26. M. P. Tropin, *Embeddability of a free lattice in the lattice of quasivarieties of pseudocomplemented distributive lattices*, Algebra i Logika **22** (1983), no. 2, 159–167; English transl. in Algebra and Logic **22** (1983), no. 2.

27. V. I. Tumanov, *Finite lattices with no independent basis of quasi-identities*, Mat. Zametki **36** (1984), no. 5, 625–634; English transl. in Math. Notes **36** (1984), no. 5–6.

28. M. E. Adams, *The Frattini sublattice of a distributive lattice*, Algebra Universalis **3** (1973), 216–228.

29. M. E. Adams, V. Koubek, and J. Sichler, *Pseudocomplemented distributive lattices with small endomorphism monoids*, Bull. Austral. Math. Soc. **28** (1983), no. 3, 305–318.

30. _____, *Homomorphisms and endomorphisms in varieties of pseudocomplemented distributive lattices (with applications to Heyting algebras)*, Trans. Amer. Math. Soc. **285** (1984), no. 1, 57–79.

31. M. E. Adams and J. Sichler, *Refinement property of reduced free products in varieties of lattices*, Contributions to Lattice Theory (Szeged, 1980), Colloq. Math. Soc. János Bolyai, vol. 33, North-Holland, Amsterdam and New York, 1983, pp. 19–53.

32. _____, *Frattini sublattices in varieties of lattices*, Colloq. Math. **44** (1981), no. 2, 181–184.

33. M. E. Adams and Věra Trnková, *Isomorphisms of sums of countable bounded distributive lattices*, Algebra Universalis **15** (1982), no. 2, 242–257.

34. K. A. Baker, *Nondefinability of projectivity in lattice varieties*, Algebra Universalis **17** (1983), no. 3, 267–274.

35. Hans-J. Bandelt, *Free objects in the variety generated by rings and distributive lattices*, Lecture Notes in Math., vol. 998, Springer-Verlag, New York and Berlin, 1983, pp. 255–260.

36. R. Beazer, *A characterization of complete bi-Brouwerian lattices*, Colloq. Math. **29** (1974), 55–59.

37. L. Beran, *Boolean and orthomodular lattices — a short characterization via commutativity*, Acta Univ. Carolin.—Math. Phys. **23** (1982), no. 1, 25–27. (Russian and Czech summaries)

38. G. M. Bergman, *Hyperidentities of groups and semigroups*, Aequationes Math. **23** (1981), no. 1, 50–65.

39. G. M. Bergman and E. Hrushovski, *Identities of cofinal sublattices*, Order **2** (1985), no. 2, 173–191.

40. T. S. Blyth and J. C. Varlet, *Subvarieties of the class of MS-algebras*, Proc. Roy. Soc. Edinburgh Sect. A **95** (1983), no. 1–2, 157–169.

41. Gabriela Bordalo, *Strongly N-normal lattices*, Proc. Ser. A Nederl. Akad. Wetensch. **87** (1984), no. 2, 113–125.

42. J. Brunner, *A finite base for M^n and maximal projective distance in M^n*, Algebra Universalis **14** (1982), no. 1, 99–106.

43. G. Bruns, *Varieties of modular ortholattices*, Houston J. Math. **9** (1983), no. 1, 1–7.

44. S. Bulman-Fleming and I. Fleischer, *One variable equational compactness in partially distributive semilattices with pseudocomplementation*, Proc. Amer. Math. Soc. **79** (1980), no. 4, 505–511.

45. Jean-Claude Carrega, *Exclusion d'algèbres*, C. R. Acad. Sci. Paris Sér. I Math. **295** (1982), no. 2, 43–46.

46. H. M. Chuong, *Remark on finitely projected modular lattices of breadth two*, Wiss. Z. Martin-Luther-Univ. Halle-Wittenberg Math.-Natur. Reihe (1981), no. 21, 127–132.

47. J. M. Cibulskis, *An algorithm for the construction of free lattices over nonfree generators*, Bull. Calcutta Math. Soc. **73** (1981), no. 1, 17–26.

48. W. H. Cornish, *A ternary variety generated by lattices*, Comment. Math. Univ. Carolin. **22** (1981), no. 4, 773–784.

49. W. H. Cornish, Teo Sturm, and Tadeusz Traczyk, *Embedding of commutative BCK-algebras into distributive lattice BCK-algebras*, Math. Japon. **29** (1984), no. 2, 309–320.

50. Gábor Czédli, *Factor lattices by tolerances*, Acta Sci. Math. (Szeged) **44** (1982), no. 1–2, 35–42.

51. A. Day, *Equational theories of projective geometries*, Contributions to Lattice Theory (Szeged, 1980), Colloq. Math. Soc. János Bolyai, vol. 33, North-Holland, Amsterdam and New York, 1983, pp. 277–316.

52. A. Day and J. Ježek, *The amalgamation property for varieties of lattices*, Trans. Amer. Math. Soc. **286** (1984), no. 1, 251–256.

53. A. Day and J. B. Nation, *A note on finite sublattices of free lattices*, Algebra Universalis **15** (1982), no. 1, 90–94.

54. R. P. Dilworth, *Aspects of distributivity*, Algebra Universalis **18** (1984), no. 1, 4–17.

55. Dietmar Dorninger, Helmut Länger, and M. Mączyński, *Zur Darstellung von Observablen auf σ-stetigen Quantenlogiken*, Österreich. Akad. Wiss. Math.-Natur. Kl. Sitzungsber. II **192** (1983), no. 4–7, 169–176.

56. Ph. Dwinger, *Characterization of the complete homomorphic images of a completely distributive complete lattice* II, Nederl. Akad. Wetensch. Indag. Math. **45** (1983), no. 1, 43–49.

57. W. Dziobiak, *Concerning axiomatizability of the quasivariety generated by a finite Heyting or topological Boolean algebra*, Studia Logica **41** (1982), no. 4, 415–428.

58. _____, *Concerning axiomatizability of the quasivariety generated by a finite Heyting or topological Boolean algebra*, Polish Acad. Sci. Inst. Philos. Sociol. Bull. Sect. Logic **10** (1981), no. 4, 177–180.

59. G. Eigenthaler, *Eine Bemerkung zur Darstellung von Polynomen über Verbänden*, Math. Nachr. **103** (1981), 299–300.

60. _____, *Über Einbettungsfragen bei frei erzeugten Algebren*, Österreich. Akad. Wiss. Math.-Natur. Kl. Sitzungsber. II **190** (1981), no. 8–10, 485–493.

61. Isidore Fleischer, *One-variable equationally compact distributive lattices*, Math. Slovaca **34** (1984), no. 4, 385–386.

62. R. Freese, *Some order-theoretic questions about free modular lattices*, Ordered Sets (Banff, Alta., 1981), NATO Adv. Study Inst. Ser. C: Math. Phys. Sci., vol. 83, Reidel, Dordrecht, Boston, 1982, pp. 355–377.

63. R. Freese and J. B. Nation, *Covers in free lattices*, Trans. Amer. Math. Soc. **288** (1985), no. 1, 1–42.

64. O. C. García and J. A. de la Peña, *Lattices with a finite Whitman cover*, Algebra Universalis **16** (1983), no. 2, 186–194.

65. M. Gavalec, *Dimensions of distributive lattices*, Mat. Časopis Sloven. Akad. Vied **21** (1971), no. 3, 177–190.

66. Michèle Jambu-Giraudet, *Interpretations d'arithmétiques dans des groupes et des treillis*, Lecture Notes in Math., vol. 890, Springer-Verlag, New York and Berlin, 1981, pp. 143–153.

67. _____, *Bi-interpretable groups and lattices*, Trans. Amer. Math. Soc. **278** (1983), no. 1, 253–269.

68. R. M. Godowski, *Varieties of orthomodular lattices with a strongly full set of states*, Demonstratio Math. **14** (1981), no. 3, 725–733.

69. P. Goralčík, V. Koubek, and P. Pröhle, *A universality condition for varieties of 0, 1-lattices*, Lectures in Universal Algebra (Szeged, 1983), Colloq. Math. Soc. János Bolyai, vol. 43, North-Holland, Amsterdam and New York, 1986, pp. 143–154.

70. G. Grätzer and A. P. Huhn, *Amalgamated free product of lattices* I. *The common refinement property*, Acta Sci. Math. (Szeged) **44** (1982), no. 1–2, 53–66.

71. _____, *Amalgamated free product of lattices* II. *Generating sets*, Studia Sci. Math. Hungar. **16** (1981), no. 1-2, 141-148.

72. _____, *Amalgamated free product of lattices* III. *Free generating sets*, Acta Sci. Math. (Szeged) **47** (1984), no. 3-4, 265-275.

73. G. Grätzer and D. Kelly, *Free m-products of lattices* I, Colloq. Math. **48** (1984), no. 2, 181-192.

74. G. Grätzer and J. Sichler, *Free decompositions of a lattice*, Canad. J. Math. **27** (1975), no. 2, 276-285.

75. Yuri Gurevich, *Decision problem for separated distributive lattices*, J. Symbolic Logic **48** (1983), no. 1, 193-196.

76. C. Herrmann, *On the word problem for the modular lattice with four free generators*, Math. Ann. **265** (1983), no. 4, 513-527.

77. _____, *On the arithmetic of projective coordinate systems*, Trans. Amer. Math. Soc. **284** (1984), no. 2, 759-785.

78. Yuang Cheh Hsueh, *Lattices of width three generate a non-finitely based variety*, Algebra Universalis **17** (1983), no. 1, 132-134.

79. A. P. Huhn, *A reduced free product of distributive lattices* I, Acta Math. Hungar. **42** (1983), no. 3-4, 349-354.

80. John Isbell, *Direct limits of meet-continuous lattices*, J. Pure Appl. Algebra **23** (1982), no. 1, 33-35.

81. Luiza Iturrioz, *Symmetrical Heyting algebras with operators*, Z. Math. Logik Grundlag. Math. **29** (1983), no. 1, 33-70.

82. B. Jónsson, *Sublattices of a free lattice*, Canad. J. Math. **13** (1961), no. 2, 256-264.

83. Morike Kamara, *Nichtdistributive, modulare Polaritätsverbände*, Arch. Math. (Basel) **39** (1982), no. 2, 126-133.

84. _____, *Spaltende modulare Polaritätsverbände*, Contributions to General Algebra, 2 (Klagenfurt, 1982), Hölder-Pichler-Tempsky, Vienna, 1983, pp. 179-190.

85. Václev Koubek, *Towards minimal binding varieties of lattices*, Canad. J. Math. **36** (1984), no. 2, 263-285.

86. V. Koubek and J. Sichler, *Universality of small lattice varieties*, Proc. Amer. Math. Soc. **91** (1984), no. 1, 19-24.

87. A. H. Lachlan, *The elementary theory of recursively enumerable sets*, Duke Math. J. **35** (1968), no. 1, 123-146.

88. Zuzana Ladzianska, *Poproduct of lattices*, Math. Slovaca **32** (1982), no. 1, 3-22.

89. _____, *M-poproduct of lattices*, Math. Slovaca **35** (1985), no. 1, 31-35.

90. _____, *Poproduct decomposition of a lattice*, Math. Slovaca **35** (1985), no. 3, 263-266.

91. K. B. Lee, *Equational classes of distributive pseudocomplemented lattices*, Canad. J. Math. **22** (1970), no. 4, 881-891.

92. Ingrid Lienkamp, *Freie Verbände über Amalgamen von Verbänden*, Mitt. Math. Sem. Giessen (1984), no. 161.

93. Z. Lomecky, *Algorithms for the computation of free lattices*, Computer Algebra (Marseille, 1982), Lecture Notes in Comput. Sci., vol. 144, Springer, Berlin-New York, 1982, pp. 223-230.

94. Jürgen Schulte Mönting, *Cut elimination and word problems for varieties of lattices*, Algebra Universalis **12** (1981), no. 3, 290-321.

95. J. B. Nation, *Finite sublattices of a free lattice*, Trans. Amer. Math. Soc. **269** (1982), no. 1, 311-337.

96. _____, *On partially ordered sets embeddable in a free lattice*, Algebra Universalis **18** (1984), no. 3, 327-333.

97. Peter Nevermann, *A note on axiomatizable order varieties*, Algebra Universalis **17** (1983), no. 1, 129-131.

98. R. Padmanabhan and P. Penner, *Bases of hyperidentities of lattices and semilattices*, C. R. Math. Rep. Acad. Sci. Canada **4** (1982), no. 1, 9-14.

99. P. Penner, *Hyperidentities of lattices and semilattices*, Algebra Universalis **13** (1981), no. 3, 307-314.

100. W. B. Powell and C. Tsinakis, *Free products of lattice ordered groups*, Algebra Universalis **18** (1984), no. 2, 178-198.

101. R. W. Quackenbush, *Non-modular varieties of semi-modular lattices with a spanning* M_3, Discrete Math. **53** (1985), 193–205.

102. I. Rival and B. Sands, *How many four-generated simple lattices?*, Universal Algebra and Applications (Warsaw, 1978), Banach Center Publ., 9, PWN, Warsaw, 1982, pp. 67–72.

103. Henry Rose, *Semidistributive lattice varieties*, Algebraic Structures and Applications (Nedlands, 1980), Lecture Notes in Pure and Appl. Math., vol. 74, Dekker, New York, 1982, pp. 149–168.

104. _____, *Nonmodular lattice varieties*, Mem. Amer. Math. Soc. **47** (1984), no. 292.

105. W. Ruckelshausen and B. Sands, *On finitely generated lattices of width four*, Algebra Universalis **16** (1983), no. 1, 17–37.

106. J. H. Schmerl, $\aleph_0$ *categorical distributive lattices of finite breadth*, Proc. Amer. Math. Soc. **87** (1983), no. 4, 707–713.

107. P. H. Schmitt, *Algebraically complete lattices*, Algebra Universalis **17** (1983), no. 2, 135–142.

108. H. K. Schuff, *Zur Darstellung von Polynomen über Verbänden*, Math. Nachr. **11** (1954), 1–4.

109. D. Schweigert, *Clones of term functions of lattices and abelian groups*, Algebra Universalis **20** (1985), no. 1, 27–33.

110. D. Schweigert and M. Szymańska, *A completeness theorem for correlation lattices*, Z. Math. Logik Grundlag. Math. **29** (1983), no. 5, 427–434.

111. Andrzej Sendlewski, *Pretabular varieties of N-lattices*, Polish Acad. Sci. Inst. Philos. Sociol. Bull. Sect. Logic **12** (1983), no. 1, 17–20.

112. _____, *Some investigations of varieties of N-lattices*, Studia Logica **43** (1984), no. 3, 257–280.

113. J. Šimon, *Opérations dérivées des treillis orthomodulaires (Part II)*, Acta Univ. Carolin.—Math. Phys. **23** (1982), no. 1, 29–36.

114. V. Slavík, *A note on the amalgamation property in lattice varieties*, Contributions to Lattice Theory (Szeged, 1980), Colloq. Math. Soc. János Bolyai, vol. 33, North-Holland, Amsterdam and New York, 1983, pp. 723–736.

115. J. D. Smith, *Universelle Algebra*, Tagungsber. Math. Forschungs. Inst. Oberwolfach (1982), no. 1–16.

116. W. Taylor, *Hyperidentities and hypervarieties*, Aequationes Math. **23** (1981), no. 1, 30–49.

117. Alasdair Urquhart, *Distributive lattices with a dual homomorphic operation* II, Studia Logica **46** (1981), 391–404.

118. _____, *Equational classes of distributive double p-algebras*, Algebra Universalis **14** (1982), no. 2, 235–243.

119. H. Werner, *A duality for the lattice variety generated by* M_3, Lectures in Universal Algebra (Szeged, 1983), Colloq. Math. Soc. János Bolyai, vol. 43, North-Holland, Amsterdam and New York, 1986, pp. 561–572.

120. Wayne B. Powell and Constantine Tsinakis, *The distributive lattice free product as a sublattice of the abelian l-group free product*, J. Austral. Math. Soc. Ser. A **34** (1983), no. 1, 92–100.

121. Ute Zimmermann and Peter Köhler, *Products of finitely based varieties of Brouwerian semilattices*, Algebra Universalis **18** (1984), 110–116.

Translated by BORIS M. SCHEIN

Chapter V
Concrete Lattices

V. I. IGOSHIN, A. V. MIKHALEV, V. N. SALIĬ, AND L. A. SKORNYAKOV

In this paper various topics are discussed pertaining to congruence lattices, subalgebras, ideals, subrings, subgroups, subsemigroups, submodules, subsets, etc. We also consider properties of lattices of varieties and of congruence varieties. See Chapter IV for lattices of lattice varieties, and Chapter III for lattices of sublattices and ideals in lattices.

§V.1. Lattices related to groups

We begin with results related to the subgroup lattice of a group. Shiryaev [521] observed that if the subgroup lattice of a group is ∧-semidistributive, it is distributive. Richter [497] found when, for every element a and every completely irreducible element v of the subgroup lattice, the join $a \vee v$ is completely ∨-irreducible in the interval $[a, 1]$. Rudolph [506] proved that an infinite group with a modular subgroup lattice with two generators a and b of finite orders and such that $\langle a \rangle \cap \langle b \rangle = 1$ contains Tarski's monster as a subgroup. He showed in [507] that a group with a modular lattice of subgroups is finite if it is generated by two elements of finite orders, one of which is a power of two (an example given by A. Yu. Ol'shanskiĭ in RZhMat 1980, 8A182 shows that this statement may fail if only finiteness of the orders is required).

Let M_n be a lattice of length two with n atoms, G a finite group, and H its subgroup. Feit [319] proved that the interval $[H, G]$ in the subgroup lattice $L(G)$ is isomorphic to M_n for $n > 0$ if and only if there exist precisely n maximal subgroups K such that $H \subset K \subset G$ (if G is solvable then, in this situation and for $n > 2$, the number $n - 1$ must be a prime

---

1991 *Mathematics Subject Classification*. Primary 06B20; Secondary 06B30, 08A30, 08A60, 08B15, 08C15, 16A21, 20M07, 22A26.

power; an example was given showing that this statement need not hold for unsolvable groups—the group A_{31} contains an interval isomorphic to M_7).

De Giovanni and Franciosi [280] proved that a locally finite group has a lower semimodular lattice of subgroups if and only if it is locally supersolvable and induces the automorphism group of a prime order in each of its noncentral principal factors.

Titov [194] established commutativity of certain groups with coinciding sets of complementable and lattice complementable subgroups.

Previato [486] proved that finite simple groups A_n, $\mathrm{Sz}(q)$, and $\mathrm{PSL}(2, q)$ have complemented lattices of subgroups. Iwahori [378] surveyed results on the subgroup lattice of Coxeter and Weil groups.

Zacher [564] investigated connections between normality of a subgroup and the subgroup lattice of a group and characterized [565] subgroups of finite index in terms of subgroup lattices. Plaumann, Strambach, and Zacher [479] studied the influence of geometric properties of various parts of the subgroup lattice of a group on those of the group. Ivanov [90] noticed connections between the subgroup lattices of a locally solvable torsion-free group and its subgroup.

Polland [481] produced tests for verifying lattice isomorphism of subgroup lattices. Koike [415] observed that the groups A_n for $n \geq 4$ and the affine Weil groups of rank ≥ 4 are determined by their subgroup lattices. Yakovlev [219] gave sufficient conditions for a lattice isomorphism to be induced by a group isomorphism. Damova [70] studied projections for lattices of maximal subgroups of a finite group that preserve the index. Jónsson [404] noted dual isomorphisms of the subgroup lattice of a group onto a certain lattice of subalgebras of its algebra of binary relations. Holmes [363] considered semidirect products of groups with coinciding subgroup lattices.

McKenzie [445] observed that the subgroup lattice of the group A_4 could not be represented as the congruence lattice of a finite algebra with a single operation.

Bulgak ([25] and [26]) considered the Möbius function on subgroup lattices of certain finite groups.

Leone and Maj ([427] and [428]) considered finite groups whose subgroup lattices are not submodular but the subgroup lattices of their nontrivial homomorphic images are submodular. Longobardi and Maj [432] studied groups whose lattice of normal subgroups is isomorphic to the lattice of normal subgroups of a free product. McKenzie [446] showed that if $\mathscr{M}$ is a variety of groups, λ a cardinal number, and for every group $G \in \mathscr{M}$ every strictly meet-irreducible element N of both the subgroup lattice and the lattice of normal subgroups of G satisfies the condition $|G: N| < \lambda$, then all finite groups in $\mathscr{M}$ are abelian. Schaller [512] studied connections between the lattices of subnormal subgroups of a finite group and its p-subgroup. Gawron and Macedońska-Nosalska [334] proved that all subgroups of a group are polynormal if and only if the normality relation in that group is transitive.

De Giovanni and Franciosi [281] studied isomorphisms between the lattices of subnormal subgroups of solvable groups that preserve order of subgroups. In this paper he analyzed images of generalized nilpotent groups under such isomorphisms. Franciosi [327] considered certain dual homomorphisms between the subgroup lattice and the lattice of normal subgroups.

Much attention was paid to studying subgroup lattices of linear groups. Zalesskiĭ [86] surveyed some sections of the theory of linear groups connected with subgroup lattices. For a wide class of rings Golubchik [62] described the structure of normal subgroups of linear and unitary groups. In a cycle [28]–[32] of his papers, Vavilov considered the structure of "parabolic" subgroups of the general linear group, unitary group, special linear group, and Chevalley groups (Vavilov and Dybkova [33] considered the symplectic group). Koĭbaev [112] described the lattice of those subgroups of a general linear group over a field that contain the group of elementary cellwise-diagonal matrices. In [113] he considered the case of a special linear group over a five-element field, and in [114] over a four-element field. In some cases Krupetskiĭ ([124]–[127]) described the lattice of those subgroups of the unitary group that contain the group of diagonal matrices. Yakovlev [220] found sufficient conditions for the existence of a lattice isomorphism of the group of elementary matrices into another group to imply the existence of a group isomorphism.

A number of papers is devoted to subgroup lattices of abelian groups. Polland's lectures [480] contain some results on lattice isomorphic abelian groups. Huhn [367] showed that for every $n > 1$ the subgroup lattice of an n-generated free abelian group is n-distributive but not $(n-1)$-distributive, and the subgroup lattice of an abelian group is n-distributive if and only if it is of rank $\leq n$. He described groups whose lattice of normal subgroups is not n-distributive. Scoppola [517] characterized the subgroup lattice of an abelian group containing independent aperiodic elements. Grinshpon [67] mentioned properties of the lattice of completely characteristic subgroups of torsion-free abelian groups. Holmes [364] considered split extensions of abelian groups with coinciding subgroup lattices, and Ivanov [89] proved the decidability of elementary theories of the subgroup lattice of a torsion-free abelian group and of the class of all such lattices. He obtained a criterion for the equivalence of such lattices.

Kutyev's monograph [134] touched upon problems of structure of the subsemigroup lattice and the determinability of a group by its subsemigroup lattice. Ovsyannilov [160] proved that a group decomposable into a direct product of two nonperiodic groups, one of which is generated by elements of infinite order, is determined by its subsemigroup lattice in the class of all semigroups.

Some topics of the theory of l-groups connected with lattices were discussed in Kopytov's monograph [117]. Kopytov [116] surveyed results on the theory of radicals in l-groups. Jakubík [381] showed that if $\mathscr{R}$ is the lattice of all torsion radicals of l-groups with the least element $\bar{0}$, then $\mathscr{R}$ is not infinitely distributive. However, if a radical $\delta \in \mathscr{R}$ is generated by a class of linearly ordered groups, then the interval $[\bar{0}, \delta]$ in $\mathscr{R}$ is completely distributive. He proved in [379] that the lattice of radical classes of linearly ordered groups is not pseudocomplemented. Jakubíková [385] found properties of the lattice $\mathscr{R}_h$ of radical classes of linearly ordered groups hereditary with respect to convex subgroups. In particular, the lattice $\mathscr{R}_h$ is Brouwerian, the hereditary radical class generated by an Archimedean linearly ordered group is an atom in $\mathscr{R}_h$, and the set of all principal hereditary classes is an ideal in $\mathscr{R}_h$. Jakubík [383] proved that there are no atoms in the lattice $\mathscr{R}_s$ of all semisimple classes of linearly ordered groups, and the lattice is not modular. He described properties of the join operation in the lattice of radical classes and relations between semisimple and radical classes.

Let G be an l-group and $K(G)$ the lattice of closed convex sub-l-groups of G. A class X of l-groups is called a K-class if there exists a class T of lattices such that $G \in X$ if and only if $K(G) \in T$ for every l-group G. A radical K-class X is called a K-radical class. Jakubík [382] studied the complete lattice $\mathscr{R}_K$ of all K-radical classes. It is a Brouwerian lattice but not a closed sublattice of the lattice $\mathscr{R}$ of all radical classes. Pringerová [488] observed that the family $\mathscr{R}_1$ of radical classes, generated by a single linearly ordered group and with no covers in the lattice $\mathscr{R}_a$ of all radical classes of abelian linearly ordered groups, is rather large. Jakubík [384] proved that $\mathscr{R}_a$ is not modular.

Medvedev [150] proved the existence of an l-group with two generators with an infinitely long lattice of radicals. Jakubík [380] studied torsion radicals of l-groups.

Anderson, Bixler, and Conrad [224] considered behavior of the lattice of convex l-groups under extensions of l-groups. Huss [369] found when the lattice of l-subgroups of an l-group is distributive. Kenoyer [407] observed that the property of an l-group to belong to a variety of a given group, the property of being Archimedean, and complete distributivity are not determined by the lattice of convex l-subgroups.

Kutyev ([132] and [133]) considered properties of the subsemigroup lattice of ordered groups.

Mukhin [158] found lattice characterizations of radical topological groups. Protasov and Tsybenko [169] described locally compact groups with operations continuous in the Chabauty topology in their lattices of closed subgroups. Yakovenko [218] characterized inductive proradical groups whose lattice of closed subgroups is of finite breadth. Komarov [115] studied the local structure of the lattice of closed subgroups. Kabenyuk [96] considered discreteness of lattices of closed subgroups of a Lie group in various topologies.

Èĭdinov [217] studied when the subformation lattice of a formation is complemented. Skiba [183] studied distributive lattices of formations of finite groups. Purdea and Both [491] considered the lattice of partial tolerances on a group. Schmidt [514] studied mappings of groups that induce isomorphisms of their coset lattices. Pivovarova [163] analyzed connections between a pair (A, Γ), where A is a set or a linear space and Γ a group, and the lattice of its subpairs.

§V.2. Lattices related to rings and modules

A number of authors continued investigating the subring lattice. Orozco and Vélez [461] considered the lattice of subfields of an extension $F(\alpha)/F$, where α is a root of the binomial $x^m - a$ over a field F of characteristic not dividing m. Beard and McConnel [238] computed orders of various groups of automorphisms of subfield lattices. Seĭtenov [181] studied elementary theory of subfield lattices of finite fields. Mináč [451] considered the distributivity property of finite intersections of valuation rings in fields.

Freĭdman [211] showed that the subring lattice of a torsion-free nil ring R is modular if and only if either $R^2 = 0$ or R is an annihilator extension of a ring with zero multiplication by a ring with zero multiplication and additive group of rank 1. Korobkov [119] proved that if an algebraic algebra without nilpotent elements is not a division ring, a direct sum of prime fields of order 2, or a field whose subfield lattice is a chain of length two, then every algebra lattice-isomorphic to it is an algebraic algebra without nilpotent elements. In [120] he described associative rings with a dense lattice of subrings (that is, 0 is a co-indecomposable element).

Geĭn [56] studied a restricted distributive law in the subalgebra lattice of a linear algebra satisfying the identity $x^2 = 0$ over a commutative ring. He proved that distributive pairs of subalgebras are dually distributive and found a criterion for distributivity of the subalgebra lattice for torsion-free algebras, as well as a criterion of decomposability of the subalgebra lattice in a direct product.

Smirnov [185] described rings with zero Jacobson radical all subrings of which have the subring lattice-anti-isomorphism to the subring lattice of a ring. In [186] he described monogenic nilpotent associative rings with additive p-group for $p \neq 2$, whose subring lattices are dually isomorphic with the subring lattice of a ring. Kiss and Ronyai [412] investigated rings for which the lattice dual to the subring lattice is algebraic.

Röhl [500] considered some conditions on the subring lattice. Kirby [410] studied the lattice of extensions for Cohen-Macaulay rings.

Vechtomov [47] obtained new characterizations of Boolean rings in terms of annihilators of elements (in particular, a ring R is Boolean if and only if left annihilators of different elements differ; this, in its own turn, is equivalent to every bijection $\varphi \colon R \to R$, for which $ab = 0$ is equivalent to $\varphi(a)\varphi(b) = 0$, being an isomorphism).

Dulatova and Pinus [81] showed that elementary equivalence of subalgebra lattices of Boolean algebras implies elementary equivalence of the Boolean algebras, and that the universal theory of the class of lattices of (finite) subalgebras of Boolean algebras is decidable. Bell [239] considered Boolean algebras with extreme cellular and compactness properties. Orhon [460] produced a criterion for the completeness of the Boolean algebra of projections in a Banach space. In [529] a study of Boolean-like rings was continued. Ramaswami [170] observed that if a triple system contains an invertible idempotent, then the class of equivalent idempotents forms a Boolean algebra.

Finkel′shteĭn [210] proved that idempotent-generated left ideals of a ring R form a sublattice in the lattice of left ideals if and only if the set of von Neumann regular elements of R is a semigroup (he obtained a more precise version of this result for Rickart rings). Yao [563] considered the lattice of ideals of tensor products of division algebras and primitive algebras.

Dubrovin [79] constructed an example of a chain ring R with one-sided zero divisors such that for all $b \in R$ either $bR \subseteq Rb$ or $Rb \subseteq bR$. He expressed properties of the Jacobson radical of a chain ring in terms of the ideal lattice. In [80] he produced an example of a prime catenary ring with nilpotent elements (this yields an example of a simple catenary radical ring all of whose elements are zero divisors; in particular, such a ring coincides with its right or left singular ideal). Bessenrodt, Brungs, and Törner [244] studied prime ideals in right chain rings.

If R is a ring with 1 and without zero divisors, let $\tilde{x}$ for $0 \neq x \in R$ denote a self-mapping of the lattice of nonzero right ideals such that $\tilde{x}(aR) = xaR$. Let $\widetilde{H}(R) = \{\tilde{x} | 0 \neq x \in R\}$ be a semigroup with the natural partial order (it is a generalization of the group of divisibility of a commutative integral domain). Brungs and Törner [254] clarified when $\widetilde{H}(R)$ is a linearly ordered semigroup. Their other paper [255] was devoted to extensions of chain rings.

Hausen [356] and Feigelstock [317] considered additive subgroups of rings with totally ordered lattices of ideals.

In recent years a systematic study of rings with the distributive lattice of one-sided ideals and of modules with the distributive submodule lattice has been initiated. Tuganbaev [198] showed that the class of right distributive right and left Noetherian rings coincides with the class of finite direct products of right catenary Artinian rings and hereditarily invariant right and left Noetherian domains. He produced an example of a right distributive right Noetherian indecomposable ring that is neither right Artinian nor semiprime. He [197] proved the equivalence of the following conditions on a right R:

(1) the ring $R[[x]]$ of power series is left distributive;

(2) R is strictly regular and injective with respect to embeddings of countably generated left ideals.

He proved that a topological space X is discrete if and only if the left $C(X, F)$-module F^X is distributive, where F is a separable topological ring which is a left Bezout ring all maximal left ideals of which are two-sided. He also proved that if the quotient ring $R/J(R)$ of a ring R over its Jacobson radical $J(R)$ is strictly regular, then the lattice $\mathscr{L}(M)$ of submodules of a left R-module M is distributive if and only if M is locally cyclic. Tuganbaev [199] proved that a right Noetherian ring is right distributive if and only if each of its right ideals is a product of prime two-sided ideals. Every right Noetherian right and left distributive ring is a finite direct product of right and left catenary right and left Artinian rings and right and left invariant domains in which every ideal is a product of maximal ideals. He proved in [200] that a semiprime ring R integral over its center is right and left distributive if and only if all right ideals of R are flat modules and for every prime ideal P in R the set $R - P$ satisfies the Ore condition. He found for which rings the ring of skew polynomials is right distributive. In [202] he proved that if R is a left distributive ring and $R/J(R)$ a strictly regular ring, then all left ideals in R are flat (if, moreover, R is left Rickart, then R is a left semihereditary ring). In [203] and [204] he described rings and their automorphisms φ for which the rings $R[[x, \varphi]]$ of skew series are left distributive.

A module is called endodistributive if it is distributive as a module over its endomorphism ring. Tuganbaev ([203] and [207]) proved that the left distributivity of R is equivalent to each of the following conditions:

(a) all injective R-modules are endodistributive;

(b) all direct sums and products of quasi-injective R-modules are endodistributive;

(c) all injective hulls of simple R-modules are endodistributive;

(d) all irreducible quasi-injective R-modules are endocatenary modules.

He proved that the class of endodistributive modules is closed under direct sums, direct products, and direct summands. Let Q be a maximal left ring of quotients of a left distributive ring R. If Q is left self-injective, then it is right distributive. If Q is right and left self-injective, then it is right and left distributive. Tuganbaev [205] proved that if R is a ring integral over its center, then the following conditions are equivalent:

(a) R is a right distributive right Noetherian ring;

(b) R is a finite direct product of right catenary right Artinian rings and right invariant domains all proper factor rings of which are decomposable in finite direct products of right catenary right Artinian rings.

A right and left distributive ring with the maximality conditions for right and left annihilators is decomposable into a finite direct product of orders in catenary Artinian rings. A right distributive ring R is a right order in a right Artinian ring if and only if R satisfies the maximality and minimality conditions for right annihilators and all of its right regular elements are regular. An example of a right distributive right Noetherian ring that is not a right order in a right Artinian ring was given. In [201] he described the structure of right semihereditary right distributive rings and distributive invariant semiprime rings. He showed that a distributive prime ring that is a finitely generated module over its center is a semihereditary order over a Prüfer domain. In [206] Tuganbaev studied, for which commutative rings R and monoids with cancellation T, the monoid ring $R[T]$ is left distributive. As a corollary, he obtained a description of left distributive group rings RG, where R is a commutative ring. He investigated completely the case when R is an arbitrary ring and T a monoid with cancellation that is not a periodic abelian group.

Mal'tsev [145] gave a criterion when every critical algebra in a variety of associative algebras over a prime field has a distributive ideal lattice.

Zamyatin [87] investigated decidability problems for elementary theories of varieties of rings, each of which has a distributive ideal lattice.

Cohn [266] proved that every finite distributive lattice is representable as a lattice of divisors of a free associative algebra $K\langle X \rangle$ for a sufficiently large set X of free generators. If R is a semifir ring, he produced conditions equivalent to R having a distributive lattice of divisors.

Let $(G, +)$ be a finite group, A its automorphism group, and $C(A; G) = \{f: G \to G | fa = af \ \forall a \in A, \ f(0) = 0\}$. Smith [525] proved that the lattice of left ideals of the near-ring $C(A; G)$ is distributive.

Nauwelaerts and Van Geel [453] studied prime Zariski central rings whose fractional ideals in a maximal symmetric ring of quotients form an abelian group.

The lattice of radicals of rings has been investigated in various papers. For example, Beĭdar and Salavova [12] for each of the lattices of N-radicals, left strict radicals, and left hereditary radicals obtained a representation of every special radical as a join of a family of supranilpotent nonspecial radicals (thus generalizing Yu. M. Ryabukhin's results about the lattice of special radicals). Vodyanyuk [48] described lattice-complementary torsions and radically semisimple classes in the class of all associative Φ-algebras. Beĭdar ([9] and [10]) showed that every maximal right ideal of a simple non-Artinian ring with zero divisors and identity is a simple ring, and the lower radical generated by it is an atom in the "lattice" of all radicals. He found that the lattice of all radicals is not atomic (connections of these results with earlier papers on the lattices of radicals are reflected in the survey [11, §1.6]). Banaschewski and Harting [233] studied properties of the partially ordered set $\mathscr{L} \operatorname{Id} R$, where $\operatorname{Id} R$ is the ideal lattice of a ring R, $\mathscr{L}$ is either the locally nilpotent radical, the Jacobson radical, or the Brown-McCoy radical, and $\mathscr{L} \operatorname{Id} R = \{I \in \operatorname{Id} R | \mathscr{L}(R/I) = 0\}$ (in particular, $\mathscr{L} \operatorname{Id} R$ is a complete lattice whose $\wedge$-irreducible elements are primary ideals).

A long cycle of papers has been devoted to lattices of subspaces and submodules. The survey [147] by Markov, Mikhalev, Skornyakov, and Tuganbaev reflected the work on lattices of submodules in the period 1973–1982. Brehm's book [253] is devoted to the lattice of submodules of a torsion-free module.

Kravchenko [123] proved that a minimal generating system of the lattice of all subspaces of a finite-dimensional linear space over a finite field with q elements has at most $\max\{q+3, 8\}$ elements (this estimate does not depend on the dimension of the space). Dzuan Tam Guè [75] showed that the minimal number of generators of the lattice of all subspaces of a finite-dimensional linear space over a field is finite if and only if the field is finitely generated over its prime subfield. He found an upper estimate for this number, not depending on the dimension of the space and linearly depending on the number of elements generating the field. He gave another estimate in [76]. Haapsalo and Niemisto [350] suggested a lattice generating program for computing in the lattice of subspaces of a linear space.

McAsey and Muhly [443] found conditions under which an isomorphism of the lattices of invariant subspaces of linear mappings is induced by a linear or adjoint linear mapping. Longstaff [433] described a diagrammatic way of picturing the lattice of invariant subspaces for a nilpotent transformation of a finite-dimensional complex linear space.

Filippov [209] proved that every isomorphism between the lattices of subgeometries of finite-dimensional projective geometries over a division ring is induced either by their isomorphism or an anti-isomorphism. Realization problems for geometric lattices were considered by Delandtsheer [282] and by Bennett [241] for Hilbert lattices. Kalinin [97] suggested an abstract characterization of U-closed subspaces, where U is a total subspace of the conjugate space. Lambrou and Longstaff [421] found properties of distributive lattices of subspaces of a Hilbert space. Mayrhofer [441] considered lattice properties of A-systems of projective geometries, and Bures [256] studied modularity in the lattice of projections of a von Neumann algebra.

Gross [342] considered when an isomorphism of sublattices in the lattice of subspaces of quadratic spaces of uncountable dimensions is induced by an isometry. Papers by Gross and Keller [343], Gross, Lomecky, and Schuppli [344], and Haapsalo [349] are related.

In a cycle of papers [289]–[291] Downey studied lattices of recursively enumerable subspaces of effective linear spaces. Guichard [346] considered the structure of automorphisms of the lattice of recursively enumerable subspaces of an effective countable-dimensional linear space over a finite or countable field. Kuzicheva ([128] and [129]) studied the lattice of recursively enumerable subspaces in connection with the description of isomorphisms of rings of recursive endomorphisms and the study of groups of recursive automorphisms of countably dimensional linear spaces.

Archinard [225] found properties of the lattice of submodules of finitely generated torsion-free modules over a Dedekind ring. Hutchinson [371] showed that the existence of an exact embedding R-Mod $\hookrightarrow$ S-Mod of categories of modules is equivalent to the fact that every universal conditional lattice identity, which holds in every lattice embedded in the lattice of an R-module, holds in every lattice embeddable in an S-module. Czédli [268] investigated connections between properties of rings and lattice identities of lattices isomorphic to the lattices of submodules.

An R-module M is said to satisfy condition AB5* if $\bigcap_{\lambda \in \Lambda}(X + X_\lambda) = X + \bigcap_{\lambda \in \Lambda} X_\lambda$ for all submodules X and downward directed families $\{X_\lambda | \lambda \in \Lambda\}$ of submodules. Brodskiĭ [21] proved that the following conditions are equivalent for a right R-module M:

(1) there exists a ring T and left T-module N such that the lattices $L_R(M)$ and $L_T(N)$ of submodules are anti-isomorphic;

(2) M satisfies AB5*.

He [22] found necessary and sufficient conditions for the existence of a canonical isomorphism of the lattice of submodules of a right R-module M and a right $\text{End}(U_R)$-module $\text{Hom}_R(U, M)$ and conditions for the existence of a canonical anti-isomorphism of the lattice of submodules $L_R(M)$ and the lattice of submodules of a left $\text{End}(U_R)$-module $\text{Hom}_R(U, M)$. In [23] he found properties of the lattice of submodules of a cogenerator, as a module over the endomorphism ring. Khuri [409] extended results on lattice isomorphisms between the lattices of submodules and ideals of rings of endomorphisms to the situation of rings with Gabriel topologies.

Dull and Saalfield [297] noted properties of projectivities of the lattices of subspaces. McDonald [444] proved an analog of the fundamental theorem of projective geometry for a projective line over a commutative ring which has "many" invertible elements. In the cases of regular weakly finite rings and of semiprime rings with the Goldie condition, this problem was solved by Sarath and Varadarajan [509]. Camillo [258] analyzed various generalizations of the fundamental theorem of projective geometry to the case of modules over rings. Dawson [278] discussed a projective geometry defined in terms of factor modules of a module, which are not sums of two proper submodules. Facchini [313] investigated situations when an isomorphism of the lattices of submodules induces an isomorphism of subfactormodules.

Bunu and Tèbyrtse [27] found a criterion of distributivity of the lattice of pre-torsions of the category of modules over a ring.

Fillmore and Longstaff [322] considered isomorphisms of lattices of closed subspaces of a topological vector space.

Bennett [240] found necessary and sufficient conditions for a lattice to be isomorphic to a lattice of convex subsets of a finite-dimensional vector space over an ordered division ring. In another paper [242] she gave separation conditions in convexity lattices. Fourneau [324] noted characterizations of subspaces and one-dimensional subspaces among the elements of the lattice of radically closed balanced convex subsets of a vector space.

Continuing research of the subalgebra lattice of a Lie algebra, Geĭn [54] proved that degrees of undecidability of lattice isomorphic Lie algebras cannot differ by more than 1 (in more detail he analyzed the effect of a change of the degree of undecidability in the case of finite-dimensional and locally supersolvable Lie algebras). He [55] described the structure of modular subalgebras of locally finite-dimensional Lie algebras in the case of characteristic zero of the ground field. He proved there that simple finite-dimensional splittable Lie algebras are lattice determinable. Geĭn [57] considered conditions for modularity, upper semimodularity and the existence of relative complements for Lie algebras.

Lashkhi [135] showed that if a Lie algebra over a principal ideal domain, which is not a field, is a wreath product of its proper subalgebras, then each of its lattice isomorphisms compatible with normalizers is induced by an isomorphism. He proved in [136] that a lattice isomorphism of a nilpotent Lie ring of degree 2 with a nonabelian proper subring is induced by an isomorphism of Lie rings; he gave examples showing that this result could not be transferred to Lie rings of higher degree of nilpotence, to normal lattice isomorphisms, and to isomorphisms of the lattices of subsemirings. Lashkhi and Zimmermann [139] found necessary and sufficient conditions for modularity or distributivity of the lattice of subideals in a finite-dimensional Lie algebra over a field of characteristic zero; they considered distributivity for Lie algebras with the transitive relation "to be an ideal". The article [137] of Lashkhi is devoted to analysis of the structure of Lie algebras with a modular lattice over a principal ideal domain, and his survey [138] is devoted to connections between modular lattices and Lie algebras.

Honda [365] considered classes of Lie algebras whose weak subideals form a sublattice in the subalgebra lattice (a subalgebra H of a Lie algebra L is called a weak subideal if there exists a chain $H = M_0 \subseteq M_1 \subseteq \cdots \subseteq M_n = L$ of subspaces such that $[M_{i+1}, H] \subseteq M_i$ for $0 \leq i < n$). Varea [555] showed that in a finite-dimensional Lie algebra L over a field F of characteristic zero all maximal subalgebras are modular elements of the subalgebra lattice if and only if the factor algebra $L/\varphi(L)$ modulo the Frattini ideal $\varphi(L)$ is isomorphic to a subalgebra of the direct product $R \times P_1 \times \cdots \times P_r$, where R is a totally solvable algebra and $P_1, \ldots, P_r$ are pairwise nonisomorphic three-dimensional simple algebras over F.

González and Elduque [339] described finite-dimensional flexible Lie-admissible algebras over a field of characteristic different from two, the subalgebra lattices of each of which coincide with the subalgebra lattice of its commutative algebra.

Herzog [361] observed that Hilbert functions of Noetherian local rings with natural order form an Artinian distributive lattice.

§V.3. Lattices related to semigroups

Johnston [396] characterized lattices isomorphic to congruence lattices of regular Rees matrix semigroups (cf. [OSL, IV, 319]).

Jones [398] proved that every band, whose congruence lattice satisfies the minimality or the maximality condition, is finite. A commutative separative semigroup G can be represented as $G = \bigcup_{\alpha \in \Gamma} G_\alpha$, where Γ is a semilattice and G_α are Archimedean semigroups. Hamilton [353] proved that the lattice Con G is modular (distributive) if and only if Γ is a tree, $\alpha, \beta \in \Gamma$, $\alpha < \beta$, $a \in G_\alpha$ and $b \in G_\beta$ imply $ab = a$, Con G_α is modular (distributive) for all $\alpha \in \Gamma$, and, among all maximal components G_α, at most one does not contain idempotents. He [352] described a fairly wide class of completely regular semigroups with modular congruence lattice. Jones [400] described hypersemisimple semigroups (this means that the semigroup itself and all of its homomorphic images are semisimple) whose congruence lattice is distributive, modular, semimodular, etc. In particular, he found a new proof of the theorem on semimodularity of the congruence lattice of a semilattice. Goberstein [338] proved that a semilattice with zero is primitive (that is, any two nonzero elements are incomparable) if and only if the lattice of its fundamental order relations is modular (or, equivalently, distributive). Sankappanavar [508] suggested a full description of semilattices with distributive congruence lattice. Trueman ([537] and [538]) proved that the congruence lattice of an ideal extension of a group by means of a nil semigroup is upper semimodular. However, for a nonmonogenic free commutative semigroup this is not so. She found certain properties of the congruence lattice of a direct product of monogenic semigroups. Tamura [532] described finite semigroups whose congruence lattice is a chain and which are extensions of congruence-free semigroups. Finite Archimedean semigroups G for which Con G is a chain are exhausted by the following list: (1) finite groups whose normal subgroups form a chain; (2) monogenic nil semigroups; and (3) semigroups consisting of two right or two left zeros [249]. Kozhukhov [111] produced a full description of semigroups whose left congruences form a chain. Eberhart [307] described commutative semigroups whose congruence lattices admit no nontrivial homomorphisms. Jones [402], Edwards [311], Koch and Madison [413] investigated congruence lattices of regular semigroups, Magill ([436] and [437]) considered the congruence lattice of semigroups of continuous transformations, and Eberhart and Williams [309] studied the congruence lattices of orthodox semigroups. Auinger's dissertation [228] was dedicated to semigroups whose congruence lattices are complemented. Tichy [536] considered the algebra $\mathfrak{F}_n$ of all functions on a semigroup G of the form $p(x_1, \ldots, x_n) = a_1 x_{i_1}^{k_1} a_2 x_{i_2}^{k_2} \cdots a_r x_{i_r}^{k_r} a_{r+1}$, where $a_i, x_i \in G$ and $\{i_1, \ldots, i_r\} = \{1, \ldots, n\}$, with respect to pointwise multiplication and the $(n+1)$-ary operation f, where $f(p_1, \ldots, p_{n+1}) = p_1(p_2, \ldots, p_{n+1})$. If G is a commutative semigroup with cancellation and $n \geq 2$, then the lattice Con $\mathfrak{F}_n$ is modular (distributive) if and only if G is a group of finite exponent (a finite cyclic group). If $n = 1$, divisible groups (the additive group of rationals and direct sums of groups of type p^∞ for different p) are added. If G is an arbitrary commutative semigroup, then Con $\mathfrak{F}_n$ is a chain if and only if G is either a semilattice, or $|G| \leq 2$, or (for $n = 1$) G is a cyclic group of prime order. Tanana and Shiryaev [193] described weakly uniform topological semigroups with a semi-Brouwerian lattice of closed subgroups (a lattice L is semi-Brouwerian if, for any a, $b \in L$, the set $\{x | x \in L, a \wedge b = a \wedge x\}$ possesses the greatest element).

Mitsch [452] and Bonzini and Cherubini [250] are surveys devoted to the congruence lattices of semigroups. Pastijn's survey [468] indicates certain properties of the congruence lattice of a regular semigroup.

Sitnikov [182] proved that every complete Boolean algebra is isomorphic to the lattice of stable tolerances of an abelian group and described all semigroups whose lattice of tolerances is Boolean in the classes of commutative semigroups, Clifford semigroups, or semigroups with cancellation. Pondělíček [482] described atoms in the lattice of stable tolerances of a commutative semigroup. He ([483] and [484]) characterized commutative regular and inverse semigroups with distributive or modular lattices of admissible tolerances. He observed that regularity and separability are equivalent conditions for commutative semigroups with admissible lattices of admissible tolerances.

Ershova [85] proved that the lattice of inverse subsemigroups of an inverse semigroup G is upper semimodular if and only if G is a chain of groups with upper semimodular lattices of subgroups. Johnston [395] suggested a description of the lattice of all inverse subsemigroups of a completely simple semigroup. She found when this lattice is modular or distributive. Shiryaev [216] suggested a description of semigroups with lower semidistributive lattice of subsemigroups. Kar Ping Shum [522] established the equivalence of the following properties of a compact topological regular semigroup G whose idempotents are central: (1) the ideals of G form a chain; (2) the idempotents of G form a chain; and (3) G is a primary (or semiprimary) semigroup. Satyanarayana [510] obtained results on semigroups with a linearly ordered set of ideals. Katsman [105] described commutative semigroups for which a semigroup with a dually isomorphic subsemigroup lattice exists. Katsman and Repnitskiĭ [106] described commutative semigroups whose subsemigroup lattice satisfies a nontrivial identity. Every such semigroup is periodic, and the indices of its elements are jointly bounded. A nil semigroup belongs to this class if and only if it is nilpotent, and a semilattice belongs to this class if and only if no free semilattice of finite rank is embeddable in it. As a corollary, it follows that every free lattice is embeddable in the subsemigroup lattice of a semilattice and of a commutative nil semigroup of index 2. Shirjaev [Shiryaev] [521] described semigroups with n-distributive subsemigroup lattice. Studies of the lattice of all regular subsemigroups of a regular semigroup that contain all idempotents of the semigroup were continued in [397] (cf. [OSL, IV, 321, 323]). Szendrei [531] characterized completely regular semigroups whose lattice of idempotent reducts is a chain. Šulka [528] considered lattices connected with nilpotent elements of a semigroup.

A projection of a semigroup S onto a semigroup S' is an isomorphism of $\mathrm{Sub}\, S$ onto $\mathrm{Sub}\, S'$. Ershova [84] proved that every projection of a Brandt semigroup with at least four idempotents is induced by a uniquely determined isomorphism of the semigroups. Jones [401] obtained the same result for a free product of inverse semigroups in the class of all inverse semigroups. Ovsyannikov [161] continued his studies of projections of nil semigroups decomposable in a free product in the variety of all semigroups (cf. [OSL, IV, 130, 133]). Sharkov ([213] and [214]) solved all principal problems connected with projections of completely simple semigroups over various classes of groups and of completely 0-simple semigroups over the one-element group. Borisov [15] refined results of Baranskiĭ [4] on projections of commutative Archimedean semigroups without idempotents. Of Ovsyannikov's results [159] we note the following: a cancellative semigroup with a nontrivial identity, which is not a group, is determined by its subsemigroup lattice if and only if its identity element (if it exists) is not externally adjoined. The same result holds for a nonperiodic cancellative semigroup that is nilpotent in Mal'tsev's sense. In both cases the lattice isomorphism is induced by an isomorphism of the semigroups. Borisov ([14]–[17]) established lattice determinability of certain classes of commutative semigroups. Ershova [84] determined when the inverse subsemigroup lattices of two inverse semigroups are isomorphic. Izbash [92] announced that if the subquasigroup lattices of two Steiner quasigroups are isomorphic, then this isomorphism is induced by an isomorphism of the quasigroups. Shevrin and Ovsyannikov [215] published a detailed survey on subsemigroup lattices.

Baranskiĭ [6] (see also [5]) published a proof of the result mentioned in the previous survey: for every group G, downwards directed set P, and a sufficiently large cardinal number $\mathfrak{m}$, there exist $2^{\mathfrak{m}}$ combinatorial inverse semigroups S such that $|S| = \mathfrak{m}$, $G \cong \operatorname{Aut} S$, and P is isomorphic to the partially ordered set of principal ideals of the semigroup [OSL, IV, 13]. It follows that, for every group G, semigroup T, and sufficiently large cardinal number $\mathfrak{m}$, there exist $2^{\mathfrak{m}}$ combinatorial inverse semigroups S such that $|S| = \mathfrak{m}$, $G \cong \operatorname{Aut} S$, and $I(T) \cong I(S)$. We note that the general concept of independence of derived structures in a given class of algebraic systems developed by the author appears here.

§V.4. Lattices related to other algebraic systems

The problem remains open whether every finite lattice is isomorphic to the congruence lattice of a finite universal algebra. Relatively long ago, Quackenbuch [492] proved that if A is a finite universal algebra with permutable congruences and $\operatorname{Con} A \cong M_n$, then $n = p^k$ for some prime p. Tůma [544] suggested another proof of this fact. In its own turn, the lattice M_{p^k+1} is realized as the subspace lattice of a linear space over the field $GF(p^k)$. Another realization of such lattices was mentioned in [527]. Ihringer [377] proved that the isomorphism $M_n \cong \operatorname{Con} A$, where $|A| < \aleph_0$ and $n \neq p^k + 1$, implies that all congruences of A are principal (cf. [414] and [511]). Ihringer observed that Feit had proved the representability of the lattices M_7 and M_{11} as the congruence lattices of a finite algebra. It follows from results of Pudlák and Tůma [490] that, for every finite lattice L with more than one element, there exists a natural number N such that if there exists an isomorphism $L \cong \mathrm{Con}\, A$ for a finite universal algebra A, then for every $n > N$ there exists a universal algebra B such that $|B| = n$ and $L \cong \mathrm{Con}\, B$. Here the main tools are the so-called regraphs which are studied in detail in a later paper [546] by Tůma. McKenzie [445] proved that a finite lattice L is isomorphic to $\mathrm{Con}\, A$, where A is a finite algebra with a single operation, if and only if L is isomorphic either to the lattice of all subspaces of a finite vector space or to the lattice of all invariant, under permutations, equivalences on a finite set. It follows from this that the subgroup lattice of the alternating group $\mathfrak{A}_4$ cannot be represented as $\mathrm{Con}\, A$, where A is a finite algebra with a single operation, thus solving a long-standing problem. Fried [330] suggested abstract and concrete characterizations of the congruence lattices of an algebra with three 4-ary operations similar to the Mal'tsev operations. Köhler [414], Palfy [466], and Tůma [545] surveyed results related to representations of finite lattices as congruence lattices.

Kogalovskiĭ ([107] and [109]) proved that every algebraic lattice L is isomorphic to $\mathrm{Con}\, G$ for a groupoid G with a new greatest element adjoined (cf. [OSL, IV, 382]), and is also isomorphic to a principal ideal of $\mathrm{Con}\, R$ for a monoid R with zero; also, algebraic lattices and only they are full endomorphic images of the congruence lattices of monoids with left zeros.

Continuing their studies [OSL, IV, 382], Kogalovskiĭ and Soldatova [110] proved:

(1) for every universal algebra A with countably many operations and a left zero there exists an algebra B with one binary and one unary operation such that $\mathrm{Con}\, B \cong \mathrm{Con}\, A$;

(2) for every unary algebra A of finite or countable signature [type] with a fixed point, there exists a 2-unary algebra B such that $\mathrm{Con}\, B \cong \mathrm{Con}\, A$, and B can be chosen finite if A is finite;

(3) for every countable (finite) algebra A there exists a countable (finite) 2-unary algebra B such that $\mathrm{Con}\, B \cong \mathrm{Con}\, A + 1$;

(4) if an algebraic lattice L with a finite or countable set of compact elements is isomorphic to $\mathrm{Con}\, A$ for a universal algebra A with zero, then L is isomorphic to $\mathrm{Con}\, B$ for a 2-unary algebra B.

Nieminen [456] established the distributivity of the congruence lattice of an algebra with a single ternary operation (x, y, z) that satisfies the identities $(x, x, y) = x$ and $((x, y, z), u, v) = ((x, u, v), (y, u, v), z)$. For each element x of such an algebra A an order $\triangleleft_x$ was defined, where $a \triangleleft_x b$ if $b = (x, a, b)$, which turns A into a lower semilattice. Here $a \wedge b = (x, a, b)$. It was found when this semilattice is a tree. In [345] the representation of a finite lattice as the lattice of admissible partitions of a finite automaton was considered. Dassow [277] (cf. [OSL, IV, 332–335]) investigated congruence lattices of the algebra of automaton mappings. Tulipani [539] proved that if A is a model of a countable positive theory T whose models possess the congruence extension property, $|A| > 2^{\aleph_0}$, and Con A has finite length, then, for every infinite cardinality $\mathfrak{m}$, there exists a model B of T such that Con B is isomorphic to a filter of Con A. Lampe's paper [422] mentioned in the previous survey [OSL, IV. 350] has been published (see also [449]). There is also a book [513] with a fairly accessible proof of the theorem on isomorphism of every algebraic lattice to the congruence lattice of a universal algebra. Leeson and Butson [426] studied the congruence lattice of an (m, n) ring.

Uroşu [552] observed that if A is a universal algebra and Con A a complemented modular lattice, then every congruence different from O_A is contained in a maximal congruence. Wroński [561] produced examples of BCK-algebras with nonmodular congruence lattices. This contrasts the distributivity of the congruence lattice of any algebra belonging to a certain variety of BCK-algebras (recall that the class of all BCK-algebras does not form a variety) ([462] and [463]). Such congruence lattices are 3-permutable. If $\wedge$ or $\vee$ appears in the signature of a BCK-algebra A, then Con A is isomorphic to the lattice of all filters of A [373]. Sin Min Lee [424] produced an example showing that, in Hashimoto's theorem on permutability of congruences of a subdirect product A of algebras with permutable congruences, the distributivity requirement for Con A cannot be replaced by the modularity requirement. In [263] properties of the distributive congruence lattice of a universal algebra decomposable in a direct product were found. Huhn [367] analyzed the n-distributivity property of the congruence lattice of a universal algebra. If A is a universal algebra and $C \subseteq \text{Con}\,A$, then a congruence $\theta \in \text{Con}\,A$ is called a C-translation if $\theta = \bigwedge_{p \in C}(\theta \vee p)$. Uroşu ([550] and [551]) found properties of C-translations and determined when all congruences of A are C-translations.

Kulikov [130] showed that a graph from a given quasivariety q of nonoriented graphs without loops is uniquely determined by its order and the lattice of those congruences whose quotient graphs belong to q. Boshchenko ([18] and [19]) described the unars whose congruence lattices are pseudocomplemented, as well as pseudocomplemented lattices isomorphic to the congruence lattices of unars. Palfy [465] proved that if a unary algebra A has proper subalgebras and the length of Con A is 2, then $|\text{Con}\,A| \leq 5$. The articles of Jakubíková-Studenovská ([386] and [387]) are connected with computation of the number of unars and partial unars defined on the same set as a given partial unar A and having the same set of congruences (cf. [OSL, IV, 202–203]). Kurinnoĭ [131] described the sublattices of the lattice Eq X which can be congruence lattices of a distributive lattice with X as the carrier. An analogous result for a semilattice was announced in [83]. Armbrust [226] discussed the language in which those properties of a sublattice L of Eq A

might be described that guarantee the existence of operations on A such that $L = \operatorname{Con} A$ for the algebra A with these operations.

Let Confin A be the set of all finitely generated congruences of a universal algebra A. Clearly, Confin A is an upper semilattice with zero. Blok and Pigozzi [247] proved that a variety $\mathfrak{V}$ had definable principal congruences if and only if Confin A is a semilattice with dual relative pseudocomplements for all $A \in \mathfrak{V}$. In particular, $\mathfrak{V}$ is congruence-distributive. If, in addition, $\mathfrak{V}$ is generated by finite algebras, then it is rationally equivalent to the variety of lattices with dual relative complements. Of Tulipani's results [540] we cite the following: if T is a first order theory, L a finite lattice, and $\mathfrak{A} = \{A | A \in \operatorname{Mod} T, \operatorname{Con} A \cong L\}$, then $\mathfrak{A}$ is closed with respect to ultraproducts if and only if all algebras from $\mathfrak{A}$ have definable compact congruences, and also if and only if there exists a sentence Φ in T such that $\mathfrak{A} = \operatorname{Mod}(T \cup \Phi)$. He ([541] and [542]) considered a theory T whose models all possess definable compact congruences and proved that the cardinality $m = \sup\{|\operatorname{Con} A| | A \in \operatorname{Mod} T\}$ can only be $2^{|A|}$, $|A|$, or a natural number. If T is complete, the third case takes place if and only if there exists a finite lattice L such that $\operatorname{Con} A \cong L$ for all $A \in \operatorname{Mod} T$. In the former two cases the criteria are given under the assumption that T is $\aleph_0$-categorical. Hutchinson [370] introduced a logic for congruences. Elementary formulas of the signature Ω of this logic are triples $\langle f, g, \theta \rangle$, where f and g are elementary formulas of signature Ω and θ is a word in the language of lattice theory. For an algebra A of signature Ω the validity of a formula $\langle f, g, \theta \rangle$ means $f \equiv g \pmod{\theta}$. He explained the basics of the logic for congruences and proved an analog of the completeness theorem (cf. [OSL, IV, 356]).

A congruence class geometry (see [OSL, IV, 286, 311, 375, 468]) is called linear if each of its lines is determined by any of its two points. Ihringer [376] proved that the class geometry of a finite non-congruence-simple algebra A with an operation of arity ≥ 2 is linear if and only if A is polynomially equivalent to a vector space over a finite field with affine operations (that is, $f(x_1, \ldots, x_n) = \lambda_0 + \lambda_1 x_1 + \cdots + \lambda_n x_n$, where λ_i are elements from the ground field). Earlier he [374] described noncongruence-simple groupoids with a linear congruence class geometry. In particular, it turned out that in the class of groupoids with identity such are the elementary abelian groups and only they, while, in the case of semigroups, semigroups of right and left zeros should be added. Smirnov and Reĭbol'd [189] considered the lattice $K(A)$ of all cosets of a universal algebra A over all congruences. They proved that $K(A)$ is simple for any finite regular algebra A with at least three elements. $K(A)$ is not simple if A is a group of type p^∞. They produced sufficient conditions for subdirect irreducibility of $K(A)$.

Skornyakov [184] considered the lattice $\operatorname{Con} A$ for a certain universal algebra A as an ordered left polygon over the monoid $\operatorname{End} A$, where

$$\varphi\theta = \{(a, b) | a, b \in A, (a\varphi, b\varphi) \in \theta\}.$$

In this respect it is natural to call a left polygon L over a monoid R realizable if there exist isomorphisms p and ω of R onto $\operatorname{End} A$ and of the meet-complete semilattice L onto $\operatorname{Con} A$, respectively, such that $\omega(\lambda x) = p(\lambda)\omega(x)$ for all $\lambda \in R$ and $x \in L$. If there exist a universal algebra A, an isomorphism p of R onto $\operatorname{End} A$, and an embedding ω of the set L into $\operatorname{Con} A$, that satisfy the conditions $\omega(\lambda x) = p(\lambda)\omega(x)$, $\omega(\inf_L \Xi) = \inf_{\operatorname{Con} A}(\omega(\Xi))$, $\omega(0) = 0_A$ and $\omega(1) = 1_A$, where $\lambda \in R$, $x \in L$, and $\varnothing \neq \Xi \subseteq L$, then the polygon L is called weakly realizable. Burmeister's results [OSL, IV.222] can be considered as necessary conditions for weak realizability. It turned out that Burmeister's conditions are not sufficient. Necessary and sufficient conditions for realizability were pointed out in the case when L is a three-element chain. The case of a two-element chain was essentially considered by Grätzer [341]. The general case is very far from being resolved.

The research of Michler [450] is conceptually close to the above. He considered a partially ordered right polygon V over a group G, which is an algebraic lattice, and a mapping $\#\colon V \to \operatorname{Sub} G$ that determines a Galois connection between the lattices V and $\operatorname{Sub} G$. If the following conditions hold:

(1) if $\lambda \in x^\#$, then $x\lambda = x$;

(2) $\lambda(x\lambda)^\# \lambda^{-1} \subseteq x^\#$;

(3) $1^\# = \{1\}$;

(4) if X is an upwards directed subset of V and $\Gamma\colon X \to G$ is an arbitrary mapping such that $x^\#\Gamma(x) \cap y^\#\Gamma(y) \neq \varnothing$ for any $x, y \in X$ implies $\bigcap\{x^\#\Gamma(x) | x \in X\} \neq \varnothing$, then there exist a universal algebra A and isomorphisms $\Phi\colon V \to \operatorname{Sub} A$ and $p\colon G \to \operatorname{Aut} A$ such that $\Phi(x)p(\lambda) = \Phi(x\lambda)$, $p(x^\#) = \{\varphi | \varphi \in \operatorname{Aut} A, U \in \Phi(x) \text{ implies } U\varphi = U\}$ for all $x \in V$ and $\lambda \in G$.

Bandelt [234] proved the distributivity of the lattice of compatible tolerances of an algebra with a single ternary operation (x, y, z), where $(x, x, y) = x$ and $((x, y, z), u, v) = (x, (y, u, v), (z, u, v))$ (observe that the class of such algebras is polynomially equivalent to a certain class of semilattices). Zelinka [568] considered when a unary operation can be recovered from the lattice of the tolerances that agree with it. Various aspects of the theory of tolerances on a universal algebra compatible with its operations were reflected in the surveys [252], [332], [439], and [440].

Switching to subalgebra lattices, we note the following results announced by Dulatova and Pinus [81]:

(1) elementary equivalence of the subalgebra lattices of two Boolean algebras implies elementary equivalence of these algebras;

(2) universal theory of subalgebra lattices of the lattices of finite subalgebras of Boolean algebras is decidable;

(3) universal theory of a class of subalgebra lattices [lattices of finite subalgebras] of algebras from a Horn class which contains a discriminator quasipri-

mal algebra is decidable. Koubek [418] clarified when, for a given algebraic lattice L and a signature Ω, there exists an algebra A of signature Ω such that $L \cong \operatorname{Sub} A$, and every subalgebra of A is congruence-simple and has a trivial semigroup of endomorphisms. Uroşu [548] studied ideal lattices of universal algebras. Palasiński [463] established infinite distributivity of the ideal lattice of any BCK-algebra. Deeba [279] found operations on the set of ideals of a BCK-algebra with respect to which this set is a Boolean algebra. Bartol [236] found when a double Brouwerian lattice is isomorphic to the subalgebra lattice of a unar. Gerstmann [336] considered idempotent algebras with n-distributive subalgebra lattices. Salii [178] proved that every complete lattice is isomorphic to the lattice of all suboperatives of a certain P-operative, where a P-operative is a set A with a fixed mapping of $2^A \setminus \{\varnothing\}$ into A as the operation; this P-operative can be chosen in such a way that all its suboperatives are monogenic. Chajda [259] used properties of the lattice of subalgebras of the algebra $A \times A$ containing the diagonal to study varieties with directly indecomposable diagonal subalgebras. Nerode and Remmel [454] is a survey of lattices of recursively enumerable subsystems of various algebraic systems (among them are vector spaces over a recursive field, fields, Boolean algebras, the chain of rational numbers, and others). Mangani and Marcja [438] considered the Boolean algebra $\mathfrak{B}(A)$ of all relatively definable subsets of an algebraic system A. They proved that a countable theory T is $\mathfrak{m}$-categorical for $\mathfrak{m} \geq \aleph_1$ if and only if the Boolean algebras $\mathfrak{B}(A)$, where $A \in \operatorname{Mod} T$ and $|A| = \mathfrak{m}$, are isomorphic. They found properties of $\mathfrak{B}(A)$ equivalent to stability, superstability, and ω-stability of T. Normann [457] proved that the lattice of recursively enumerable degrees of continuous functionals is distributive and contains all finite distributive lattices. Szendrei [530] considered chains in the lattice of clones of operations on a finite set. Danil'chenko ([71] and [72]) considered the lattice of operations on a three-element set. Hwang [372] investigated lattices connected with Latin squares.

§V.5. Lattices related to sets, topologies, categories, etc.

Ježek [393] proved that if X is a countable set, then the equality $|X| = |Y|$ is equivalent to the elementary equivalence of the partition lattices on X and Y. Obviously, every permutation of X induces an automorphism of the partition lattice on X. Hanlon [355] investigated the lattice of fixed points of this automorphism. McNulty [447] studied partition lattices embeddable in lattices of varieties connected with a given variety. Peele [471] suggested estimates for the number of elements of a set of the least cardinality, such that a given finite lattice is representable by its partitions. If X is a subset of the lattice P_n of partitions of the set $\underline{n} = \{1, \ldots, n\}$, then there exists the least lattice among the sublattices of P_n that contain X and coincide with

Con A for some universal algebra A with n as the carrier. This lattice is said to be cl-generated by X. Zádori [566] proved that if $n = 3$ or $n \geq 5$, then P_n is cl-generated by three elements (it is known that, for ordinary generation, four generators are needed). Tůma [543] obtained the following result: if $\varphi_i: L_i \to \Pi$ ($i = 1, 2$) are mappings of finite lattices L_i into the lattice Π of partitions of a finite set X and $f: L_1 \to L_2$ is one-to-one, then an automorphism $\Phi \in \text{Aut}\,\Pi$ such that $\varphi_1(a) = \Phi(\varphi_2(f(a)))$ for every $a \in L_1$ exists if and only if $a < b$, where $a, b \in L_1$, implies $f(a) \leq f(b)$ or $f(a) \| f(b)$. Partition n-lattices (that is, for systems Σ of subsets of a set X such that every n elements of X are contained in precisely one subset in Σ) were investigated by Hsü [366]. The papers of Rousseau [504] and Sha and Kleitman [520] are connected with partition lattices.

Korec [417] made more precise results of his paper [OSL, IV, 346] on distributive sublattices of the lattice $\text{Eq}\,X$ that consists of pairwise permutable equivalences. Kaarli [94] considered analogous topics. Šešelja and Ušan ([519] and [554]) investigated the lattice of $(n+1)$-ary relations, and Kalyuzhnaya and Sinyavskiĭ ([98] and [99]) the lattice of abstract functional relations. It was found in [308] when the lattice of all closed equivalences is complete on a countable or containing a Cantor perfect set separable topological space. Haiman [351] considered essentially sublattices of the lattice $\text{Eq}\,X$ consisting of pairwise permutable congruences (he called such lattices linear) and worked out a general method for proving sentences of the form

$$\forall x_1, \ldots, x_n \left(\bigwedge_{i \in I} (P_i \leq Q_i) \Rightarrow (P \leq Q) \right)$$

in such lattices. New proofs of Desargues' theorem and related statements are among the applications.

Kung [419] defined an elementary event (A, m) on a set X as a system of finite subsets of the set X, whose intersections with a fixed subset A contain precisely m elements; he described fields and rings of sets generated by various elementary events.

A family Σ of subsets of a set X is called a mod-family if $A, B \in \Sigma$ implies $|A \cap B| < \aleph_0$. It was clarified in [237] when a Boolean algebra is isomorphic to a factor algebra of $2^X - I$, where I is the ideal generated by a mod-family that contains all finite subsets. In particular, all countable Boolean algebras have this property and, if Martin's axiom holds, all algebras of cardinality less than $2^{\aleph_0}$. To some extent, [405] is also devoted to the same topic.

The papers [229], [230], and [231] consider lattices of subsets of a set endowed with a measure. Essentially, Seitz [518] considered a partial lattice of subsets. Edelman and Klingsberg [310] considered the lattice of suborders of a given order.

Herrmann ([358] and [360]) established the undecidability of the elementary theory of the lattice of recursively enumerable sets. He [359] found sublattices elementarily definable in the language of the lattice of recursively enumerable sets. Maas [434] characterized those recursively enumerable sets whose lattice of supersets is effectively isomorphic to the lattice of all recursively enumerable sets. The lattice of recursively enumerable sets was studied in [177], [190], [435], and [526].

Cohen and Rubin [265] generalized a known result of Kaplansky on the determinability of a topological space by the lattice of functions on it to the case of ordered topological spaces. Fourneau [325] proved, inter alia, that every maximal ideal of the lattice of continuous functions on a topological space contains all constants. It follows that the lattice of bounded continuous functions has no maximal elements. Lochan and Strauss [431] connected a function $\varphi: Y \to X$ with a lattice homomorphism $\Phi: C(X) \to C(Y)$, where $C(X)$ and $C(Y)$ are lattices of functions on topological spaces X and Y. This function φ is a homomorphism if and only if Φ is an isomorphism. It was determined when Φ is continuous. Muravitskii ([156] and [157]) investigated the pseudo-Boolean algebra of open sets of a topological space. In particular, he found properties of a topological space implying the local finiteness of this pseudo-Boolean algebra. If X is an infinite compact topological space and B a set of its regular open subsets which forms a σ-complete and δ-complemented base for X, then $|B|^\omega = |B|$ [246]. Komiya [416] suggested axioms for convexity based on consideration of a certain closure operator on the lattice of all subsets of a topological space. The article [454] is a survey of results related to the lattice of recursively enumerable open subsets. Veksler ([35] and [36]) considered the lattice of nets of a topological space (that is, complements to the union of a maximal system of pairwise disjoint clopen sets). Mazzocca [442] characterized the lattice of faces of the n-dimensional cube endowed with an additional binary operation of reflection of x with respect to y. Šmarda [523] characterized sets with closure for which the lattice of closed subsets is algebraic and distributive. The lattice of topologies on a given set was considered in [286]–[288], [429], and [535]; the lattice of compactifications in [493]; the lattice of the so-called para-H-closed extensions in [567]; the lattice of topologies in a topos in [251]; and the lattice of ring topologies on the space $C(X)$ in [212].

Hušek [368] proved that for every partially ordered set P without zero in the category $\mathfrak{K}$ of topological spaces there exists a partially ordered set of (mono) coreflexions whose intersection is not a (mono) coreflexion, and which is isomorphic to P. As $\mathfrak{K}$ one can take the category of uniform or proximity spaces. The dual result holds as well. Korotenkov [121] established the modularity of the lattice of so-called I-ideals of a category satisfying certain conditions. Giuli and Tozzi [337] investigated the lattice of subcategories closed under products and extremal subobjects of the topological category.

Weispfenning ([557] and [558]) considered a lattice E consisting of all equivalence classes of existential formulas with respect to a theory T. We note these among this result: (1) E is a Boolean algebra if and only if T

is model complete; and (2) E is a Stone algebra if and only if T possesses a model completion. Analogous results were obtained for other p-algebras. Golunkov and Savel'ev [63] studied subalgebra lattices of algebras of single-place partially recursive functions and predicate algebras, and systems of related algorithmic algebras. Dong Ping Yang [562] clarified when an α-recursive presentable lattice is embeddable in the lattice of so-called recursive degrees. Lattices of degrees were considered in [73], [74], [143], [162], [222], [223], [314], [315], [320], [321], [430], [459], [515]. Cooper [267] introduced a new order on the set of complete theories. Mikheeva [155] investigated lattices connected with multivalued logic. Various aspects of lattices of logics were considered in [78], [298], [302], [316], and [326]. Mal'tsev [142] and Goncharov [64] considered semilattices of numerations. The lattices considered in [100], [101], [102], [196], [257], [354], [423], [489], [534] are also connected with logic.

§V.6. Lattices of varieties

Artamonov [3] analyzed results of investigations of lattices of varieties in the period 1973–1977; and the period from 1978 to June, 1982 was reflected in §7 of the survey "Concrete lattices" of Mikhalev, Salii, and Skornyakov [OSL, IV].

A long paper by Ježek [389]–[391] was devoted to the lattice $L(\Omega)$ of equational theories of type Ω. He described modular elements in this lattice and showed that the set of finitely based equational theories is definable in $L(\Omega)$. If Ω is type (1) or $(1, 0)$, then the automorphism group of $L(\Omega)$ is a symmetric group on a countable set. For every other type $L(\Omega)$ has no "nonobvious" automorphisms. In [392] he investigated $\vee$-irreducible elements of $L(\Omega)$. McNulty [447] studied conditions for embeddability of partition lattices as intervals of the lattice of equational theories. In another paper [448] he showed that, for every finite algebra A that is not finitely based, there exists a finite algebra B with the same property, whose equational theory $Eq(B)$ is properly contained in $Eq(A)$. Let $\mathfrak{M}$ be a minimal variety of algebras of a certain type. Dudek and Płonka [296] found a sufficient condition for the join $\mathfrak{V} \vee \mathfrak{M}$, where $\mathfrak{V}$ is a variety that does not contain $\mathfrak{M}$, to cover $\mathfrak{V}$. Volkov [51] established that every element of the lattice of subvarieties of an arbitrary variety which is a union of the so-called nilpotent varieties of algebras of finite type Ω has a cover (it is assumed that Ω has neither nullary, nor unary operations).

Let $\mathfrak{V}$ be a (quasi)variety of algebras such that every nonidentity congruence on every $\mathfrak{V}$-algebra has a nontrivial class which is a subalgebra. Vernikov and Volkov [42] proved the equivalence of the following conditions for the lattice $\mathfrak{L}(\mathfrak{V})$ of sub(quasi)varieties of such a (quasi)variety:

(1) $\mathfrak{L}(\mathfrak{V})$ is complemented;
(2) $\mathfrak{L}(\mathfrak{V})$ is Boolean; and
(3) $\mathfrak{V}$ is a join of finitely many minimal sub(quasi)varieties.

A (quasi)variety is called hereditarily selfdual if each of its sub(quasi)varieties has the selfdual lattice of sub(quasi)varieties. Vernikov [39] described hereditarily selfdual permutation varieties and hereditarily selfdual (quasi) varieties whose lattice of sub(quasi)varieties satisfies the quasi-identity $((y \wedge z = 0) \Rightarrow ((x \vee y) \wedge z = x \wedge z))$. Gorbunov and Tumanov [66] found a representation of lattices of quasivarieties as lattices of algebraic sets of free systems of countable rank. Wroński [559] investigated splitting of lattices of subquasivarieties.

Gil'man [59] represented the lattice of all subquasivarieties of a locally finite quasivariety of algebraic systems of finite type as the lattice of all closed subsets with respect to a certain algebraic closure operator. Tumanov [208] proved that for every finite distributive lattice L there exists a finitely based locally finite quasivariety $\mathfrak{Q}$ of finite type such that L is isomorphic to the lattice of all $\mathfrak{Q}$-subquasivarieties. McKenzie [445] produced a finite lattice not embeddable in the lattice of subvarieties of a locally finite variety.

Dziobiak [305] found conditions under which the lattice of subquasivarieties of a locally finite semisimple variety is distributive. He [306] constructed a countable family of semiprimal varieties whose lattices of subquasivarieties generate the variety of all lattices. Gil'man ([60] and [61]) characterized ∨-irreducible and irreducible elements in the lattices of universal classes of algebraic systems and announced that every distributive lattice is isomorphic to the lattice of universal subclasses of a universal class generated by a finite set of finite commutative groupoids. García and Taylor [333] considered the following quasi-order on the class of all varieties of algebras of a given type: $\mathfrak{V}_1 \leq \mathfrak{V}_2$ if and only if there exists a functor from $\mathfrak{V}_2$ into $\mathfrak{V}_1$ that commutes with the forgetful functor onto the carrier. The kernel of this quasi-order is an equivalence of varieties in the sense of Mal'tsev. Identifying equivalent varieties we arrive at a complete lattice. It was proved to be nonmodular. Images of the following varieties are ∧-irreducible in it: (1) groups; (2) abelian groups; (3) rings with identity; (4) modular lattices. They studied, in addition, sublattice intervals and abelian groups. Smirnov [188] established that Mal'tsev theories form a continual lattice with respect to inclusion, which is dual to the lattice of nonempty special Mal'tsev classes. Kogalovskiĭ [108] characterized lattices of varieties of algebras with nullary and unary operations and showed that the local structure of these lattices might be arbitrarily complex. He [109] established that, for every irregular variety $\mathfrak{V}$ of unary algebras, whose cardinal number of operations does not exceed $\mathfrak{m}$, there exists an $\mathfrak{m}$-generated monoid R with left zero such that the lattice $\operatorname{Con} R$ is anti-isomorphic to $\mathfrak{L}(\mathfrak{V})$. The converse statement holds too. Ignatov [91] produced an exhaustive description of the lattice of quasivarieties of convexors. Galanter [53] found irreducible elements of the lattice of varieties of pseudo-Boolean algebras. Vazhenin and Sizyĭ [34] described atoms of the lattice of varieties of endographs (that is, of algebraic systems $(A|\varphi, \rho)$, where φ is a unary operation, ρ a unary predicate, and φ is an endomorphism of the graph $(A|p))$. Egorova, Kartashev, and Skornyakov [82] proved that varieties determined by identities of the form $f^i(x) = f^j(x)$ constitute a distributive sublattice in the lattice of varieties of n-layered heterogeneous unars. Kartashov [103] described completely ∨-irreducible elements in the lattice of all quasivarieties of unars. He found criteria for the lattice of subquasivarieties of a quasivariety of unars to be algebraic, distributive, semidistributive, Boolean, Stone, complemented, pseudocomplemented, or a chain. Gil'man [60] established that there are no dual atoms in the lattice of all universal classes of unars. In [58] he investigated coverings in this lattice and described all its chain elements, while in his report [60] he characterized distributive lattices isomorphic to the lattices of subquasivarieties of a suitable quasivariety of unars.

Pigozzi [478] discovered a sublattice isomorphic to the lattice of all finite subsets of a countable set in the ordered set of varieties generated by finite groupoids which are not finitely based. In the lattice of all varieties of groupoids he found intervals consisting entirely of varieties of the above-mentioned form and described the structure of such intervals. Oates-Williams [458] showed the lattice of subvarieties of the variety generated by the Murskiĭ algebra to be continual. Dziobiak [299] constructed a seven-element groupoid such that there exists a continuum of subvarieties containing the Murskii algebra in the variety it generates. Ježek and Kepka [394] clarified the structure of the lattice of varieties of medial cancellative groupoids and the lattice of varieties of commutative medial groupoids. Dudek [295] established that the variety of commutative medial groupoids with the identity $x + ny = x$ is equationally complete if and only if the Mersenne number M_n is prime. Ježek [388] showed that the lattice of varieties of commutative idempotent groupoids has a continual set of atoms.

Sukhanov [192] characterized those varieties of semigroups whose every proper subvariety is catenary. Vernikov and Volkov ([41] and [46]) announced the following theorems:

(1) if $\mathfrak{V}$ is a variety of semigroups and the lattice $\mathfrak{L}(\mathfrak{V})$ is modular, then either $\mathfrak{V}$ is of index ≤ 2 or $\mathfrak{V} = \mathfrak{G} \vee \mathfrak{U}' \vee \mathfrak{U}''$, where $\mathfrak{G}$ is the variety of periodic abelian groups, $\mathfrak{U}' \subseteq [x^2 = x^3, xy = yx]$ and $\mathfrak{U}'' \subseteq [x^2y = xyx = yx^2 = 0]$, or $\mathfrak{V} \subseteq \mathfrak{N} \vee \mathfrak{W}$, where $\mathfrak{N}$ is a nil variety and $\mathfrak{W} = [x = x^2, xy = yx]$;

(2) if $\mathfrak{V}$ is a variety of commutative semigroups, then the lattice $\mathfrak{L}(\mathfrak{V})$ is distributive if and only if $\mathfrak{V}$ satisfies at least one of the identities $x^n y = x^2 y$ ($n \geq 3$), $x^2 y^2 = x^3 y$, $x^2 y^2 = x^4 y$, and $x^3 y = x^3 y^2$;

(3) the following are equivalent for a variety $\mathfrak{V}$ of semigroups:

(1) $\mathfrak{L}(\mathfrak{V})$ is hereditarily selfdual;

(2) $\mathfrak{L}(\mathfrak{U}) \cong C_1 \times \cdots \times C_n$, where C_i are finite chains;

(3) $\mathfrak{L}(\mathfrak{V})$ is modular and $\mathfrak{V}$ is semicatenary (that is, $\mathfrak{U} = \bigvee_{i=1}^{n} \mathfrak{C}_i$, where $\mathfrak{L}(\mathfrak{C}_i)$ is a finite chain and $\mathfrak{C}_i \wedge \mathfrak{C}_j$ is trivial if $i \neq j$);

(4) $\mathfrak{V}$ is contained in a variety of either the form

$$\mathfrak{G}_c \vee [xy = x] \vee [xy = y] \vee \mathfrak{W} \vee [xy = t^2, xy = yx],$$

or the form

$$[x^3y = y, xy = yx] \vee [xy_z = t^3, xy = yx] \vee [xyz = t^2],$$

where $\mathfrak{G}_c$ is a semicatenary group variety.

Biryukov [13] described those varieties $\mathfrak{V}$ of semigroups for which $\mathfrak{L}(\mathfrak{V})$ is a lattice of finite breadth. Gurchenkov [69] pointed out that the lattice of subquasivarieties of nilpotent of degree 3 lattice ordered groups is not distributive. Koryakov [122] studied filters in the lattice of varieties of commutative nilpotent semigroups. Kepka [408] described the structure of an 88-element lattice of varieties of left distributive (that is, satisfying the identity $xyz = xyxz$) semigroups. Torlopova [195] noted certain properties of the lattice of subvarieties of the variety generated by semigroups of rank 2. Goralčík and Koubek [340] showed that a variety $\mathfrak{V}$ is $\vee$-irreducible in the lattice of varieties of idempotent semigroups if and only if every $\mathfrak{V}$-algebra is embeddable in a subdirectly irreducible $\mathfrak{V}$-algebra. Adair [221] constructed the lattice of varieties of idempotent semigroups with involution $x \mapsto x^{-1}$ such that $xx^{-1}x = x$. It is obtained from the four-element Boolean lattice in which the greatest element is replaced by the completed chain of natural numbers.

Petrich and Reilly [477], continuing their investigations of homomorphisms of the lattice $\mathfrak{L}(\mathrm{CS})$ of varieties of completely simple semigroups, showed that the mapping $\mathfrak{V} \mapsto (\mathfrak{V} \wedge \mathfrak{C}, \mathfrak{V} \vee \mathfrak{C})$ where $\mathfrak{C}$ is the variety of all central completely simple semigroups, is an isomorphism of $\mathfrak{L}(\mathrm{CS})$ onto a precisely described subdirect product of $\mathfrak{L}(\mathfrak{C})$ and the interval $[\mathfrak{L}, \mathrm{CS}]$ of $\mathfrak{L}(\mathrm{CS})$. They [476] clarified the structure of certain intervals of the lattice $\mathfrak{L}(\mathfrak{C})$. In their third paper [474] they studied the kernel of a certain homomorphism of $\mathfrak{L}(\mathrm{CS})$ onto the lattice of near-varieties of idempotent generated completely simple semigroups. Rasin [171] clarified the structure of the lattice of subvarieties of the join of the variety of all completely simple semigroups and that of all idempotent semigroups. Pastijn and Trotter [470] obtained general properties of the lattice of varieties of completely regular semigroups. Jones [399] investigated neutral elements of that lattice, while Petrich [473] studied the structure of its "lower floors". Gerhard and Petrich [335] described the lattice of varieties of regular orthogroups. Reilly [495] showed that the lattice of varieties of inverse semigroups has continual $\wedge$-breadth. He and Petrich [475] proved that the lattice of varieties of strictly inverse semigroups is decomposable into a direct product of three lattices of varieties: of rectangular bands, Brandt combinatorial semigroups, and groups.

Rasin [172] showed how the lattice of subvarieties of the variety of orthodox Clifford semigroups, all of whose subgroups belong to a periodic group variety $\mathfrak{V}$, could be constructed from the lattices $\mathfrak{L}(\mathfrak{V})$ and $\mathfrak{L}(J)$, where J is the variety of all idempotent semigroups. Makaridina [141] established that the lattice of varieties of ordered semigroups contains a continuum of elements without covers. On the other hand, she [140] showed that every variety of ordered semigroups determined by semigroup identities has a cover in this lattice. Repnitskiĭ [174] proved that the variety of all l-semigroups is $\vee$-irreducible in the lattice $\mathfrak{L}$ of all varieties of l-semigroups. There are no dual atoms in $\mathfrak{L}$ and every element, except the greatest one, has a cover. He established [173] that the set of covers for every Cross variety of commutative dld-semigroups is finite and is strict. Aĭvazyan [1] considered certain infinite chains of semigroup varieties. Newman [455] described the lattice of varieties of all metabelian groups of exponent 4. Fitzpatrick and Kovacs [323] showed that the variety of nilpotent groups of class 4 whose free groups do not contain elements of even order, form a distributive lattice which, however, is not a sublattice in the lattice of all group varieties. On the other hand, varieties of nilpotent groups of class 4, whose free groups contain elements of even order, form a nondistributive sublattice in the lattice of all group varieties. Sapir [179] showed that the lattice of all quasivarieties of groups is isomorphic to a quotient lattice of a certain sublattice of the lattice of quasivarieties of commutative 3-nilpotent idempotent semigroups. Budkin [24] found a condition under which a quasivariety of groups possesses a cover in the lattice of all quasivarieties of groups. Let $\mathfrak{V}$ be one of the following varieties: all groups, nilpotent groups of class 2, and metabelian groups. He proved that the set of quasivarieties of groups from $\mathfrak{V}$ that contain free $\mathfrak{V}$-groups and have no covers in the lattice of $\mathfrak{V}$-quasivarieties is continual.

Feil [318] constructed a linearly ordered set of varieties of l-groups isomorphic to the segment $[0, 1]$ of the real line. Powell and Tsinakis [485] found a simple proof of the fact that only the variety of all l-groups and the variety of l-groups with subnormal jumps are $\wedge$-irreducible in the lattice Λ of all varieties of l-groups. Reilly [494] considered a subsemilattice of Λ. Repnitskiĭ [175] established that Λ contains a continual interval each element of which has a continuum of covers. Gurchenkov [68] proved that every proper variety of l-groups possesses a cover in Λ. Moreover, all nontrivial proper varieties of l-groups, different from the variety of normal-valued l-groups determined by the identity $(x \vee e)(y \vee e) \leq (y \vee e)^2(x \vee e)^2$, have an infinite number of covers. Holland [362] mentioned varieties of l-groups, found by Scrimger, Medvedev, and Bergman, which cover in Λ the variety of abelian l-groups. Kopytov [118] found another such variety. Medvedev ([151] and [152]) investigated the lattice Λ_0 of 0-residually finite varieties of l-groups. He showed that this lattice is not Brouwerian, contains nonidentity elements without covers and, on the other hand, contains a continuum of elements with an uncountable number of covers. Medvedev

[153] also proved that the lattice Λ^* of quasivarieties of l-groups is not modular. Arora [227] observed that the lattice Λ of all varieties of l-groups is not a complete sublattice of Λ^*. Pringerová [487] established that the ordered set of all semisimple classes of linearly ordered groups is a nonmodular complete lattice without dual atoms; she studied topics connected with coverings in this lattice. Jakubíková [385] showed that the lattice of hereditarily radical classes of linearly ordered groups is Brouwerian, that the class generated by the Archimedean linearly ordered group is an atom in it, and that the set of all principal classes forms an ideal.

Volkov [49] investigated "lower floors" of the lattice of varieties of (associative) rings. In [50] he proved the distributivity of the lattice of subvarieties of the variety of rings determined by an identity of the form $x^2 = x^3 f(x)$, where $f(x) \in \mathbb{Z}[x]$. Volkov and Vernikov ([40], [43], and [44]) considered varieties of associative rings admitting a duality, that is, varieties $\mathfrak{V}$ such that $\mathfrak{L}(\mathfrak{V})$ is dual to $\mathfrak{L}(\mathfrak{V}^*)$ for some variety $\mathfrak{V}^*$. They proved that every such variety is generated by a finite ring. They described varieties whose every subvariety admits a duality, as well as admitting duality varieties consisting of semisimple (in the sense of Jacobson) rings. They used a description of ring varieties whose lattice of subvarieties contains exactly one pair of incomparable elements. Vernikov [38] proved that every ring variety is representable as a finite disjoint join of varieties whose lattice of subvarieties is not decomposable as a direct product of its proper sublattices. He [37] characterized nilpotent ring varieties whose lattice of subvarieties has $\wedge$-breadth 2. Vernikov and Volkov [45] announced the following theorem: if $\mathfrak{V}$ is a variety of associative rings of prime characteristic, and if all nilpotent rings from $\mathfrak{V}$ are commutative, then $\mathfrak{L}(\mathfrak{V})$ is distributive if and only if $\mathfrak{L}(\mathfrak{V}')$ is distributive for every nilpotent variety $\mathfrak{V}' \subseteq \mathfrak{V}$. Romanowska [501] constructed a 16-element lattice of varieties of idempotent distributive semirings (ID-semirings) with a semilattice reduct. An ID-semiring is called normal (rectangular) if both of its reducts are normal (rectangular) bands. Pastijn and Romanowska [469] described the lattice of varieties of normal and the lattice of varieties of rectangular ID-semirings. Bandelt [235] proved that the lattice join of the variety $\mathfrak{R}$ of all rings and $\mathfrak{D}$ of distributive lattices, which are considered as varieties of inverse semirings, coincide with the Mal'tsev product $\mathfrak{R} \cdot \mathfrak{D}$. Mal'tsev [146] described varieties of associative algebras over a finite field with a distributive lattice of subvarieties. Volkov [52] announced that he had constructed two series of almost distributive varieties of associative rings, that is, varieties $\mathfrak{V}$ such that $\mathfrak{L}(\mathfrak{V})$ is not distributive but $\mathfrak{L}(\mathfrak{V}')$ is distributive for all $\mathfrak{V}' \subsetneq \mathfrak{V}$. Zakharova [88] characterized varieties of associative algebras over a Noetherian ring, whose lattice of subvarieties satisfies the maximality condition. Drenski [77] and Stoyanova-Venkova [191] represented certain lattices of varieties of associative algebras. Popov [168] established the distributivity of the lattice of subvarieties of Grassmann algebras with identity of the variety determined by the identities $[x, y, [u, v], z] = 0$ and $[[x, y]^2, y] = 0$. A variety of associative algebras is called monomial if it can be defined by a system of polylinear monomial identities. Mironov [154] proved that finitely based varieties do not form a sublattice in the lattice of monomial varieties, and that varieties without a finite basis for identities do not form a sublattice either.

Let M be a fixed specializing submodule of a free associative algebra. Sverchkov [180] proved that M-separated varieties of Jordan algebras form a lattice with the greatest element. Il'tyakov [93] found a complete description of the lattice of subvarieties of the variety of two-step solvable alternative algebras over a field of characteristic 0 which do not contain a primitive third root of 1. Martirosyan ([148] and [149]) found necessary and sufficient conditions for a variety of alternative (right alternative) algebras to have the distributive lattice of subvarieties. Bakhturin [7] investigated completely the problem of closedness of a set of special varieties of Lie algebras of characteristic 0 with respect to the lattice operations. Ruckelshausen and Sands [505] proved that an infinite finitely generated lattice of width 4 generates a variety which has infinite height as an element of the lattice of lattice varieties. Vas de Carvalho [556] constructed a seven-element lattice of varieties of bounded distributive lattices with a unary operation f satisfying the identities $f(x \wedge y) = f(x) \vee f(y)$ and $f^4(x) = x$, while Blyth and Varlet [248] described a 20-element lattice of varieties of algebras of the same type but with the identities $f(x \wedge y) = f(x) \vee f(y)$, $x \leq f^2(x)$, and $f(1) = 0$. Urquhart [553] established continuality of the lattice of varieties of distributive double p-algebras. Kamara [406] listed all covers of the variety of distributive polarity lattices in the lattice of all varieties of polarity lattices. Palasiński [464] observed that the lattice of varieties of BCK-algebras is distributive, and Romanowska and Traczyk [502] clarified the structure of the completely distributive lattice of subvarieties of the variety of commutative BCK-algebras described by the identity $xy^{n+1} = xy^n$. Wroński [560] found three Heyting algebras generating a quasivariety whose lattice of subquasivarieties is isomorphic to the pentagon N_5. Dziobiak [303] stated a criterion for distributivity of the lattice of subquasivarieties of the variety $\mathfrak{V}$ of Heyting algebras in terms of belonging or not belonging to $\mathfrak{V}$ of certain subsets of a set of 17 finite lattices. He [302] showed that the lattice join of two finitely axiomatizable quasivarieties generated by finite Heyting algebras (or by finite Boolean algebras) need not be finitely axiomatizable. Here is another result of the same author [304]: a quasivariety generated by a finite Sugihara matrix possesses finitely many subquasivarieties. Gil'man [59] studied lattices of subquasivarieties of certain varieties of lattices with a closure operator. Penner [472] showed that the image of the variety of semilattices in the lattice of closed sets of superidentities is the join of the images of the variety of distributive lattices and the variety of abelian groups and is covered by the image of the variety of diagonal semigroups. Jónsson [403] proved a number of theorems on the lattice of varieties of relation algebras in the sense of Tarski.

§V.7. Congruence conditions

A variety $\mathfrak{V}$ of universal algebras is called congruence-distributive [congruence-modular] if the lattice Con A is distributive [modular] for every algebra $A \in \mathfrak{V}$.

Fried and Kiss [331] characterized congruence-distributive varieties in terms of existence of certain polynomials and proved that every variety whose algebras possess restricted equationally definable principal congruences is congruence-distributive (this solves a problem raised by Fried, Grätzer, and Quackenbush [OSL, III, 88]). They introduced a so-called Pixley condition which generalizes the concept of a Mal'tsev condition and proved that congruence-distributivity is a Pixley condition. A special congruence-distributive variety was considered in [330]. Rosenberg [503] found a criterion for functional completeness of an algebra in a congruence-distributive variety. Pinus [165] applied Boolean degrees to studying congruence-distributive varieties of algebraic systems and proved the undecidability of certain theories connected with such varieties and the existence of nonisomorphic but, in some sense equivalent, algebraic systems in such varieties. He [167] announced that the set of isomorphism types of algebras of countable cardinality in a nontrivial congruence-distributive variety is linearly ordered if and only if the variety is generated by a quasiprimal algebra without proper subalgebras. In [166] quasi-orders "...is of the isomorphism type of the subalgebra of..." and "...is an isomorphism type of a homomorphic image of..." are studied on the set of isomorphism types of a congruence-distributive variety.

Lee [425] constructed a simple nonassociative ring that generates a congruence-distributive variety and contains arbitrarily large finite simple algebras. Denecke [285] described finite nontrivial preprimal algebras that have no nontrivial subalgebras and generate congruence-distributive subalgebras. See also Zlatoš [569]. Freese [329] surveyed results connected with Jónsson's theorems on generation of congruence-distributive varieties of universal algebras.

A set Σ of lattice identities is said to imply a lattice congruence-identity ε if in any variety whose congruence lattices of algebras satisfy Σ, all these lattices satisfy ε as well. Czédli [273] showed that if a set Σ of lattice identities implies congruence-distributivity, then there exists a finite subset Σ' of Σ that implies congruence-distributivity as well. He announced that Day and Freese had established an analogous fact for the identity of modularity. Czédli and Freese [275] established that the class of identities, each of which implies congruence-distributivity, as well as the class of identities, each of which implies congruence-modularity, are both recursively defined.

Blok and Pigozzi [247] considered the upper semilattice $Cp(A)$ of all finitely generated congruences on an algebra A and proved that all algebras of a variety $\mathfrak{V}$ have definable principal congruences if and only if $Cp(A)$ possesses dual relative pseudocomplements for every A in $\mathfrak{V}$. They proved that the variety generated by the class $\{\mathrm{Cp}(A) | A \in \mathfrak{V}\}$, for a variety $\mathfrak{V}$, is a variety of distributive lattices with dual relative pseudocomplements if and only if $\mathfrak{V}$ is generated by finite algebras and the principal congruences of all algebras in $\mathfrak{V}$ are definable.

Gumm's monograph [347] is devoted to various topics connected with Mal'tsev-like theorems on congruences of universal algebras. It develops a geometric approach to algebras from congruence-modular varieties, that is, varieties all of whose algebras have modular congruence lattices. The article [348] is close to the book. Freese [329] gave a proof for a version of a generalization of Jónsson's theorem to the case of congruence-modularity which is due to Hrushovsky (unpublished), as well as results on properties of two subdirectly irreducible algebras that generate the same congruence-modular variety. Gumm [347] described a study of the commutation operation for congruences in congruence-modular varieties. Earlier [OSL, III, 111] he gave a Mal'tsev characterization of congruence-modularity in terms of ternary polynomials. This latter result received a new proof by Lakser, Taylor, and Tschantz [420] in which the commutation operation for congruences was not used. Herrmann [359] showed that every congruence-modular variety of algebras whose congruence lattices are complemented for all algebras is polynomially equivalent to a variety of modules over a ring. Kiss [411] described varieties of algebras whose congruence lattices of algebras are modular with complemented principal congruences. Freese [328] introduced and investigated a notion of similarity for subdirectly irreducible algebras of a congruence-modular variety. Repnitskiĭ [176] gave a sufficient condition for every nilpotent locally finite subvariety of a congruence-modular variety of universal algebras of finite type to be finitely based and generated by a finite algebra. Zlatoš [570] associated a ring with every congruence-modular variety of universal algebras and studied its properties. He [569] gave some properties of congruence-modular varieties.

Taylor [533] introduced a concept of term condition and, as an application, produced a new proof of Herrmann's theorem about every abelian algebra in a congruence-modular variety being affine. Biculo and Lazari [245] observed that a concept of a commutator of congruences in a Mal'tsev variety, introduced by Smith, and a corresponding concept for congruence-modular varieties, introduced by Hagemann and Herrmann, coincide for Mal'tsev varieties. McNulty's article [449] contains results on congruence-modular varieties.

Varieties that are both congruence-distributive and congruence-permutable are called arithmetic. Bredikhin [20] proved that for algebras from such varieties the concept of R-isomorphism is reduced to that of isomorphism of semigroups of correspondences which preserves the universal correspondence. Chajda [261] showed that congruence-principal congruence-permutable varieties might be characterized by a sufficiently simple strong Mal'tsev condition which was given in the paper explicitly. His article [260] is close to this topic. Chajda and Duda [264] found necessary and sufficient conditions for permutability and n-permutability of congruences on every algebra of a variety in terms of the least congruences, quasi-orders, tolerances, and subalgebras containing certain ordered pairs. Chajda [260] described varieties with permutable and transferrable congruences.

Uroşu [549] used a ternary Mal'tsev polynomial in an algebra A from a congruence-permutable variety to define a congruence on A^2 and produced elementary properties of this congruence.

Varieties with permutable congruences on their algebras or congruence-permutable varieties are a particular case of congruence-modular varieties. Chajda [262] found conditions of Mal'tsev type that characterize regular arithmetic varieties. Denecke ([283] and [284]) studied groups of automorphisms and weak automorphisms of preprimal algebras which generate arithmetic varieties. Kaarli [95] announced that an arithmetic variety of algebras of finite type is affine complete if and only if it is generated by a single finite algebra that has no proper subalgebras or if and only if it is generated by a finite number of finite algebras that have no proper subalgebras. Schweigert and Szymańska [516] showed that the variety of correlation lattices is arithmetic, while Duda [294] described arithmetic varieties of rings with identity.

Czédli [269] found a weak condition of Mal'tsev type that characterizes varieties all of whose algebras have the congruence lattice satisfying the ∨-semidistributivity condition (thus solving one of Jónsson's problems [OSL, III, 126]) and in [271] found a weak condition of Mal'tsev type that characterizes varieties with ∧-semidistributive congruence lattices. He [272], answering a question asked by Jónsson [OSL, III, 126], gave an algorithm which, for every Horn sentence that holds in congruence lattices of algebras from an n-permutable variety, produces a corresponding Mal'tsev condition. Czédli and Day [274] found an algorithm which associates with every universal Horn sentence that satisfies Whitman's condition a suitable continuously weak Mal'tsev condition.

Duda ([292] and [293]) investigated a variety of algebras possessing the property of direct decomposition for congruences and found sufficiently simple conditions of Mal'tsev type characterizing them.

Czédli [270] showed that certain varieties satisfy the following condition: if the variety Con $\mathfrak{V}$ of lattices generated by congruence lattices of all algebras from $\mathfrak{V}$ does not coincide with the class of all lattices, then Con $\mathfrak{V}$ is modular. Riedel [499] called an algebra congruence uniform if the classes of each of its congruences have the same cardinality. Here are some of his results: if $A_1, \ldots, A_r$ are algebras from a variety all of whose algebras are congruence uniform, $A = A_1 \times \cdots \times A_r$, and θ is the intersection of all maximal congruences of A, then A is n-generated if and only if the algebras $A_1, \ldots, A_r$ and A/θ are n-generated. More precise results are given for congruence-modular varieties.

Smirnov [187] generalized to universal congruence-formulas the known theorem of Pixley-Wille on congruence-quasi-identities. He proved that every disjunctive congruence-identity determines an M_δ-class that does not necessarily have M_δ type. Gorbunov [65] called a quasivariety $\mathfrak{Q}$ of algebras distributive if the lattice of $\mathfrak{Q}$-congruences is distributive for every algebra from $\mathfrak{Q}$; he announced that every minimal distributive locally finite quasivariety of algebras of finite type possesses a finite basis of quasi-identities. Pinus [164] proved that every congruence-distributive Horn class which contains a rigid nontrivial system with a rigid linearly ordered set of principal congruences contains rigid systems of arbitrarily large uncountable cardinality.

Czédli and Lenkehegyi [276] introduced a concept of ordered congruence on an ordered universal algebra and found conditions of Mal′tsev type that characterize classes of ordered algebras closed with respect to subalgebras and direct products, all of whose algebras possess distributive lattices of ordered congruences. Evans and Ganter [312], in the language of derived terms and for an arbitrary variety of algebras, found necessary and sufficient conditions for all algebras to have the modular lattice of subalgebras. Richter [498] introduced concepts of an equivalence-permutable (equivalence-distributive, equivalence-modular) category as a category all of whose objects have permutable equivalences (respectively, form a distributive or modular lattice), and transferred theorems to such categories and described such categories by conditions of Mal′tsev type, thus generalizing theorems of A. I. Mal′tsev, Jónsson, and Day.

Tulipani ([541] and [542]) considered cardinal numbers $L_T(\lambda)$ and $C_T(\lambda)$ defined as the supremum of lengths and cardinalities of congruence lattices of models of cardinality λ of a theory T. Assuming that T possesses only functional symbols (not counting the equality) he proved that only three possibilities can hold for these numbers:

(1) $L_T(\lambda) = C_T(\lambda) = 2^\lambda$;
(2) $\exists n < \aleph_0 \ \forall \lambda \geq \aleph_0 \ ((L_T(\lambda) = n) \ \& \ (C_T(\lambda) = \lambda))$;
(3) $\exists m, n < \aleph_0 \ \forall \lambda \geq \aleph_0 \ ((L_T(\lambda) = m) \ \& \ (C_T(\lambda) = n))$.

If all models of T have definable principal congruences, then case (3) holds if and only if the lattices Con A are isomorphic to one and the same finite lattice L for every model A of T. A criterion for cases (1) and (2) was suggested under assumption that T is $\aleph_0$-categorical. Stability of T is sufficient for case (3). In a sense, the following result from [540] is connected with the above: if $\mathfrak{K}(L, T) = \{A | A \in \text{Mod } T, \text{Con } A \cong L\}$, where L is a fixed finite lattice, then the following are equivalent:

(1) the class $\mathfrak{K}$ is closed under ultraproducts;
(2) all models from $\mathfrak{K}(L, T)$ have definable compact congruences;
(3) there exists a sentence Φ in the language of T such that $\mathfrak{K}(L, T) = \text{Mod}(T \cup \Phi)$.

Bibliography

1. S. V. Aĭvazyan, *Some infinite chains of varieties of groups*, Akad. Nauk Armyan. SSR Dokl. **76** (1983), no. 5, 198–200. (Russian)

2. L. A. Al'shanskiĭ, *On the lattice of varieties of representations of semigroups*, Abstract of the XVIII All-Union Algebraic Conference (Kishinev, 1985), Part I, 1985, p. 15. (Russian)

3. V. A. Artamonov, *Lattices of varieties*, Ordered Sets and Lattices, no. 7, Izdat. Saratov Univ., 1983, pp. 97–106. (Russian)

4. V. A. Baranskiĭ, *Lattice isomorphisms of commutative Archimedean semigroups without idempotents*, Ural. Gos. Univ. Mat. Zap. **9** (1975), no. 3, 8–13. (Russian)

5. _____, *On independence of related structures of algebraic systems*, Izv. Vyssh. Uchebn. Zaved. Mat. **1982**, no. 11, 75–77; English transl. in Soviet Math. (Iz. VUZ) **26** (1982), no. 11.

6. _____, *Independence of automorphism groups and ideal lattices of semigroups*, Mat. Sb. **123** (1984), no. 3, 348–368; English transl. in Math. USSR-Sb. **51** (1985), no. 2.

7. Yu. A. Bakhturin, *Special varieties of Lie algebras*, Algebra i Logika **20** (1981), no. 5, 522–530; English transl. in Algebra and Logic **20** (1981), no. 5.

8. E. L. Bashkirov, *On linear groups over a field that contain a special unitary group of nonzero index*, Abstract of the XVIII All-Union Algebraic Conference (Kishinev, 1985), Part I, 1985, p. 38. (Russian)

9. K. I. Beĭdar, *On atoms in the lattice of radicals*, Abstract of the XVII All-Union Algebraic Conference (Minsk, 1983), Part II, 1983, p. 18. (Russian)

10. _____, *Atoms in the "lattice" of radicals*, Mat. Issled. (1985), no. 85, 21–31. (Russian)

11. K. I. Beĭdar et al., *Associative rings*, Itogi Nauki i Tekhniki: Algebra, Topologiya, Geometriya, vol. 22, VINITI, Moscow, 1984, pp. 3–115; English transl. in J. Soviet Math., vol. 38, 1987.

12. K. I. Beĭdar and K. Salavova, *The lattices of N radicals, left strong radicals, and left hereditary radicals*, Acta Math. Hungar. **42** (1983), no. 1–2, 81–95. (Russian)

13. A. P. Biryukov, *On semigroup varieties of finite width*, Abstract of the XVIII All-Union Algebraic Conference (Kishinev, 1985), Part I, 1985, p. 54. (Russian)

14. A. A. Borisov, *Lattice determinacy of a class of commutative semigroups without idempotents*, Izv. Vyssh. Uchebn. Zaved. Mat. **1982**, no. 12, 11–16; English transl. in Soviet Math. (Iz. VUZ) **26** (1982), no. 12.

15. _____, *Commutative semigroups that are lattice-isomorphic to commutative Archimedean semigroups without idempotents*, Izv. Vyssh. Uchebn. Zaved. Mat. **1984**, no. 9, 14–20; English transl. in Soviet Math. (Iz. VUZ) **28** (1984), no. 9.

16. _____, *Lattice closedness of the class of all commutative separative semigroups without idempotents*, Preprint No. 4021-84, deposited at VINITI by the editors of Sibirsk. Mat. Zh., 1984. (Russian)

17. _____, *Lattice isomorphisms of commutative semigroups without idempotents*, Izv. Vyssh. Uchebn. Zaved. Mat. **1986**, no. 2, 54–56; English transl. in Soviet Math. (Iz. VUZ) **30** (1986), no. 2.

18. A. P. Boshchenko, *On the congruence lattice of unars*, Abstract of the XVII All-Union Algebraic Conference (Minsk, 1983), Part II, 1983, p. 28. (Russian)

19. _____, *A characterization of pseudocomplemented lattices isomorphic to the congruence lattices of unars*, Abstract of the XVIII All-Union Algebraic Conference (Kishinev, 1985), Part I, 1985, p. 66. (Russian)

20. D. A. Bredikhin, *Bands of correspondences and R-isomorphisms of algebras with permutable congruences*, Teoriya Polugrupp i ee Prilozheniya, No. 7, Izdat. Saratov Univ., Saratov, 1984, pp. 4–9. (Russian)

21. G. M. Brodskiĭ, *Dualism in modules and the $AB5^*$ condition*, Uspekhi Mat. Nauk **38** (1983), no. 2, 201–202; English transl. in Russian Math. Surveys **38** (1983), no. 2.

22. _____, *Hom functors and lattices of submodules*, Trudy Moskov. Mat. Obshch. **46** (1983), 164–186; English transl. in Trans. Moscow Math. Soc. **1984**, no. 2.

23. _____, *On the least cogenerator as a module over its ring of endomorphisms*, Abstract of the XVIII All-Union Algebraic Conference (Kishinev, 1985), Part I, 1985, p. 66. (Russian)

24. A. I. Budkin, *Quasivarieties of groups without coverings*, Mat. Zametki **37** (1985), no. 5, 609–616; English transl. in Math. Notes **37** (1985), no. 5–6.

25. A. S. Bulgak, *The Möbius function of the lattice of subgroups of a finite group*, Moskov. Inst. Inzh. Zheleznodorozh. Transporta Trudy (1982), no. 653, 103–109. (Russian)

26. _____, *On the Möbius function on the lattices of subgroups in certain finite groups*, Abstract of the IX All-Union Theory of Groups Symposium (Moscow, 1984), 1984, pp. 84–85. (Russian)

27. I. D. Bunu and E. I. Tebyrce [Tèbyrtse], *The distributivity of a lattice of pretorsions*, Abelian Groups and Modules, Tomsk. Gos. Univ., Tomsk, 1984, pp. 3–10. (Russian)

28. N. A. Vavilov, *Subgroups of the general linear group over a ring that contains the group of block triangular matrices*. II, Vestnik Leningrad. Univ. Mat. Mekh. Astronom. **1982**, no. 13, 5–9; English transl. in Vestnik Leningrad Univ. Math. **15** (1983).

29. _____, *On subgroups of the unitary group over a semilocal ring*, Uspekhi Mat. Nauk **37** (1982), no. 4, 147–148; English transl. in Russian Math. Surveys **37** (1982), no. 4.

30. _____, *Parabolic subgroups of the Chevalley group over a commutative ring*, Zap. Nauchn. Sem. Leningrad. Otdel. Mat. Inst. Steklov. (LOMI) **116** (1982), 20–43; English transl. in J. Soviet Math. **26** (1984), no. 3.

31. _____, *Subgroups of a special linear group containing the group of diagonal matrices*, Abstract of the IX All-Union Theory of Groups Symposium (Moscow, 1984), 1984, pp. 182–183. (Russian)

32. _____, *Subgroups of the special linear group which contain the group of diagonal matrices*. I, Vestnik Leningrad. Univ. Mat. Mekh. Astronom. **1985**, no. 4, 3–7; English transl. in Vestnik Leningrad Univ. Math. **18** (1985).

33. N. A. Vavilov and E. V. Dybkova, *Subgroups of the general symplectic group containing the group of diagonal matrices*. II, Zap. Nauchn. Sem. Leningrad. Otdel. Mat. Inst. Steklov. (LOMI) **132** (1983), 44–56; English transl. in J. Soviet Math. **30** (1985), no. 1.

34. Yu. M. Vazhenin and S. V. Sizyĭ, *Atoms of the lattice of quasivarieties of primitive endographs*, Abstract of the XVIII All-Union Algebraic Conference (Kishinev, 1985), Part I, 1985, p. 77. (Russian)

35. A. I. Veksler, *Nets in topological spaces*, Izv. Vyssh. Uchebn. Zaved. Mat. **1984**, no. 3, 25–33; English transl. in Soviet Math. (Iz. VUZ) **28** (1984), no. 3.

36. _____, *Nets in zero-dimensional spaces*, Izv. Vyssh. Uchebn. Zaved. Mat. **1985**, no. 1, 17–25; English transl. in Soviet Math. (Iz. VUZ) **29** (1985), no. 1.

37. B. M. Vernikov, *On varieties of associative rings of d-width* 2, Ural. Gos. Univ. Mat. Zap. **13** (1983), no. 3, 16–26. (Russian)

38. _____, *Varieties of associative rings whose lattice of subvarieties is indecomposable into a direct product*, Ural. Gos. Univ. Mat. Zap. **13** (1984), no. 4, 3–10. (Russian)

39. _____, *Hereditarily self-dual varieties and quasivarieties*, Izv. Vyssh. Uchebn. Zaved. Mat. **1984**, no. 9, 25–29; English transl. in Soviet Math. (Iz. VUZ) **28** (1984), no. 9.

40. _____, *Self-dual varieties of associative rings*, Ural. Gos. Univ. Mat. Zap. **14** (1985), no. 1, 31–37. (Russian)

41. _____, *Hereditarily selfdual varieties of semigroups*, Abstract of the XVIII All-Union Algebraic Conference (Kishinev, 1985), Part I, 1985, p. 90. (Russian)

42. B. M. Vernikov and M. V. Volkov, *Complements in lattices of varieties and quasivarieties*, Izv. Vyssh. Uchebn. Zaved. Mat. **1982**, no. 11, 17–20; English transl. in Soviet Math. (Iz. VUZ) **26** (1982), no. 11.

43. _____, *Dualities in lattices of varieties of associative rings*, Izv. Vyssh. Uchebn. Zaved. Mat. **1984**, no. 9, 66–69; English transl. in Soviet Math. (Iz. VUZ) **28** (1984), no. 9.

44. _____, *Dualisms in lattices of varieties of associative rings*, Ural. Gos. Univ. Mat. Zap. **13** (1984), no. 4, 11–38. (Russian)

45. _____, *On distributive varieties of associative rings*, Report No. 516, Algebraic Systems Seminar (Sverdlovsk, 1985). (Russian)

46. _____, *On varieties of semigroups with the modular lattice of subvarieties*, Report No. 535, Algebraic Systems Seminar (Sverdlovsk, 1985). (Russian)

47. E. M. Vechtomov, *Boolean rings*, Mat. Zametki **39** (1986), no. 2, 182–185; English transl. in Math. Notes **39** (1986), no. 1–2.

48. E. A. Vodyanyuk, *Lattice-complementary torsions and radical-semisimple classes*, Preprint No. 40462-85, deposited at VINITI, 1985. (Russian)

49. M. V. Volkov, *Lower levels of a lattice of varieties of associative rings*, Ural. Gos. Univ. Mat. Zap. **13** (1983), no. 3, 27–38. (Russian)

50. _____, *Distributivity of certain lattices of varieties of associative rings*, Sibirsk. Mat. Zh. **25** (1984), no. 6, 23–30; English transl. in Siberian Math. J. **25** (1984), no. 6.

51. _____, *On the join of varieties*, Simon Stevin **58** (1984), no. 4, 311–317.

52. _____, *On distributive varieties of associative rings*, Report No. 529, Algebraic Systems Seminar (Sverdlovsk, 1985). (Russian)

53. G. I. Galanter, *On irreducible elements of the lattice of varieties of pseudo-Boolean algebras*, Abstract of the XVIII All-Union Algebraic Conference (Kishinev, 1985), Part I, 1985, p. 106. (Russian)

54. A. G. Geĭn, *Levels of solvability of lattice-isomorphic Lie algebras*, Ural. Gos. Univ. Mat. Zap. **13** (1982), no. 1, 7–15. (Russian)

55. _____, *Modular subalgebras and projections of locally finite-dimensional Lie algebras of characteristic* 0, Ural. Gos. Univ. Mat. Zap. **13** (1983), no. 3, 39–51. (Russian)

56. _____, *The distributive law in a lattice of subalgebras*, Serdica **11** (1985), no. 2, 171–179. (Russian)

57. _____, *On Lie algebras with a restriction on subalgebras*, Abstract of the XVIII All-Union Algebraic Conference (Kishinev, 1985), Part I, 1985, p. 111. (Russian)

58. E. A. Gil'man, *On the lattice of universal classes of unars*, Preprint No. 6186-82, deposited at VINITI, 1982. (Russian)

59. _____, *Lattices of subquasivarieties of locally finite quasivarieties of algebraic systems*, Ordered Sets and Lattices, No. 8, Izdat. Saratov Univ., Saratov, 1982, pp. 22–32. (Russian)

60. _____, *Indecomposable elements of lattices of universal classes*, Izv. Vyssh. Uchebn. Zaved. Mat. **1983**, no. 12, 58–60; English transl. in Soviet Math. (Iz. VUZ) **27** (1983), no. 12.

61. _____, *An abstract characterization of finite lattices of universal algebras*, Abstract of the XVIII All-Union Algebraic Conference (Kishinev, 1985), Part I, 1985, p. 114. (Russian)

62. I. Z. Golubchik, *Normal subgroups of the linear and unitary groups over associative rings*, Spaces over Algebras, and Some Problems in the Theory of Nets, Bashkir. Gos. Ped. Inst., Ufa, 1985, pp. 122–142. (Russian)

63. Yu. V. Golunkov and A. A. Savel'ev, *A lattice of systems of algorithmic algebras of partial recursive functions and predicates*, Izv. Vyssh. Uchebn. Zaved. Mat. **1984**, no. 11, 57–59; English transl. in Soviet Math. (Iz. VUZ) **28** (1984), no. 11.

64. S. S. Goncharov, *Positive numerations of families with single-valued numerations*, Algebra i Logika **22** (1983), no. 5, 481–488; English transl. in Algebra and Logic **22** (1983), no. 5.

65. V. A. Gorbunov, *On the finite basedness of minimal distributive quasivarieties*, Abstract of the XVIII All-Union Algebraic Conference (Kishinev, 1985), Part I, 1985, p. 127. (Russian)

66. V. A. Gorbunov and V. I. Tumanov, *Construction of lattices of quasivarieties*, Mathematical Logic and the Theory of Algorithms, Trudy Inst. Mat., vol. 2, "Nauka" Sibirsk. Otdel., Novosibirsk, 1982, pp. 12–44. (Russian)

67. S. Ya. Grinshpon, *Structure of completely characteristic subgroups of torsion-free groups*, Abelian Groups and Modules, Tomsk. Gos. Univ., Tomsk, 1981, pp. 56–92. (Russian)

68. S. A. Gurchenkov, *Coverings in the lattice of l-varieties*, Mat. Zametki **35** (1984), no. 5, 677–684; English transl. in Math. Notes **35** (1984), no. 5–6.

69. _____, *On theory of l-groups*, Abstract of the XVIII All-Union Algebraic Conference (Kishinev, 1985), Part I, 1985, p. 152. (Russian)

70. O. V. Damova, *The lattice of maximal subgroups of a finite group*, Sibirsk. Mat. Zh. **23** (1982), no. 6, 74–79; English transl. in Siberian Math. J. **23** (1982), no. 6.

71. A. F. Danil'chenko, *Some properties of the lattice of parametrically closed classes of functions of three-valued logic*, Abstract of the VII All-Union Conference on Math. Logic (Novosibirsk, 1984), 1984, p. 53. (Russian)

72. _____, *On Boolean sublattices of the lattice of bicentralizers of algebras of cardinality* 3, Abstract of the XVIII All-Union Algebraic Conference (Kishinev, 1985), Part I, 1985, p. 156. (Russian)

73. A. N. Dëgtev, *Semilattices of disjunctive and linear degrees*, Mat. Zametki **38** (1985), no. 2, 310–316, 350; English transl. in Math. Notes **38** (1985), no. 1–2.

74. _____, *Relations between reducibilities of tabular type*, Algebra i Logika **22** (1983), no. 3, 829-840; English transl. in Algebra and Logic **22** (1983), no. 3.

75. Dzuan Tam Guè, *The minimum number of generators of a lattice of submodules of a semisimple module*, Modules and Algebraic Groups, vol. 2, Zap. Nauchn. Sem. Leningrad. Otdel. Mat. Inst. Steklov. (LOMI) **132** (1983), 110-113; English transl. in J. Soviet Math. **30** (1985), no. 1.

76. _____, *The best estimate for the minimal number of generators of the lattice* $L(Fq, n)$, Preprint No. 6271-85, deposited at VINITI, 1985. (Russian)

77. V. Drenski, *Lattices of varieties of associative algebras*, Serdica **8** (1982), no. 1, 20-31. (Russian)

78. Ya. M. Drugush, *Union of logics modelled by finite trees*, Algebra i Logika **21** (1982), no. 2, 149-161; English transl. in Algebra and Logic **21** (1982), no. 2.

79. N. I. Dubrovin, *Chain rings*, Uspekhi Mat. Nauk **37** (1982), no. 4, 139-140; English transl. in Russian Math. Surveys **37** (1982), no. 4.

80. _____, *An example of a chain primitive with nilpotent elements*, Mat. Sb. **120** (1983), no. 3, 441-447; English transl. in Math. USSR-Sb. **48** (1984), no. 2.

81. Z. A. Dulatova and A. G. Pinus, *On the subalgebra lattice of certain classes of algebras*, Abstract of the XVIII All-Union Algebraic Conference (Kishinev, 1985), Part I, 1985, p. 178. (Russian)

82. D. P. Egorova, V. K. Kartashev, and L. A. Skornyakov, *The variety of heterogeneous unars*, Algebraic Systems, Ivanov. Gos. Univ., Ivanovo, 1981, pp. 122-133. (Russian)

83. V. D. Ermakova and G. Ch. Kurinnoĭ, *On representations of equivalences by congruences of a semilattice*, Abstract of the XVIII All-Union Algebraic Conference (Kishinev, 1985), Part I, 1985, p. 187. (Russian)

84. T. I. Ershova, *Projection of Brandt semigroups*, Ural. Gos. Univ. Mat. Zap. **13** (1982), no. 1, 27-39. (Russian)

85. _____, *Inverse semigroups with upper semimodular lattice of inverse subsemigroups*, Preprint No. 3544-83, deposited at VINITI, 1983. (Russian)

86. A. E. Zalesskiĭ, *Linear groups*, Itogi Nauki i Tekhniki: Algebra, Topologiya, Geometriya, vol. 21, VINITI, Moscow, 1983, pp. 135-182; English transl. in J. Soviet Math., vol. 31, 1985.

87. A. P. Zamyatin, *Decidability of the elementary theories of certain varieties of rings*, Ural. Gos. Univ. Mat. Zap. **13** (1983), no. 3, 52-74. (Russian)

88. E. N. Zakharova, *Varieties of associative algebras with the maximality condition for subvarieties*, Izv. Akad. Nauk Moldav. SSR Ser. Fiz.-Tekhn. Mat. Nauk **1983**, no. 2, 10-15. (Russian)

89. A. A. Ivanov, *Theory of subgroup lattices of torsion-free Abelian groups of rank 1*, Preprint No. 4932-83, deposited at VINITI by the editors of Sibirsk. Mat. Zh., 1983. (Russian)

90. S. G. Ivanov, *L-homomorphisms of locally solvable torsion-free groups*, Mat. Zametki **37** (1985), no. 5, 627-635; English transl. in Math. Notes **37** (1985), no. 5-6.

91. V. V. Ignatov, *Quasivarieties of convexors*, Izv. Vyssh. Uchebn. Zaved. Mat. **1985**, no. 7, 12-14; English transl. in Soviet Math. (Iz. VUZ) **29** (1985), no. 7.

92. V. I. Izbash, *On lattice isomorphic quasigroups*, Abstract of the XVIII All-Union Algebraic Conference (Kishinev, 1985), Part I, 1985, p. 213. (Russian)

93. A. V. Il'tyakov, *Lattice of subvarieties of a variety of two-set solvable alternative algebras*, Algebra i Logika **21** (1982), no. 2, 170-177; English transl. in Algebra and Logic **21** (1982), no. 2.

94. K. K. Kaarli, *Compatible function extension property*, Algebra Universalis **17** (1983), no. 2, 200-207.

95. _____, *On affine completeness of varieties*, Abstract of the XVIII All-Union Algebraic Conference (Kishinev, 1985), Part I, 1985, p. 221. (Russian)

96. M. I. Kabenyuk, *Discreteness of lattices of closed subgroups of Lie groups*, Abstract of the IX All-Union Theory of Groups Symposium (Moscow, 1984), 1984, pp. 205-206. (Russian)

97. V. V. Kalinin, *Subspaces of a vector space*, Probabilistic Methods and Cybernetics, no. 19, Kazan. Gos. Univ., Kazan, 1983, pp. 26-29. (Russian)

98. S. A. Kalyuzhnaya, *Lattice of abstract functional dependency relations*, Vestnik Khar'kov. Univ. (1985), no. 277, 91-101. (Russian)

99. S. A. Kalyuzhnaya and V. V. Sinyavskiĭ, *Lattices of sets of right parts of the relation of abstract multivalued dependence*, Mat. Metody Analiz. Dinamicheskikh Sistem (1984), no. 8, Kharkov, 166–171. (Russian)

100. M. I. Kanovich, *Implicativity of the lattice of truth-table degrees of algorithmic problems*, Dokl. Akad. Nauk SSSR **270** (1983), no. 5, 1045–1050; English transl. in Soviet Math. Dokl. **27** (1983), no. 3.

101. M. I. Kanovich, *Complexity and convergence of algorithmic mass problems*, Dokl. Akad. Nauk SSSR **272** (1983), no. 2, 289–293; English transl. in Soviet Math. Dokl. **28** (1983), no. 2.

102. _____, *Reducibility via general recursive operators*, Dokl. Akad. Nauk SSSR **273** (1983), no. 4, 793–796; English transl. in Soviet Math. Dokl. **28** (1983), no. 3.

103. V. K. Kartashov, *Lattices of quasivarieties of unars*, Sibirsk. Mat. Zh. **26** (1985), no. 3, 49–62; English transl. in Siberian Math. J. **26** (1985), no. 3.

104. _____, *A characterization of distributive lattices of varieties of unars*, Abstract of the XVIII All-Union Algebraic Conference (Kishinev, 1985), Part I, 1985, p. 234. (Russian)

105. S. I. Katsman, *Commutative semigroups with self-dual lattice of subsemigroups*, Semigroup Forum **18** (1979), no. 2, 119–161.

106. S. I. Katsman and V. V. Repnitskiĭ, *Commutative semigroups the lattice of whose subsemigroups satisfies a nontrivial identity*, Report No. 533, Algebraic Systems Seminar (Sverdlovsk, 1985). (Russian)

107. S. R. Kogalovskiĭ, *A remark on congruence lattices of universal algebras*, Algebraic Systems, Ivanov. Gos. Univ., Ivanovo, 1981, pp. 153–157. (Russian)

108. _____, *On lattices of varieties of unary algebras*, Preprint No. 5280-84, deposited at VINITI, 1984. (Russian)

109. _____, *Lattices of equational theories of unary algebras*, Preprint, Ivanov. Gos. Univ., Ivanovo, 1986. (Russian)

110. S. R. Kogalovskiĭ and V. V. Soldatova, *Remarks on congruence lattices of universal algebras*, Studia Sci. Math. Hungar. **25** (1990), no. 1–2, 33–43. (Russian)

111. I. V. Kozhukhov, *Left chain semigroups*, Semigroup Forum **22** (1981), no. 1, 1–8.

112. V. A. Koĭbaev, *Subgroups of the general linear group containing a group of elementary block-diagonal matrices*, Vestnik Leningrad. Univ. Mat. Mekh. Astronom. **1982**, no. 13, 33–40; English transl. in Vestnik Leningrad Univ. Math. **15** (1983).

113. _____, *Subgroups of a special linear group over the field of five elements containing the group of diagonal matrices*, Abstract of the IX All-Union Theory of Groups Symposium (Moscow, 1984), 1984, pp. 210–211. (Russian)

114. _____, *Subgroups of a special linear group over the field of four elements containing the group of diagonal matrices*, Abstract of the XVIII All-Union Algebraic Conference (Kishinev, 1985), Part I, 1985, p. 264. (Russian)

115. Yu. A. Komarov, *Local structure of the lattice of subgroups of a topological group*, Dokl. Akad. Nauk Ukrain. SSR Ser. A **1983**, no. 1, 18–20. (Russian)

116. V. M. Kopytov, *Ordered groups*, Itogi Nauki i Tekhniki: Algebra, Topologiya, Geometriya, vol. 19, VINITI, Moscow, 1981, pp. 3–29; English transl. in J. Soviet Math., vol. 23, 1983.

117. _____, *Lattice-ordered groups*, "Nauka", Moscow, 1984. (Russian)

118. _____, *A nonabelian variety of lattice-ordered groups in which every solvable l-group is abelian*, Mat. Sb. **126** (1985), no. 2, 247–266; English transl. in Math. USSR-Sb. **126** (1985), no. 1.

119. S. S. Korobkov, *Lattice isomorphisms of algebraic algebras without nilpotent elements*, Ural Gos. Univ. Mat. Zap. **14** (1985), no. 1, 75–84. (Russian)

120. _____, *Associative rings with a dense lattice of subrings*, Abstract of the XVIII All-Union Algebraic Conference (Kishinev, 1985), Part I, 1985, p. 277. (Russian)

121. Yu. G. Korotenkov, *Universally normal categories*, Preprint No. 1386-85, deposited at VINITI, 1985. (Russian)

122. I. O. Koryakov, *A sketch of the lattice of commutative nilpotent semigroup varieties*, Semigroup Forum **24** (1982), no. 4, 285–317.

123. A. A. Kravchenko, *The minimal number of generators of the lattice of subspaces of a finite-dimensional linear space*, Zap. Nauchn. Sem. Leningrad. Otdel. Mat. Inst. Steklov. (LOMI) **114** (1982), 148–149; English transl. in J. Soviet Math. **27** (1984), no. 4.

124. S. L. Krupetskiĭ, *On certain subgroups of a unitary group over a quadratic extension of an ordered Euclidean field*, Algebra and Number Theory, Nal'chik, 1979, pp. 39–48.

125. ____, *Intermediate subgroups in the unitary group over the quaternion algebra*, Zap. Nauchn. Sem. Leningrad. Otdel. Mat. Inst. Steklov. (LOMI) **116** (1982), 96–101; English transl. in J. Soviet Math. **26** (1984), no. 3.

126. ____, *On linear groups containing the unitary group*, Abstract of the IX All-Union Theory of Groups Symposium (Moscow, 1984), 1984, pp. 214–215. (Russian)

127. ____, *On subgroups of a unitary group containing a maximal nonsplittable torus*, Abstract of the XVIII All-Union Algebraic Conference (Kishinev, 1985), Part I, 1985, p. 287. (Russian)

128. V. A. Kuzicheva, *Recursive endomorphisms of countable vector spaces with recursive operations*, Preprint No. 4175-84, deposited at VINITI, 1984. (Russian)

129. ____, *Groups of recursive linear transformations of countable vector spaces*, Preprint No. 6353-84, depostied at VINITI, 1984. (Russian)

130. N. A. Kulikov, *Definability of graphs by lattices of congruence*, Algebra i Logika **24** (1985), no. 1, 13–25; English transl. in Algebra and Logic **24** (1985), no. 1.

131. G. Ch. Kurinnoĭ, *Which equivalence systems are systems of congruences of distributive lattices*, Izv. Vyssh. Uchebn. Zaved. Mat. **1985**, no. 12, 67–68; English transl. in Soviet Math. (Iz. VUZ) **29** (1985), no. 12.

132. K. M. Kutyev, *A subsemigroup lattice characterization of a linearly ordered group isomorphic to a subgroup of real numbers*, Preprint No. 3617-82, deposited at VINITI, 1982. (Russian)

133. ____, *Certain SL-criteria for partially ordered groups*, Preprint No. 3618-82, deposited at VINITI, 1982. (Russian)

134. ____, *Subsemigroups in a group*, Preprint No. 1384-84, deposited at VINITI, 1984. (Russian)

135. A. A. Lashkhi, *Projection of wreath products of Lie algebras*, Atti Accad. Naz. Lincei Rend. Cl. Sci. Fis. Mat. Natur. (8) **69** (1980), no. 6, 313–316. (Italian summary)

136. ____, *Projections of mixed Lie rings*, Universal Algebra and Applications (Warsaw, 1978), Banach Center Publ., 9, PWN, Warsaw, 1982, pp. 57–66.

137. ____, *Lie algebras with a modular lattice of subalgebras*, Soobshch. Akad. Nauk Gruzin. SSR **118** (1985), no. 2, 277–280. (Russian)

138. ____, *Lattices with a modular identity and Lie algebras*, Itogi Nauki i Tekhniki: Sovremennye Problemy Mat.: Noveĭshie Dostizheniya, vol. 26, VINITI, Moscow, 1985, pp. 213–257; English transl. in J. Soviet Math., vol. 38, 1987.

139. A. A. Lashkhi and I. A. M. Zimmermann, *Modularität und Distributivität im Subidealverband einer Lie-Algebra*, Rend. Sem. Mat. Univ. Padova (1985), no. 73, 169–177.

140. V. A. Makaridina, *Coverings of varieties of ordered semigroups of semigroup identities*, Associative Actions, Leningrad. Gos. Ped. Inst., Leningrad, 1983, pp. 76–81. (Russian)

141. ____, *Coverings in the lattice varieties of ordered semigroups*, Algebraic Systems with One Action and Relation, Leningrad. Gos. Ped. Inst., Leningrad, 1985, pp. 69–73. (Russian)

142. An. A. Mal'tsev, *Upper semilattices of numerations*, Sibirsk. Mat. Zh. **23** (1982), no. 4, 122–136; English transl. in Siberian Math. J. **23** (1982), no. 4.

143. ____, *Structure of the semilattice of tt1-degrees*, Sibirsk. Mat. Zh. **26** (1985), no. 2, 132–139; English transl. in Siberian Math. J. **26** (1985), no. 2.

144. Yu. N. Mal'tsev, *Basis rank of varieties of associative algebras*, Serdica **10** (1984), no. 4, 442–448. (Russian)

145. ____, *Varieties of algebras whose critical algebras are arithmetic*, Sibirsk. Mat. Zh. **25** (1984), no. 1, 204–207; English transl. in Siberian Math. J. **25** (1984), no. 1.

146. ____, *Distributive varieties of associative algebras*, Mat. Issled. (1984), no. 76, 73–98. (Russian)

147. V. T. Markov et al., *Rings of endomorphisms and lattice of submodules*, Itogi Nauki i Tekhniki: Algebra, Topologiya, Geometriya, vol. 21, VINITI, Moscow, 1983, pp. 183–254; English transl. in J. Soviet Math., vol. 31, 1985.

148. V. D. Martirosyan, *On the distributivity of lattices of subvarieties of varieties of alternative algebras*, Mat. Sb. **118** (1982), no. 1, 118–137; English transl. in Math. USSR-Sb. **46** (1983), no. 1.

149. _____, *Distributivity of lattices of subvarieties of varieties of right-alternative algebras*, Akad. Nauk Armyan. SSR Dokl. **78** (1984), no. 5, 199–202. (Russian)

150. N. Ya. Medvedev, *The lattice of radicals of finitely generated l-groups*, Math. Slovaca **33** (1983), no. 2, 185–188. (Russian)

151. _____, *Coverings in a lattice of l-varieties*, Algebra i Logika **22** (1983), no. 1, 53–60; English transl. in Algebra and Logic **22** (1983), no. 1.

152. _____, *The lattice of o-approximable l-varieties*, Czechoslovak Math. J. **34(109)** (1984), no. 1, 6–17. (Russian)

153. _____, *Free products of l-groups*, Algebra i Logika **23** (1984), no. 5, 493–511; English transl. in Algebra and Logic **23** (1984), no. 5.

154. V. V. Mironov, *Union of monomial varieties of algebras*, Mat. Zametki **35** (1984), no. 6, 789–794; English transl. in Math. Notes **35** (1984), no. 5–6.

155. E. A. Mikheeva, *On atoms of the lattice of closed classes of many-valued logic*, Abstract of the XVIII All-Union Algebraic Conference (Kishinev, 1985), Part II, 1985, p. 35. (Russian)

156. A. Yu. Mutavitskiĭ, *On pseudo Boolean algebras of topological space*, Abstract of the XVII All-Union Algebraic Conference (Minsk, 1983), Part II, 1983, pp. 164–165. (Russian)

157. _____, *On a local representation of pseudo Boolean algebras of topological spaces*, Abstract of the XVIII All-Union Algebraic Conference (Kishinev, 1985), Part II, 1985, p. 48. (Russian)

158. Yu. N. Mukhin, *Subgroup lattices of topological radical groups*, Abstract of the XVIII All-Union Algebraic Conference (Kishinev, 1985), Part II, 1985, p. 53. (Russian)

159. A. Ya. Ovsyannikov, *Lattice isomorphisms of nonperiodic semigroups with cancellation laws*, Izv. Vyssh. Uchebn. Zaved. Mat. **1980**, no. 3, 32–44; English transl. in Soviet Math. (Iz. VUZ) **24** (1980), no. 3.

160. _____, *Determinability by the lattice of semigroups of nonperiodic groups decomposable into a direct product*, Ural. Gos. Univ. Mat. Zap. **13** (1982), no. 1, 98–101. (Russian)

161. _____, *Structural isomorphisms of semigroups decomposable into a free product in certain varieties of nilsemigroups*, Ural. Gos. Univ. Mat. Zap. **13** (1984), no. 4, 92–103. (Russian)

162. R. Sh. Omanadze, *The upper semilattice of recursively enumerable Q degrees*, Algebra i Logika **23** (1984), no. 2, 175–184; English transl. in Algebra and Logic **23** (1984), no. 2.

163. G. V. Pivovarova, *Lattice pairs*, Algebra and Discrete Mathematics, Latv. Gos. Univ, Riga, 1984, pp. 81–95.

164. A. G. Pinus, *The spectrum of rigid systems of Horn classes*, Sibirsk. Mat. Zh. **22** (1981), no. 5, 153–157; English transl. in Siberian Math. J. **22** (1981), no. 5.

165. _____, *Applications of Boolean powers of algebraic systems*, Sibirsk. Mat. Zh. **26** (1985), no. 3, 117–125, 225; English transl. in Siberian Math. J. **26** (1985), no. 3.

166. _____, *Imbeddability and epimorphism relations on congruence-distributive varieties*, Algebra i Logika **24** (1985), no. 5, 588–607, 621–622; English transl. in Algebra and Logic **24** (1985), no. 5.

167. _____, *On epimorphism relations on countable algebras of a congruences-distributive variety*, Abstract of the XVIII All-Union Algebraic Conference (Kishinev, 1985), Part II, 1985, p. 96. (Russian)

168. A. P. Popov, *Identities of the tensor square of a Grassmann algebra*, Algebra i Logika **21** (1982), no. 4, 442–471; English transl. in Algebra and Logic **21** (1982), no. 4.

169. I. V. Protasov and Yu. V. Tsybenko, *Chabauty's topology in the lattice of closed subgroups*, Ukrain. Mat. Zh. **36** (1984), no. 2, 207–213; English transl. in Ukrainian Math. J. **36** (1984), no. 2.

170. V. Ramaswami, *Idempotent elements in associative triple systems*, Publ. Math. Debrecen **31** (1984), no. 3–4, 265–270. (Russian)

171. V. V. Rasin, *On the varieties of Cliffordian semigroups*, Semigroup Forum **23** (1981), no. 3, 201–220.

172. _____, *Varieties of orthodox Clifford semigroups*, Izv. Vyssh. Uchebn. Zaved. Mat. **1982**, no. 11, 82–85; English transl. in Soviet Math. (Iz. VUZ) **26** (1982), no. 11.

173. V. B. Repnitskiĭ, *Cross and almost Cross varieties of commutative dld-semigroups*, Ural. Gos. Univ. Mat. Zap. **13** (1982), no. 1, 102–116. (Russian)

174. _____, *On varieties of l-semigroups*, Izv. Vyssh. Uchebn. Zaved. Mat. **1982**, no. 11, 54–58; English transl. in Soviet Math. (Iz. VUZ) **26** (1982), no. 11.

175. _____, *On covering elements in the lattice of varieties of l-groups*, Preprint No. 5944-83, deposited at VINITI, 1983. (Russian)

176. _____, *On nilpotent varieties of algebras*, Abstract of the XVIII All-Union Algebraic Conference (Kishinev, 1985), Part II, 1985, p. 127. (Russian)

177. A. A. Savel′ev, *A means of description of maximal subsystems of algebras*, Probabilistic Methods and Cybernetics, no. 18, Kazan. Gos. Univ., Kazan, 1982, pp. 77–83. (Russian)

178. V. N. Salii, *P-operatives with monogenic P-suboperatives*, Teoriya Polugrupp i ee Prilozheniya (1983), Izdat. Saratov Univ., Saratov, 45–48. (Russian)

179. M. V. Sapir, *On lattices of quasivarieties of semigroups and groups*, Ural. Gos. Univ. Mat. Zap. **13** (1984), no. 4, 124–133. (Russian)

180. S. R. Sverchkov, *Special varieties of Jordan algebras*, Preprint No. 34, Inst. Mat. Sibirsk. Otdel. Akad. Nauk SSSR, 1983. (Russian)

181. S. M. Seĭtenov, *Elementary theory of lattices of subfields of finite fields*, Topics in Algebraic Number Theory and Constructive Models, Alma-Ata, 1985, pp. 71–80. (Russian)

182. V. M. Sitnikov, *On semigroups with a Boolean algebra of stable tolerances*, Ural. Gos. Univ. Mat. Zap. **13** (1983), no. 3, 146–158. (Russian)

183. A. N. Skiba, *On distributive lattices of formations of finite groups*, Abstract of the IX All-Union Theory of Groups Symposium (Moscow, 1984), 1984, p. 122. (Russian)

184. L. A. Skornyakov, *The congruence lattice as an act over the endomorphism monoid*, Lectures in Universal Algebra (Szeged, 1983), Colloq. Math. Soc. János Bolyai, vol. 43, North-Holland, Amsterdam and New York, 1986, pp. 469–496.

185. V. D. Smirnov, *Semisimple rings, all of whose subrings have duals*, Ural. Gos. Univ. Mat. Zap. **13** (1983), no. 3, 159–171. (Russian)

186. _____, *Monogenic nilpotent rings possessing duals*, Preprint No. 5312-85, deposited at VINITI, 1985. (Russian)

187. D. M. Smirnov, *On universal determinability of Mal′cev classes*, Algebra i Logika **21** (1982), no. 5, 721–738; English transl. in Algebra and Logic **21** (1982), no. 5.

188. _____, *The lattice of Mal′cev theories*, Algebra i Logika **23** (1984), no. 3, 296–304; English transl. in Algebra and Logic **23** (1984), no. 3.

189. D. M. Smirnov and A. V. Reĭbol′d, *Lattices of congruence classes of algebras*, Algebra i Logika **23** (1984), no. 6, 684–701; English transl. in Algebra and Logic **23** (1984), no. 6.

190. B. Solon, *pc-degrees inside an e-degree of a hyperimmune retractable set*, Algebraic Systems, Ivanov. Gos. Univ., Ivanovo, 1981, pp. 203–217. (Russian)

191. A. N. Stoyanova-Venkova, *Some lattices of varieties of associative algebras defined by identities of fifth degree*, C. R. Acad. Bulgare Sci. **35** (1982), no. 7, 867–868. (Russian)

192. E. V. Sukhanov, *Almost-linear varieties of semigroups*, Mat. Zametki **32** (1982), no. 4, 469–476; English transl. in Math. Notes **32** (1982), no. 3–4.

193. B. P. Tanana and V. M. Shiryaev, *On topological semigroups with semi-Brouwerian lattices of closed semigroups*, Abstract of the XVIII All-Union Algebraic Conference (Kishinev, 1985), Part II, 1985, p. 197. (Russian)

194. G. N. Titov, *Commutativity of certain torsion-free FS-groups*, Abstract of the IX All-Union Theory of Groups Symposium (Moscow, 1984), 1984, p. 69. (Russian)

195. N. G. Torlopova, *On semigroups of rank* 2, Preprint No. 5590-82, deposited at VINITI, 1982. (Russian)

196. P. I. Trofimov, *Predicates, lattices, and Boolean rings*, Preprint No. 3389-85 deposited at VINITI, 1985. (Russian)

197. A. A. Tuganbaev, *Distributive modules and rings*, Uspekhi Mat. Nauk **39** (1984), no. 1, 157–158; English transl. in Russian Math. Surveys **39** (1984), no. 1.

198. _____, *Distributive rings*, Mat. Zametki **35** (1984), no. 3, 329–332; English transl. in Math. Notes **35** (1984), no. 3–4.

199. _____, *Right distributive rings*, Izv. Vyssh. Uchebn. Zaved. Mat. **1985**, no. 1, 46–51; English transl. in Soviet Math. (Iz. VUZ) **29** (1985), no. 1.

200. _____, *Rings with plane right ideals and distributive rings*, Mat. Zametki **38** (1985), no. 2, 218–228; English transl. in Math. Notes **38** (1985), no. 1–2.

201. _____, *Rings with the distributive ideal*, Abelian Groups and Modules, Tomsk. Gos. Univ., Tomsk, 1985, pp. 88–104. (Russian)

202. _____, *Distributive Bezout rings*, Vīsnik Kiïv. Unīv. Ser. Mat. Mekh. (1985), no. 27, 109–110. Ukrainian

203. _____, *Distributive rings*, Abstract of the XVIII All-Union Algebraic Conference (Kishinev, 1985), Part II, 1985, p. 218. (Russian)

204. _____, *Distributive rings of series*, Mat. Zametki **39** (1986), no. 4, 518–528; English transl. in Math. Notes **39** (1986), no. 3–4.

205. _____, *Rings in which the lattice of right ideals is distributive*, Izv. Vyssh. Uchebn. Zaved. Mat. **1986**, no. 2, 44–49; English transl. in Soviet Math. (Iz. VUZ) **30** (1986), no. 2.

206. _____, *Rings with a distributive lattice of right ideals*, Uspekhi Mat. Nauk **41** (1986), no. 3, 203–204; English transl. in Russian Math. Surveys **41** (1986), no. 3.

207. _____, *Distributive rings and endodistributive modules*, Ukrain. Mat. Zh. **38** (1986), no. 1, 63–67; English transl. in Ukrainian Math. J. **38** (1986), no. 1.

208. V. I. Tumanov, *Finite distributive lattices of quasivarieties*, Algebra i Logika **22** (1983), no. 2, 168–181; English transl. in Algebra and Logic **22** (1983), no. 2.

209. N. D. Filippov, *Projections of projective geometries*, Colloq. on Ordered Sets (Szeged, 1985), p. 10.

210. M. Ya. Finkel'shteĭn, *Rings in which the annihilators make up a sublattice of a lattice of ideals*, Sibirsk. Mat. Zh. **24** (1983), no. 6, 160–167; English transl. in Siberian Math. J. **24** (1983), no. 6.

211. P. A. Freidman, *Nilrings without torsion with a modular lattice of subrings*, Ural. Gos. Univ. Mat. Zap. **13** (1982), no. 1, 133–137. (Russian)

212. N. I. Shakenko, *Topological rings of continuous real-valued functions*, Uspekhi Mat. Nauk **37** (1982), no. 5, 207–208; English transl. in Russian Math. Surveys **37** (1982), no. 5.

213. Yu. A. Sharkov, *Lattice isomorphisms of completely simple semigroups*, Preprint No. 3413-81, deposited at VINITI, 1981. (Russian)

214. _____, *Lattice isomorphisms of completely 0-simple semigroups over a one-element group*, Preprint No. 4673-83, deposited at VINITI, 1983. (Russian)

215. L. N. Shevrin and A. Ya. Ovsyannikov, *Semigroups and their subsemigroup lattices*, Semigroup Forum **27** (1983), no. 1, 1–154.

216. V. M. Shiryaev, *Semilattices with semidistributive lattices of subsemilattices*, Vestnik Beloruss. Gos. Univ. Ser. I Fiz. Mat. Mekh. (1985), no. 1, 61–64. (Russian)

217. M. I. Èĭdinov, *On the lattice of subformations*, Abstract of the IX All-Union Theory of Groups Symposium (Moscow, 1984), 1984, p. 136. (Russian)

218. E. M. Yakovenko, *Topological groups of finite width*, Ural. Gos. Univ. Mat. Zap. **13** (1982), no. 1, 138–146. (Russian)

219. B. V. Yakovlev, *Lattice determinability of groups from a certain class*, Abstract of the IX All-Union Theory of Groups Symposium (Moscow, 1984), 1984, pp. 256–257. (Russian)

220. _____, *On lattice determinability of projective special groups over a ring*, Abstract of the XVIII All-Union Algebraic Conference (Kishinev, 1985), Part II, 1985, p. 305. (Russian)

221. C. L. Adair, *Bands with evolution*, J. Algebra **75** (1982), no. 2, 297–314.

222. K. Ambos-Spies, *On the structure of polynomial time degrees*, Lecture Notes in Comput. Sci., vol. 166, Springer-Verlag, Berlin and New York, 1984, pp. 198–208.

223. _____, *An extension of the nondiamond theorem in classical and α-recursion theory*, J. Symbolic Logic **49** (1984), no. 2, 586–607.

224. M. Anderson, P. Bixler, and P. Conrad, *Vector lattices with no proper a-subspaces*, Arch. Math. (Basel) **41** (1983), no. 5, 427–433.

225. G. Archinard, *Submodules of a torsion free and finitely generated module over a Dedekind ring*, Colloq. Math. **48** (1984), no. 2, 193–204.

226. M. Armbrust, *Equivalence relations versus unary operations*, Z. Math. Logik Grundlag. Math. **29** (1983), no. 6, 569–571.

227. A. K. Arora, *Quasivarieties of lattice ordered groups*, Algebra Universalis **20** (1985), no. 1, 34–50.

228. K. Auinger, *Halbgruppen mit komplementärem Kongruenzverband*, Dissertation, Univ. Wien, 1982.

229. G. Bachman and P. D. Stratigos, *Criteria for σ-smoothness, τ-smoothness and tightness of lattice regular measures with applications*, Canad. J. Math. **33** (1981), no. 6, 1498–1525.

230. _____, *Lattice repleteness and some of its applications to topology*, J. Math. Anal. Appl. **99** (1984), no. 2, 472–493.

231. _____, *A general measure decomposition theorem by means of the generalized Wallman remainder*, J. Austral. Math. Soc. Ser. A **36** (1984), no. 1, 87–105.

232. K. A. Baker, *Nondefinability of projectivity in lattice varieties*, Algebra Universalis **17** (1983), no. 3, 267–274.

233. B. Banaschewski and R. Harting, *Lattice aspects of radical ideals and choice principles*, Proc. London Math. Soc. **50** (1985), no. 3, 385–404.

234. H.-J. Bandelt, *Tolerances on median algebras*, Czechoslovak Math. J. **33(108)** (1983), no. 3, 344–347.

235. _____, *Free objects in the variety generated by rings and distributive lattices*, Lecture Notes in Math., vol. 998, Springer-Verlag, New York and Berlin, 1983, pp. 255–260.

236. W. Bartol, *Subalgebra lattices of monounary algebras*, Algebra Universalis **12** (1981), no. 1, 66–69.

237. J. E. Baumgartner and M. Weese, *Partition algebras for almost disjoint families*, Trans. Amer. Math. Soc. **274** (1982), no. 2, 619–630.

238. J. T. B. Beard and R. M. McConnel, *Matrix field extensions*, Acta Arith. **41** (1982), no. 3, 213–221.

239. M. Bell, *Two Boolean algebras with extreme cellular and compactness properties*, Canad. J. Math. **35** (1983), no. 5, 824–838.

240. M. K. Bennett, *Lattices of convex sets*, Trans. Amer. Math. Soc. **234** (1977), no. 1, 279–288.

241. _____, *Affine geometry: A lattice characterization*, Proc. Amer. Math. Soc. **88** (1983), no. 1, 21–26.

242. _____, *Separation conditions on convexity lattices*, Lecture Notes in Math., vol. 1149, Springer-Verlag, New York and Berlin, 1985, pp. 22–36.

243. Cl. Bernardi and G. Mazzanti, *Different types of congruences in direct products*, J. Algebra **74** (1982), no. 1, 96–111.

244. C. Bessenrodt, H.-H. Brungs, and G. Törner, *Prime ideals in right chain rings*, Mitt. Math. Sem. Giessen (1984), no. 163, 141–167.

245. I. Bicudo and H. Lazari, *The commutator in universal algebra*, Rev. Mat. Estatist **1** (1983), 1–5. (Portuguese. English summary)

246. A. Błaszczyk, *On the power of lattices of regular open sets*, Bull. Polish Acad. Sci. Math. **32** (1984), no. 11, 635–642.

247. W. J. Blok and D. Pigozzi, *On the structure of varieties with equationally definable principal congruences*. I, Algebra Universalis **15** (1982), no. 2, 195–227.

248. T. S. Blyth and J. C. Varlet, *Subvarieties of the class of MS-algebras*, Proc. Roy. Soc. Edinburgh Sect. A **95** (1983), no. 1–2, 157–169.

249. C. Bonzini and A. Cherubini, *Sui Δ-semigruppi di Putcha*, Istit. Lombardo Accad. Sci. Lett. Rend. A **114** (1980), 179–194.

250. _____, *On the lattice of congruences of a semigroup*, Proc. Conf. on Near-Rings and Near-Fields, San Benedetto del Tronto, 1981, Parma, 1982, pp. 135–141.

251. F. Borceaux and G. Van den Bossche, *Structure des topologies d'un topos*, Cahiers Topologie Géom. Différentielle **25** (1984), no. 1, 37–39.

252. N. Both and I. Purdea, *Tolerances*, Preprint, No. 2, Babes-Bolyai Univ. Fac. Math. Res. Semin., 1982, pp. 8-38.

253. U. Brehm, *Untermodulverbände torsionfreier Moduln*, Freiburg, 1983.

254. H. H. Brungs and G. Törner, *Right chain rings and the generalized semigroup of divisibility*, Pacific J. Math. **97** (1981), no. 2, 293–305.

255. _____, *Extensions of chain rings*, Math. Z. **185** (1984), no. 1, 93–104.

256. D. Bures, *Modularity in the lattice of projections of a von Neumann algebra*, Canad. J. Math. **36** (1984), no. 6, 1021–1030.

257. G. Burosch et al., *On subalgebra of predicates*, Elektron. Informationsverarb. Kybernet. **21** (1985), no. 1–2, 9–22.

258. V. P. Camillo, *Inducing lattice map by semilinear isomorphisms*, Rocky Mountain J. Math. **14** (1984), no. 2, 475–486.

259. I. Chajda, *Varieties with directly decomposable diagonal subalgebras*, Ann. Univ. Sci. Budapest. Eötvös Sec. Math. **25** (1982), 193–201.

260. _____, *Coherence, regularity and permutability of congruences*, Algebra Universalis **17** (1983), no. 2, 170–173.

261. _____, *A Mal'cev condition for congruence principal permutable varieties*, Algebra Universalis **19** (1984), no. 3, 337–340.

262. _____, *Regularity in arithmetical varieties*, Arch. Math. (ČSSR) **20** (1984), no. 4, 177–182.

263. _____, *Transferable tolerances and weakly tolerance regular lattices*, Lectures in Universal Algebra (Szeged, 1983), Colloq. Math. Soc. János Bolyai, vol. 43, North-Holland, Amsterdam and New York, 1986, pp. 27–40.

264. I. Chajda and J. Duda, *Finitely generated relations and their applications to permutable and n-permutable varieties*, Comment. Math. Univ. Carolin. **23** (1982), no. 1, 41–54.

265. M. Cohen and M. Rubin, *Lattices of continuous monotonic functions*, Proc. Amer. Math. Soc. **86** (1982), no. 4, 685–691.

266. P. M. Cohn, *Ringe mit distributivem Faktorverband*, Abh. Braunschweig. Wiss. Ges. **33** (1982), 35–40.

267. G. R. Cooper, *On complexity of complete first-order theories*, Z. Math. Logik Grundlag. Math. **28** (1982), no. 2, 93–136.

268. G. Czédli, *On properties of rings that can be characterized by infinite lattice identities*, Studia Sci. Math. Hungar. **16** (1981), no. 1–2, 45–60.

269. _____, *A Mal'cev-type condition for the semidistributivity of congruence lattices*, Acta Sci. Math. (Szeged) **43** (1981), no. 3/4, 267–272.

270. _____, *An application of Mal'cev type theorems to congruence varieties*, Universal Algebra (Esztergom, 1977), Colloq. Math. Soc. János Bolyai, vol. 29, North-Holland, Amsterdam and New York, 1982, pp. 169–171.

271. _____, *A characterization for congruence semidistributivity*, Lecture Notes in Math., vol. 1004, Springer-Verlag, Berlin and New York, 1983, pp. 104–110.

272. _____, *Mal'cev conditions for Horn sentences with congruence permutability*, Acta Math. Hungar. **44** (1984), no. 1–2, 115–124.

273. _____, *A note on the compactness of the consequence relation for congruence varieties*, Algebra Universalis **15** (1982), no. 1, 142–143.

274. G. Czédli and A. Day, *Horn sentence with (W) and weak Mal'cev conditions*, Algebra Universalis **19** (1984), no. 2, 217–230.

275. G. Czédli and R. Freese, *On congruence distributivity and modularity*, Algebra Universalis **17** (1983), no. 2, 216–219.

276. G. Czédli and A. Lenkehegyi, *On congruence n-distributivity of ordered algebras*, Acta Math. Hungar. **41** (1983), no. 1–2, 17–26.

277. J. Dassow, *On the congruence lattice of algebras of automation mappings*, Finite Algebra and Multiple-Valued Logic (Szeged, 1979), Colloq. Math. Soc. János Bolyai, vol. 28, North-Holland, Amsterdam and New York, 1981, pp. 161–182.

278. J. E. Dawson, *The projective geometry arising from a hollow module*, J. Austral. Math. Soc. Ser. A **37** (3), no. 1984, 351–357.

279. E. Y. Deeba, *On the algebra of ideals of BCK-algebra*, Math. Japon. **30** (1985), no. 3, 383–391.

280. F. de Giovanni and S. Franciosi, *On submodular infinite groups*, Rend. Circ. Mat. Palermo (2) **31** (1982), no. 2, 257–266. (Italian. English summary)

281. _____, *Isomorphisms between subnormal structures of groups*, Ann. Mat. Pura Appl. (4) **137** (1984), 123–138. (Italian. English summary)

282. A. Delandtsheer, *Finite geometric lattices with highly transitive automorphism groups*, Arch. Math. (Basel) **42** (1984), no. 4, 376–383.

283. K.-D. Denecke, *Schwache Automorphismen präprimaler Algebren, die arithmetische Varietäten erzeugen*, Diskrete Mathematik und ihre Anwendungen in der Mathematischen Kybernetik, Rostock. Math. Kolloq. (1978), no. 10, 23–35.

284. _____, *Präprimale Algebren, die arithmetische Varietäten erzeugen*, Universal Algebra and Applications (Warsaw, 1978), Banach Center Publ., 9, PWN, Warsaw, 1982, pp. 391–398.

285. _____, *Eine algebraische Charakterisierung einer Klasse präprimaler Algebren*, Rostock. Math. Kolloq. (1983), no. 23, 43–53.

286. G. Di Maio and A. Russo, *On the lattice of topologies over a set and over some of its sublattices*, Ricerche Mat. **30** (1981), no. 1, 133–144. (Italian)

287. _____, *Some lattice properties of T_1^α-topologies and of one of their generalizations*, Rend. Accad. Sci. Fis. Mat. Napoli (4) **48** (1980/81), 565–573. (Italian. English summary)

288. _____, *Lattice of T_1-topologies weaker than a given T_1-topology and their T_1-complementarity*, Rend. Mat. (7) **3** (1983), no. 4, 775–786. (Italian. English summary)

289. R. G. Downey, *Abstract dependence, recursion theory, and the lattice of recursively enumerable filters*, Bull. Austral. Math. Soc. **27** (1983), no. 3, 461–464.

290. _____, *Some remarks on a theorem of Iraj Kalantari concerning convexity and recursion theory*, Z. Math. Logik Grundlag. Math. **30** (1984), no. 4, 295–302.

291. _____, *Co-immune subspaces and complementation in V_∞*, J. Symbolic Logic **49** (1984), no. 2, 528–538.

292. J. Duda, *A Mal'cev characterization of n-permutable varieties with directly decomposable congruences*, Algebra Universalis **16** (1983), no. 3, 269–274.

293. _____, *Directly decomposable compatible relations*, Glas. Mat. Ser. III **19** (1984), no. 2, 225–229.

294. _____, *Polynomial pairs characterizing principality*, Lectures in Universal Algebra (Szeged, 1983), Colloq. Math. Soc. János Bolyai, vol. 43, North-Holland, Amsterdam and New York, 1986, pp. 109–122.

295. J. Dudek, *Medial groupoids and Mersenne numbers*, Fund. Math. **114** (1981), no. 2, 109–112.

296. J. Dudek and J. Płonka, *On covering in lattices of varieties of algebras*, Bull. Polish Acad. Sci. Math. **31** (1983), no. 1–2, 1–4.

297. M. R. Dull and A. J. Saalfeld, *A note on projectivities*, J. Algebra **96** (1985), no. 2, 603–607.

298. W. Dzik, *On the content of lattices of logics. I*, Rep. Math. Logic (1981), no. 13, 17–27.

299. W. Dziobiak, *A variety generated by a finite algebra with $2^{\aleph_0}$ subvarieties*, Polish Acad. Sci. Inst. Philos. Sociol. Bull. Sect. Logic **9** (1980), no. 1, 2–9.

300. _____, *A variety generated by a finite algebra with $2^{\aleph_0}$ subvarieties*, Algebra Universalis **13** (1981), no. 2, 148–156.

301. _____, *On infinite subdirectly irreducible algebras in locally finite equational classes*, Algebra Universalis **13** (1981), no. 3, 393–394.

302. _____, *Cardinalities of proper ideals in some lattices of strengthenings of the intuitionistic propositional logic*, Proceedings of the Finnish-Polish-Soviet logic conference (Polanica Zdrój, 1981), Studia Logica **42** (1983), no. 2–3, 173–177.

303. _____, *On distributivity of the lattice of subquasivarieties of a variety of Heyting algebras*, Polish Acad. Sci. Inst. Philos. Sociol. Bull. Sect. Logic **12** (1983), no. 1, 37–40.

304. _____, *Quasivariety generated by a finite Sugihara structure has finitely many subquasivarieties*, Polish Acad. Sci. Inst. Philos. Sociol. Bull. Sect. Logic **12** (1983), no. 1, 27–31.

305. _____, *On distributivity of the lattice of subquasivarieties of a locally finite semisimple arithmetical variety*, Algebra Universalis **19** (1984), no. 1, 130–132.

306. _____, *On subquasivariety lattices of semiprimal varieties*, Algebra Universalis **20** (1985), no. 1, 127–129.

307. C. Eberhart, *On Abelian semigroups whose congruences form a simple lattice*, Houston J. Math. **8** (1982), no. 3, 323–332.

308. C. Eberhart and J. W. Stepp, *Lattices of closed equivalences: Simplicity results*, Bull. Polish Acad. Sci. Math. **30** (1982), no. 3–4, 171–177.

309. C. Eberhart and W. Williams, *Elementary orthodox semigroups*, Semigroup Forum **29** (1984), no. 3, 351–364.

310. P. H. Edelman and P. Klingsberg, *The subposet lattice and the order polynomials*, European J. Combin. **3** (1982), no. 4, 341–346.

311. P. M. Edwards, *On the lattice of congruences on an eventually regular semigroup*, J. Austral. Math. Soc. Ser. A **38** (1985), no. 2, 281–286.

312. T. Evans and B. Ganter, *Varieties with modular subalgebra lattices*, Bull. Austral. Math. Soc. **28** (1983), no. 2, 247–254.

313. A. Facchini, *Lattice of submodules and isomorphism of subquotients*, Abelian Groups and Modules (Udine, 1984), CISM Courses and Lectures, vol. 287, Springer, Vienna, 1984, pp. 491–501.

314. P. Farrington, *The first-order theory of the c-degrees with the #-operation*, Z. Math. Logik Grundlag. Math. **28** (1982), no. 6, 487–493.

315. ____, *Constructible lattices of c-degrees*, J. Symbolic Logic **47** (1982), no. 4, 739–754.

316. D. H. Faust, *The Boolean algebra of formulas of first-order logic*, Ann. Math. Logic **23** (1982), no. 1, 27–53.

317. Sh. Feigelstock, *The additive groups of rings with totally ordered lattice of ideals*, Quaestiones Math. **4** (1980/81), no. 4, 331–335.

318. T. Feil, *An uncountable tower of l-group varieties*, Algebra Universalis **14** (1982), no. 1, 129–131.

319. W. Feit, *An interval in the subgroup lattice of a finite group which is isomorphic to M_7*, Algebra Universalis **17** (1983), no. 2, 220–221.

320. P. A. Fejer, *Branching degrees above low degrees*, Trans. Amer. Math. Soc. **273** (1982), no. 1, 157–180.

321. ____, *The density of the nonbranching degrees*, Ann. Pure Appl. Logic **24** (1983), no. 2, 113–130.

322. P. A. Fillmore and W. E. Longstaff, *On isomorphisms of lattices of closed subspaces*, Canad. J. Math. **36** (1984), no. 5, 820–829.

323. P. Fitzpatrick and L. G. Kovacs, *Varieties of nilpotent groups of class four.* I, J. Austral. Math. Soc. Ser. A **35** (1983), no. 1, 59–73; II, J. Austral. Math. Soc. Ser. A **35** (1983), no. 1, 74–108; III, J. Austral. Math. Soc. Ser. A **35** (1983), no. 1, 109–122.

324. R. Fourneau, *Caractérisations de certain sous-lattis du lattis des convexes équilibrés radialiment fermés*, Bull. Soc. Roy. Sci. Liège **50** (1981), no. 9–10, 310–312.

325. ____, *Idéaux des lattis $C(X)$ et $C^*(X)$*, Bull. Soc. Roy. Sci. Liège **51** (1982), no. 5–8, 167–169.

326. B. C. van Fraasen, *Quantification as an act of mind*, J. Philos. Logic **11** (1982), no. 3, 343–369.

327. S. Franciosi, *Some dual homomorphisms between lattices of subgroups and lattices of normal subgroups*, Ricerche Mat. **30** (1981), no. 2, 179–193. (Italian)

328. R. Freese, *Subdirectly irreducible algebras in modular varieties*, Lecture Notes in Math., vol. 1004, Springer-Verlag, Berlin and New York, 1983, pp. 142–152.

329. ____, *On Jónsson's theorem*, Algebra Universalis **18** (1984), no. 1, 70–76.

330. E. Fried, *Congruence-lattices of discrete RUCS varieties*, Algebra Universalis **19** (1984), no. 2, 177–196.

331. E. Fried and E. W. Kiss, *Connections between congruence-lattices and polynomial properties*, Algebra Universalis **17** (1983), no. 3, 227–262.

332. M. Froda-Schechter, *Remarks on tolerances*, Preprint, No. 2, Babes-Bolyai Univ. Fac. Math. Res. Semin., 1982, pp. 76–88.

333. O. C. García and W. Taylor, *The lattice of interpretability types of varieties*, Mem. Amer. Math. Soc. **50** (1984), no. 305.

334. P. W. Gawron and O. Macedońska-Nosalska, *Pewne wlasnosci struktury podgrup dany grupy*, Zeszyty Nauk. Polytech. Slask. Mat. Fiz. (1984), no. 42, 5–13.

335. J. A. Gerhard and M. Petrich, *All varieties of regular orthogroups*, Semigroup Forum **31** (1985), no. 3, 311–351.

336. H. Gerstmann, *Über die schwache Distributivität des Verbandes der subalgebren idempotenter Algebren*, Contributions to General Algebra, 2 (Klagenfurt, 1982), Hölder-Pichler-Tempsky, Vienna, 1983, pp. 359–364.

337. E. Guili and A. Tozzi, *On the lattice of epidense subcategories of a topological category*, Lecture Notes in Math., vol. 1060, Springer-Verlag, Berlin and New York, 1984, pp. 271–277.

338. S. M. Goberstein, *On the modularity of the lattice of fundamental orders on a semilattice*, Semigroup Forum **24** (1982), no. 1, 83–86.

339. S. González and A. Elduque, *Flexible Lie-admissible algebras with A and A^- having the same lattice of subalgebras*, Algebras Groups Geom. **1** (1984), no. 1, 137–143.

340. P. Goralčík and V. Koubek, *There are too many subdirectly irreducible bands*, Algebra Universalis **15** (1982), no. 2, 187–194.

341. G. Grätzer, *On the endomorphism semigroup of simple algebras*, Math. Ann. **170** (1967), no. 4, 334–338.

342. H. Gross, *The lattice method in the theory of quadratic spaces of non-denumerable dimensions*, J. Algebra **75** (1982), no. 1, 23–42.

343. H. Gross and H. A. Keller, *On the problem of classifying infinite chains in projective and orthogonal geometry*, Ann. Acad. Sci. Fenn. Ser. A I Math. **8** (1983), no. 1, 67–86.

344. H. Gross, Z. Lomecky, and R. Schuppli, *Lattice problems originating in quadratic space theory*, Algebra Universalis **20** (1985), no. 3, 267–291.

345. J. W. Grzymala-Busse, *On the representation of finite lattices in the class of finite automata*, MTA Számitástechn. És Autom. Kut. Intéz. Tanul. (1982), no. 137, 199–204.

346. D. R. Guichard, *Automorphisms of substructure lattices in recursive algebras*, Ann. Pure Appl. Logic **25** (1983), no. 1, 47–58.

347. H.-P. Gumm, *Geometrical methods in congruence modular algebras*, Mem. Amer. Math. Soc. **45** (1983), no. 286.

348. _____, *Geometrical reasoning and analogy in universal algebra*, Universal Algebra and its Links with Logic, Algebra, Combinatorics and Computer Science (Darmstadt, 1983), R & E Res. Exp. Math., 4, Heldermann, Berlin, 1984, pp. 14–28.

349. L. Haapasalo, *Von Vektorraumisometrien induzierte Verbandsisomorphismen zwischen nicht orthostabilen und nicht distributiven Vektorraumverbänden*, Ann. Acad. Sci. Fenn. Ser. A I Math. Dissertationes (1981), no. 37.

350. L. Haapasalo and P. Niemistö, *Lattice generation program for computing a vector space lattice*, Beiträge Algebra Geom. **14** (1983), 15–21.

351. M. Haiman, *Proof theory for linear lattices*, Adv. in Math. **58** (1985), no. 3, 209–242.

352. H. Hamilton, *Modular permutation composed completely regular semigroups*, Abstract Spec. Sess. Semigroups and Connect. Other Fields, 777th Meet. Amer. Math. Soc. (Davis, CA, 1980), 1980, pp. 9-12.

353. _____, *Modularity and distributivity of the congruence lattice of a commutative separative semigroup*, Math. Japon. **27** (1982), no. 5, 581–589.

354. W. P. Hanf and D. Myers, *Boolean sentence algebras: Isomorphism construction*, J. Symbolic Logic **48** (1983), no. 2, 329–338.

355. Ph. Hanlon, *The fixed-point partition lattices*, Pacific J. Math. **96** (1981), no. 2, 319–341.

356. J. Hausen, *The additive group of rings with totally ordered ideal lattices*, Quaestiones Math. **6** (1983), no. 4, 323–332.

357. _____, *On varieties of algebras having complemented modular lattices of congruences*, Algebra Universalis **16** (1983), no. 1, 129–130.

358. E. Herrmann, *Definable Boolean pairs in the lattice of recursively enumerable sets*, Proc. First Easter Conference on Model Theory (Diedrich Shagen, 1983), Seminarberichte, vol. 49, Humboldt Universität, Berlin, 1983, pp. 42–67.

359. _____, *Definable structures in the lattice of recursively enumerable sets*, J. Symbolic Logic **49** (1984), no. 4, 1190–1197.

360. _____, *The undecidability of the elementary theory of the lattice of recursively enumerable sets*, Math. Res. **20** (1984), 66–72.

361. B. Herzog, *Die Hilbertfunktionen bilden einen Artinschen Verband*, Beiträge Algebra Geom. (1981), no. 11, 123–126.

362. W. C. Holland, *A survey of varieties of lattice ordered groups*, Lecture Notes in Math., vol. 1004, Springer-Verlag, Berlin and New York, 1983, pp. 153–158.

363. Ch. Holmes, *Generalized Rottlander, Honda, Yff groups*, Houston J. Math. **10** (1984), no. 3, 405–414.

364. _____, *Split extensions of abelian groups with identical subgroup structures*, Contemp. Math. **33** (1984), Amer. Math. Soc., Providence, RI, 265–273.

365. M. Honda, *Joins of weak subideals of Lie algebras*, Hiroshima Math. J. **12** (1982), no. 3, 657–673.

366. Chen Jung Hsü, *On the characterization of the partition lattice $LP_n(s)$ as a geometric lattice*, Chinese J. Math. **9** (1981), no. 1, 37–46.

367. A. P. Huhn, *Schwach distributive Verbände* II, Acta Sci. Math. (Szeged) **46** (1983), no. 1–4, 85–98.

368. M. Hušek, *Applications of category theory to uniform structures*, Lecture Notes in Math., vol. 962, Springer-Verlag, Berlin and New York, 1982, pp. 138–144.

369. M. E. Huss, *The lattice of lattice ordered subgroups of a lattice ordered group*, Houston J. Math. **10** (1984), no. 4, 500–505.

370. G. Hutchinson, *A complete logic for n-permutable congruence lattices*, Algebra Universalis **13** (1981), no. 2, 206–224.

371. _____, *Exact embedding functors between categories of modules*, J. Pure Appl. Algebra **25** (1982), no. 1, 107–111.

372. J. S. Hwang, *The lattice of stable marriages and permutations*, J. Austral. Math. Soc. Ser. A **33** (1982), no. 3, 401–410.

373. P. M. Idziak, *Filters and congruence relations in BCK-semilattices*, Math. Japon. **29** (1984), no. 6, 975–980.

374. T. Ihringer, *On groupoids having a linear congruence class*, Math. Z. **180** (1982), no. 3, 395–411.

375. _____, *On certain linear congruence class geometries*, Ann. Discrete Math. **18** (1983), 481–492.

376. _____, *On finite algebras having a linear congruence class geometry*, Algebra Universalis **19** (1984), no. 1, 1–10.

377. _____, *A property of finite algebras having M_n's as congruence lattices*, Algebra Universalis **19** (1984), no. 2, 269–271.

378. N. Iwahori, *Some topics on Coxeter groups and Weil groups*, Proc. Internat. Math. Conf. (Singapore, 1981), North-Holland, Amsterdam and New York, 1982, pp. 27–34.

379. J. Jakubík, *On the lattice of radical classes of linearly ordered groups*, Studia Sci. Math. Hungar. **16** (1981), no. 1–2, 77–86.

380. _____, *Torsion radicals of lattice ordered groups*, Czechoslovak Math. J. **32(107)** (1982), no. 3, 347–363.

381. _____, *Distributivity of intervals of torsion radicals*, Czechoslovak Math. J. **32(107)** (1982), no. 4, 548–555.

382. _____, *On K-radical classes of lattice ordered groups*, Czechoslovak Math. J. **33(108)** (1983), no. 1, 149–163.

383. _____, *On the lattice of semisimple classes of linearly ordered groups*, Časopis Pěst. Mat. **108** (1983), no. 2, 183–190.

384. _____, *On radical classes of abelian linearly ordered groups*, Math. Slovaca **35** (1985), no. 2, 141–154.

385. M. Jakubíková, *Hereditary radical classes of linearly ordered groups*, Časopis Pěst. Mat. **108** (1983), no. 2, 199–207.

386. D. Jakubíková-Studenovská, *Partial monounary algebras with common congruence relations*, Czechoslovak Math. J. **32(107)** (1982), no. 2, 307–326.

387. _____, *On congruence relations of monounary algebras*. I, Czechoslovak Math. J. **32(107)** (1982), no. 3, 437–459; II, Czechoslovak Math. J. **33** (1983), no. 3, 448–466.

388. J. Ježek, *The number of minimal varieties of idempotent groupoids*, Comment. Math. Univ. Carolin. **23** (1982), no. 1, 199–205.

389. _____, *The lattice of equational theories*. I. *Modular elements*, Czechoslovak Math. J. **31(106)** (1981), no. 1, 127–152.

390. _____, *The lattice of equational theories*. II. *The lattice of full sets of terms*, Czechoslovak Math. J. **31(106)** (1981), no. 4, 573–603.

391. _____, *The lattice of equational theories*. III. *Definability and automorphisms*, Czechoslovak Math. J. **32(107)** (1982), no. 1, 129–164.

392. _____, *On join-irreducible equational theories*, Lecture Notes in Math., vol. 1004, Springer-Verlag, Berlin and New York, 1983, pp. 159–165.

393. _____, *Elementarily non-equivalent infinite partition lattices*, Algebra Universalis **20** (1985), no. 1, 132–133.

394. J. Ježek and T. Kepka, *Equational theories of medial groupoids*, Algebra Universalis **17** (1983), no. 2, 174–190.

395. K. G. Johnston, *Subalgebra lattices of completely simple semigroups*, Semigroup Forum **29** (1984), no. 1–2, 109–121.

396. _____, *Non-modular congruence lattices of Rees matrix semigroups*, Czechoslovak Math. J. **35(110)** (1985), no. 3, 429–433.

397. J. Ježek and P. K. Jones, *The lattice of full regular subsemigroups of a regular semigroup*, Proc. Roy. Soc. Edinburgh Sect. A **98** (1984), no. 3–4, 203–214.

398. P. R. Jones, *A band whose congruence lattice has ACC or DCC is finite*, J. Algebra **64** (1980), no. 2, 336–339.

399. _____, *On the lattice of varieties of completely regular semigroups*, J. Austral. Math. Soc. Ser. A **35** (1983), no. 2, 227–235.

400. _____, *On congruence lattices of regular semigroups*, J. Algebra **82** (1983), no. 1, 18–39.

401. _____, *Lattice isomorphisms of free products of inverse semigroups*, J. Algebra **89** (1984), no. 2, 280–290.

402. _____, *Joins and meets of congruences on a regular semigroup*, Semigroup Forum **30** (1984), no. 1, 1–16.

403. B. Jónsson, *Varieties of relation algebras*, Algebra Universalis **15** (1982), no. 3, 273–298.

404. _____, *Maximal algebras of binary relations*, Contemp. Math. **33** (1984), Amer. Math. Soc., Providence, RI, 299–307.

405. W. Just and A. Krawczyk, *On certain Boolean algebras $\mathscr{P}(\omega)/I$*, Trans. Amer. Math. Soc. **285** (1984), no. 1, 411–429.

406. M. Kamara, *Nichtdistributive modulare Polaritätsverbände*, Arch. Math. (Basel) **39** (1982), no. 2, 126–133.

407. D. Kenoyer, *Recognizability in the lattice of convex l-subgroups of a lattice-ordered group*, Czechoslovak Math. J. **34(109)** (1984), no. 3, 411–416.

408. T. Kepka, *Varieties of left distributive semigroups*, Acta Univ. Carolin.–Math. Phys. **25** (1984), no. 1, 3–18.

409. S. M. Khuri, *Endomorphism rings and Gabriel topologies*, Canad. J. Math. **36** (1984), no. 2, 193–205.

410. D. Kirby, *Subrings of the first neighbourhood ring. II*, Math. Proc. Cambridge Philos. Soc. **92** (1982), no. 1, 35–39.

411. E. W. Kiss, *Complemented and skew congruences*, Ann. Univ. Ferrara Sez. VII **29** (1983), 111–127.

412. E. W. Kiss and L. Ronyai, *On rings having a special type of subring lattice*, Acta Math. Acad. Sci. Hungar. **37** (1981), no. 1–3, 223–234.

413. R. J. Koch and B. L. Madison, *Relations on the lattice of congruences of a regular semigroup*, Semigroup Forum **31** (1985), no. 2, 227–233.

414. P. Köhler, *M_7 as an interval in a subgroup lattice*, Algebra Universalis **17** (1983), no. 3, 263–266.

415. K. Koike, *On some groups which are determined by their subgroup lattices*, Tokyo J. Math. **6** (1983), no. 2, 413–421.

416. H. Komiya, *Convexity on a topological space*, Fund. Math. **111** (1981), no. 2, 107–113.

417. I. Korec, *Arithmetical equivalence lattice for which no Pixley function exists*, Acta Math. Univ. Comenian. (1982), no. 40–41, 267–274.

418. V. Koubek, *Subalgebra lattices, simplicity and rigidity*, Acta Sci. Math. (Szeged) **47** (1984), no. 1–2, 71–83.

419. J. P. S. Kung, *On algebraic structures underlying the Poisson process*, Algebra Universalis **13** (1981), no. 2, 137–147.

420. H. Lakser, W. Taylor, and S. T. Tschantz, *A new proof of Gumm's theorem*, Algebra Universalis **20** (1985), no. 1, 115–122.

421. M. S. Lambrou and W. E. Longstaff, *Abelian algebras and reflexive lattices*, Bull. London Math. Soc. **12** (1980), no. 3, 165–168.

422. W. A. Lampe, *Congruence lattices of algebras of fixed similarity type*. II, Pacific J. Math. **103** (1982), no. 2, 475–508.

423. D. Lascar, *Ordre de Rudin-Keisler et poids dans les théories stables*, Z. Math. Logik Grundlag. Math. **28** (1982), no. 5, 413–430.

424. Sin Min Lee, *Note on congruences of a direct product of algebras*, Algebra Universalis **16** (1983), no. 1, 126–128.

425. _____, *A construction of simple non-associative Boolean rings*, Bull. Malaysian Math. Soc. (2) **7** (1984), no. 1, 35–37.

426. J. J. Leeson and A. T. Butson, *On the general theory of (m, n) rings*, Algebra Universalis **11** (1980), no. 1, 42–76.

427. A. Leone and M. Maj, *Minimal nonsubmodular finite groups*, Ricerche Mat. **31** (1982), no. 2, 377–388. (Italian)

428. _____, *Nonsubmodular finite groups with submodular proper quotients*, Rend. Accad. Sci. Fis. Mat. Napoli (4) **50** (1982-1983), 185–193. (Italian. English summary)

429. J. Lihová, *On the lattice of convexly compatible topologies on a partially ordered set*, Contributions to Lattice Theory (Szeged, 1980), Colloq. Math. Soc. János Bolyai, vol. 33, North-Holland, Amsterdam and New York, 1983, pp. 609–625.

430. P. Lindström, *On certain lattices of degrees of interpretability*, Notre Dame J. Formal Logic **25** (1984), no. 2, 127–140.

431. R. Lochan and D. Strauss, *Lattice homomorphisms of spaces of continuous functions*, J. London Math. Soc. (2) **25** (1982), no. 2, 379–384.

432. P. Longobardi and M. Maj, *Some groups whose lattice of normal subgroups is isomorphic to the lattice of normal subgroups of a free product*, Rend. Mat. (7) **3** (1983), no. 4, 725–734. (Italian. English summary)

433. W. E. Longstaff, *Picturing the lattice of invariant subspaces of a nilpotent complex matrix*, Linear Algebra Appl. **56** (1984), 161–168.

434. W. Maas, *Characterization of recursively enumerable sets with supersets effectively isomorphic to all recursively enumerable sets*, Trans. Amer. Math. Soc. **279** (1983), no. 1, 311–336.

435. W. Maas and M. Stob, *The intervals of the lattice of recursively enumerable sets determined by major subsets*, Ann. Pure Appl. Logic **24** (1983), no. 2, 189–212.

436. K. D. Magill, *Congruences on semigroups of continuous self-maps*, Semigroup Forum **29** (1984), no. 1–2, 159–182.

437. _____, *Semigroups for which the continuum congruences form finite chains*, Semigroup Forum **30** (1984), no. 2, 221–230.

438. P. Mangani and A. Marcja, $\aleph_1$-*Boolean spectrum and stability*, Atti Accad. Naz. Lincei Rend. Cl. Sci. Fis. Mat. Natur. (8) **72** (1982), no. 5, 269–272.

439. I. Maurer, *Topics and bibliography of tolerance relations*, Preprint, No. 2, Babes-Bolyai Univ. Fac. Math. Res. Semin., 1982, pp. 2–7.

440. I. Maurer, I. Purdea, and I. Virag, *Tolerances on algebras*, Preprint, No. 2, Babes-Bolyai Univ. Fac. Math. Res. Semin., 1982, pp. 39–75.

441. P. Mayrhofer, *Der Verband der A-Primärsysteme eines A-Systems*, Sitzber. Österr. Akad. Wiss. Math.-Natur. Kl. **191** (1982), no. 1–3, 23–33.

442. F. Mazzocca, *On a class of lattices associated with n-cubes*, Discrete Math. **49** (1984), no. 2, 133–138.

443. M. J. McAsey and P. S. Muhly, *On projective equivalence of invariant subspace lattices*, Linear Algebra Appl. **43** (1982), 167–175.

444. B. R. McDonald, *Projectivities over rings with many units*, Comm. Algebra **9** (1981), no. 2, 195–204.

445. R. McKenzie, *Finite forbidden lattices*, Lectures Notes in Math., vol. 1004, Springer-Verlag, New York and Berlin, 1983, pp. 176–205.

446. _____, *A note on residually small varieties of semigroups*, Algebra Universalis **17** (1983), no. 2, 143–149.

447. G. F. McNulty, *Structural diversity in the lattice of equational theories*, Algebra Universalis **13** (1981), no. 3, 271–292.

448. _____, *Infinite chains of nonfinitely based equational theories of finite algebras*, Algebra Universalis **13** (1981), no. 3, 373–378.

449. _____, *Fifteen possible previews in equational logic*, Lectures in Universal Algebra (Szeged, 1983), Colloq. Math. Soc. János Bolyai, vol. 43, North-Holland, Amsterdam and New York, 1986, pp. 307–331.

450. L. Michler, *Über das Operieren der Automorphismengruppe auf dem Unteralgebrenverband einer Algebra*, Theory of Semigroups (Conf. on Theory and Appl. of Semigroups, Greifswald, 1984), 1984, pp. 92–99.

451. J. Mináč, *The distributivity property of finite intersections of valuation rings*, Math. Slovaca **34** (1984), no. 3, 277–279.

452. H. Mitsch, *Semigroups and their lattice of congruences*, Semigroup Forum **26** (1983), no. 1–2, 1–63.

453. E. Nauwelaerts and J. Van Geel, *Arithmetical Zariski central rings*, Lecture Notes in Math., vol. 951, Springer-Verlag, Berlin and New York, 1982, pp. 132–142.

454. A. Nerode and J. Remmel, *A survey of r.e. substructures*, Recursion Theory (Ithaca, NY, 1982), Proc. Sympos. Pure Math., vol. 42, Amer. Math. Soc., Providence, RI, 1985, pp. 323–375.

455. M. F. Newman, *Metabelian groups of prime power exponent*, Lecture Notes in Math., vol. 1098, Springer-Verlag, Berlin and New York, 1984, pp. 87–98.

456. J. Nieminen, *The congruence lattice of simple ternary algebras*, Serdica **8** (1982), no. 1, 115–122.

457. D. Normann, *R.e. degrees of continuous functionals*, Arch. Math. Logik Grundlag. **23** (1983), no. 1–2, 79–98.

458. Sh. Oates-Williams, *On the variety generated by Murskiĭ algebra*, Algebra Universalis **18** (1984), no. 2, 175–177.

459. P. Odifreddi, *Global properties (automorphisms and definability) of m-degrees*, Boll. Un. Mat. Ital. A (6) **4** (1985), no. 1, 71–76. (Italian. English summary)

460. M. Orhon, *Boolean algebras of commuting projections*, Math. Z. **183** (1983), no. 4, 531–537.

461. M. A. Orozco and W. Y. Vélez, *The lattice of subfields of a radical extension*, J. Number Theory **15** (1982), no. 3, 388–405.

462. M. Palasiński, *On a problem on BCK-algebras*, Math. Japon. **26** (1981), no. 5, 545–546.

463. _____, *On ideals and congruence lattices of BCK-algebras*, Math. Sem. Notes Kobe Univ. **9** (1981), no. 2, 441–443.

464. _____, *The distributivity of the lattice of varieties of BCK-algebras*, Math. Sem. Notes Kobe Univ. **10** (1982), no. 2/2, 747–748.

465. P. P. Palfy, *On certain congruence lattices of finite unary algebras*, Comment. Math. Univ. Carolin. **19** (1978), no. 1, 89–95.

466. _____, *Distributive congruence lattices of finite algebras*, Colloq. on Ordered Sets (Szeged, 1985), p. 21.

467. F. Pastijn, *Idempotent distributive semirings. II*, Semigroup Forum **26** (1983), no. 1–2, 151–166.

468. _____, *Congruences on regular semigroups—a survey*, Proc. 1984 Marquette Conf. on Semigroups (Milwaukee, WI, 1984), Marquette University, Milwaukee, WI, 1985, pp. 159–175.

469. F. Pastijn and A. Romanowska, *Idempotent distributive semirings. I*, Acta Sci. Math. (Szeged) **44** (1982), no. 3–4, 239–253.

470. F. Pastijn and P. G. Trotter, *Lattices of completely regular semigroup varieties*, Pacific J. Math. **119** (1985), no. 1, 191–214.

471. R. Peele, *On finite partition representation of lattices*, Discrete Math. **42** (1982), no. 2–3, 267–280.

472. P. Penner, *Hyperidentities of semilattices*, Houston J. Math. **10** (1984), no. 1, 81–108.

473. M. Petrich, *On varieties of completely regular semigroups*, Semigroup Forum **25** (1982), no. 1–2, 153–169.

474. M. Petrich and N. R. Reilly, *Near varieties of idempotent generated completely simple semigroups*, Algebra Universalis **16** (1983), no. 1, 83–104.

475. _____, *The join of the varieties of strict inverse semigroups and rectangular bands*, Glasgow Math. J. **25** (1984), no. 1, 59–74.

476. _____, *All varieties of central completely simple semigroups*, Trans. Amer. Math. Soc. **280** (1983), no. 2, 623–636.

477. _____, *Certain homomorphisms of the lattice of varieties of completely simple semigroups*, J. Austral. Math. Soc. Ser. A **37** (1984), no. 3, 298–306.

478. D. Pigozzi, *Finite groupoids without finite bases for their identities*, Algebra Universalis **13** (1981), no. 3, 329–354.

479. P. Plaumann, K. Strambach, and G. Zacher, *Gruppen mit geometrischen Abschnitten im Untergruppenverband*, Monatsh. Math. **99** (1985), no. 2, 117–145.

480. J. Polland, *On lattice-isomorphic abelian groups*, Carleton Math. Lecture Note (1984), no. 49.

481. _____, *On verifying lattice isomorphisms between groups*, Arch. Math. (Basel) **44** (1985), no. 4, 309–310.

482. B. Pondělíček, *Atomicity of tolerance lattices of commutative semigroups*, Czechoslovak Math. J. **33(108)** (1983), no. 3, 485–498.

483. _____, *Modularity and distributivity of tolerance lattices of commutative inverse semigroups*, Czechoslovak Math. J. **35(110)** (1985), no. 1, 146–157.

484. _____, *Modularity and distributivity of tolerance lattices of commutative separative semigroups*, Czechoslovak Math. J. **35(110)** (1985), no. 2, 333–337.

485. W. B. Powell and C. Tsinakis, *Meet-irreducible varieties of lattice-ordered groups*, Algebra Universalis **20** (1985), no. 2, 262–263.

486. E. Previato, *Some families of simple groups whose lattices are complemented*, Boll. Un. Mat. Ital. B (7) **1** (1982), no. 3, 1003–1014.

487. G. Pringerová, *On semisimple classes of abelian linearly ordered groups*, Časopis Pěst. Mat. **108** (1983), no. 1, 40–52.

488. _____, *Covering condition in the lattice of radical classes of linearly ordered groups*, Math. Slovaca **33** (1983), no. 4, 363–369.

489. P. Pudlák, *Some prime elements in the lattice of interpretability types*, Trans. Amer. Math. Soc. **280** (1983), no. 1, 255–275.

490. P. Pudlák and J. Tůma, *Regraphs and congruence lattices*, Algebra Universalis **17** (1983), no. 3, 339–343.

491. I. Purdea and N. Both, *Partial tolerances of groups*, Mathematica **24** (1982), no. 1–2, 139–142.

492. R. W. Quackenbush, *A note on a problem of Goralčík*, Contributions to Universal Algebra (Szeged, 1975), Colloq. Math. Soc. János Bolyai, vol. 17, North-Holland, Amsterdam, 1977, pp. 363–364.

493. T. G. Raghavan, *On the lattice of one-point near-compactification*, Indian J. Pure Appl. Math. **16** (1985), no. 4, 357–364.

494. N. R. Reilly, *A subsemilattice of the lattice of varieties of lattice ordered groups*, Canad. J. Math. **33** (1981), no. 6, 1309–1318.

495. _____, *The breadth of the lattice of those varieties of inverse semigroups which contain the variety of groups*, Proc. Roy. Soc. Edinburgh Sect. A **93** (1983), no. 3–4, 319–325.

496. B. Rendi, *On the lattice structure of universal algebras with an infinitary operation*, Bul. Ştiinţ. Tehn. Inst. Politehn. "Traian-Vuia" Timişoara **(23)(37)** (1978), no. 2, 104–105. (Romanian. French summary)

497. G. Richter, *Application of some lattice theoretic results in group theory*, Contributions to General Algebra, 2 (Klagenfurt, 1982), Hölder-Pichler-Tempsky, Vienna, 1983, pp. 305–317.

498. G. Richter, *Mal'cev conditions for categories*, Categorical Topology (Toledo, Ohio, 1983), Sigma Ser. Pure Math., vol. 5, Heldermann, Berlin, 1984, pp. 453–469.

499. H. H. J. Riedel, *Growth sequences of finite algebras*, Algebra Universalis **20** (1985), no. 1, 90–95.

500. F. Röhl, *Über eine Endlichkeitsbedingung an den Unterringverband eines Ringes*, Abh. Math. Sem. Univ. Hamburg **51** (1981), 210–225.

501. A. Romanowska, *Idempotent distributive semirings with a semilattice reduct*, Math. Japon. **27** (1982), no. 4, 483–493.

502. A. Romanowska and T. Traczyk, *Commutative BCK-algebras. Subdirectly irreducible algebras and varieties*, Math. Japon. **27** (1982), no. 1, 35–48.

503. I. G. Rosenberg, *Functionally complete algebras in congruence distributive varieties*, Acta Sci. Math. (Szeged) **43** (1981), no. 3–4, 347–352.

504. R. Rousseau, *Representations of a sublattice of the partition lattice on a lattice*, Discrete Math. **47** (1983), no. 2–3, 307–314.

505. W. Ruckelshausen and B. Sands, *On finitely generated lattices of width four*, Algebra Universalis **16** (1983), no. 1, 17–37.

506. R. Rudolph, *Ein Untergruppensatz für modulare Gruppen*, Monatsh. Math. **94** (1982), no. 2, 149–153.

507. _____, *Die Endlichkeit einigen modularer Gruppen*, Arch. Math. (Basel) **38** (1982), no. 6, 506–510.

508. H. P. Sankappanavar, *Congruence-distributivity and join-irreducible congruences on a semilattice*, Math. Japon. **30** (1985), no. 4, 495–502.

509. B. Sarath and K. Varadarajan, *Fundamental theorem of projective geometry*, Comm. Algebra **12** (1984), no. 7–8, 937–952.

510. M. Satyanarayana, *Structure and ideal theory of commutative semigroups*, Czechoslovak Math. J. **28(103)** (1978), no. 2, 171–180; *Correction*, Czechoslovak Math. J. **29(104)** (1979), no. 4, 662–663.

511. N. Sauer, M. G. Stone, and R. Weedmark, *Every finite algebra with congruence lattice M_7 has principal congruences*, Lecture Notes in Math., vol. 1004, Springer-Verlag, Berlin and New York, 1983, pp. 273–292.

512. K.-U. Schaller, *Über den Verband der Subnormalteiler einer endlichen, p-auflösbaren Gruppe*, Arch. Math. (Basel) **38** (1982), no. 6, 496–500.

513. _____, *A survey on congruence lattice representations*, Teubner Texte zur Math., 42, Teubner, Leipzig, 1982.

514. R. Schmidt, *Affinities of groups*, Rend. Sem. Mat. Univ. Padova **72** (1984), 163–190.

515. S. Schwarz, *The quotient semilattice of the recursively enumerable degrees modulo the cappable degrees*, Trans. Amer. Math. Soc. **283** (1984), no. 1, 315–328.

516. D. Schweigert and M. Szymańska, *A completeness theorem for correlation lattices*, Z. Math. Logik Grundlag. Math. **29** (1983), no. 5, 427–434.

517. C. M. Scoppola, *A lattice-theoretic characterization of the lattice of subgroups of an abelian group containing two independent aperiodic elements*, Rend. Sem. Mat. Univ. Padova **73** (1985), 191–207. (Italian)

518. K. Seitz, *A lattice theoretical characterization of network systems*, Mathematical Theory of Networks and Systems (Beer Sheva, 1983), Lecture Notes in Comput. Sci., vol. 58, Springer, Berlin and New York, 1984, pp. 796–803.

519. B. Šešelja and J. Ušan, *Structure of generalized equivalences contained in $(2, n\bar{A}_1)$-RT relations*, Univ. u Novom Sadu Zb. Rad. Prirod.-Mat. Fak. Ser. Mat. **11** (1981), 275–286.

520. Ji Chang Sha and D. J. Kleitman, *Superantichains in the lattice of partitions of a set*, Stud. Appl. Math. **71** (1984), no. 3, 207–241.

521. V. M. Shiryaev, *Semigroups with $\wedge$-semidistributive subsemigroup lattices*, Semigroup Forum **31** (1985), no. 1, 47–68.

522. Kar Ping Shum, *Topological semiprimary semigroups*, Studia Sci. Math. Hungar. **15** (1980), no. 1–3, 15–18.

523. B. Šmarda, *Polarity compatible with a closure system*, Czechoslovak Math. J. **29(104)** (1979), no. 1, 13–20.

524. J. D. Smith, *Universelle Algebra*, Tagungsber. Math. Forschungs. Inst. Oberwolfach (1982), no. 1–16.

525. K. C. Smith, *The lattice of left ideals in a centralizer nearring is distributive*, Proc. Amer. Math. Soc. **85** (1982), no. 3, 313–317.

526. M. Stob, *Invariance of properties under automorphisms of the lattice of recursively enumerable sets*, Pacific J. Math. **100** (1982), no. 2, 445–471.

527. M. G. Stone and R. H. Weedmark, *On representing $M_n S$ by congruence lattices of finite algebras*, Discrete Math. **44** (1983), no. 3, 299–308.

528. R. Šulka, *On three lattices that belong to every semigroup*, Math. Slovaca **34** (1984), no. 2, 217–228.

529. V. Swaminathan, *Structure of Boolean-like rings*, Math. Sem. Notes Kobe Univ. **9** (1981), no. 2, 471–493.

530. A. Szendrei, *Short maximal chains in the lattice of clones over a finite set*, Math. Nachr. **110** (1983), 43–58.

531. M. B. Szendrei, *Completely regular semigroups whose idempotent reducts form a chain*, Studia Sci. Math. Hungar. **16** (1981), no. 1–2, 125–139.

532. T. Tamura, *Semigroups whose congruences form a chain and which are extensions of congruence-free semigroups*, Semigroups (Proc. Conf. Clayton, Australia, 1979), 1979, pp. 85–102.

533. W. Taylor, *Some applications of the term condition*, Algebra Universalis **14** (1982), no. 1, 11–24.

534. F. Thieffine, *Compatible complement in Piron's system and ordinary modal logic*, Lett. Nuovo Cimento (4) **36** (1983), no. 12, 377–381.

535. W. J. Thron and R. A. Valent, *A class of maximal ideals in the lattice of topologies*, Proc. Amer. Math. Soc. **87** (1983), no. 2, 330–334.

536. R. F. Tichy, *Über den Kongruenzrelationenverband der Algebra der Polynomfunktion einer Halbgruppe*, Österreich. Akad. Wiss. Math. Natur. Kl. Sitzungsber. II **191** (1982), no. 1–3, 35–51.

537. D. C. Trueman, *The lattice of congruences on direct products of cyclic semigroups and certain other semigroups*, Proc. Roy. Soc. Edinburgh Sect. A **95** (1983), no. 3–4, 203–214.

538. _____, *Lattices of congruences on free finitely generated commutative semigroups and direct products of cyclic monoids*, Proc. Roy. Soc. Edinburgh Sect. A **100** (1985), no. 1–2, 175–179.

539. S. Tulipani, *Congruence lattices of algebraic theories: Models of positive universal theory*, Note Mat. **2** (1982), no. 1, 57–71.

540. _____, *On classes of algebras with the definability of congruences*, Algebra Universalis **14** (1982), no. 3, 269–279.

541. _____, *On the congruence lattice of stable algebras with definability of compact congruences*, Czechoslovak Math. J. **33(108)** (1983), no. 2, 286–291.

542. _____, *On the size of congruence lattices for models of theories with definability of congruences*, Algebra Universalis **17** (1983), no. 3, 346–359.

543. J. Tůma, *On a question of K. Leeb*, Comment. Math. Univ. Carolin. **23** (1982), no. 3, 589–591.

544. _____, *A simple geometric proof of a theorem on N_n's*, Comment. Math. Univ. Carolin. **26** (1985), no. 2, 233–239.

545. _____, *Incidence structures and lattice representations*, Preprint, Math. Inst. Czechosl. Acad. Sci., 1984.

546. _____, *Perfect chamber systems*, Lectures in Universal Algebra (Szeged, 1983), Colloq. Math. Soc. János Bolyai, vol. 43, North-Holland, Amsterdam and New York, 1986, pp. 533–548.

547. C. Uroşu, *Congruences décomposables sur produits directs des algèbres universelles*, An. Univ. Timişoara Ser. Ştiinţ. Mat. **18** (1980), no. 2, 177–185.

548. _____, *Ideals of universal algebras*, Bul. Ştiinţ. Tehn. Inst. Politehn. "Traian-Vuia" Timişoara **26(40)** (1981), no. 1, 27–29. (Romanian. French summary)

549. _____, *On a condition of Mal'tsev type for universal algebras*, Bul. Ştiinţ. Tehn. Inst. Politehn. "Traian-Vuia" Timişoara **26(40)** (1981), no. 2, 21–22. (Romanian. French summary)

550. _____, *C-translate congruences*, Bul. Ştiinţ. Tehn. Inst. Politehn. "Traian-Vuia" Timişoara **27(41)** (1982), no. 1, 9–12. (Romanian. French summary)

551. _____, *C-translated congruences on universal algebras*, Stud. Cerc. Mat. **35** (1983), no. 5, 452–457. (Romanian. French summary)

552. _____, *Sur les propriétés laticeales des congruences*, Inst. Politehn. "Traian-Vuia" Timişoara. Lucrǎr. Sem. Mat. Fiz. **1983**, 76–78.

553. A. Urquhart, *Equational classes of distributive double p-algebras*, Algebra Universalis **14** (1982), no. 2, 235–243.

554. J. Ušan and B. Šešelja, *Transitive n-ary relations and characterization of generalized equivalences*, Univ. u Novom Sadu Zb. Rad. Prirod.-Mat. Fak. Ser. Mat. **11** (1981), 231–245.

555. V. R. Varea, *Lie algebras whose maximal subalgebras are modular*, Proc. Roy. Soc. Edinburgh Sect. A **94** (1983), no. 1–2, 943.

556. J. Vaz de Carvalho, *The subvariety $K_{2,0}$ of Ockham algebras*, Bull. Soc. Roy. Sci. Liège **53** (1984), no. 6, 393–400.

557. V. Weispfenning, *The model-theoretic significance of complemented existential formulas*, J. Symbolic Logic **46** (1981), no. 4, 843–850.

558. _____, *Model theory and lattice of formulas*, Patras Logic Symposion (Patras, 1980), Stud. Logic. Found. Math., vol. 109, North-Holland, Amsterdam and New York, 1982, pp. 261–295.

559. A. Wroński, *Splittings of lattices of quasivarieties (abstract)*, Polish Acad. Sci. Inst. Philos. Sociol. Bull. Sect. Logic **10** (1981), no. 3, 128–129.

560. _____, *Quasivarieties of Heyting algebras (abstract)*, Polish Acad. Sci. Inst. Philos. Sociol. Bull. Sect. Logic **10** (1981), no. 3, 130–134.

561. _____, *Reflections and distensions of BCK-algebras*, Math. Japon. **28** (1983), no. 2, 215–225.

562. Dong Ping Yang, *On the embedding of α-recursive presentable lattices into the α-recursive degrees below $0'$*, J. Symbolic Logic **49** (1984), no. 2, 488–502.

563. M. Yao, *Ideals of tensor products of division algebras and primitive algebras*, Chinese Ann. Math. Ser. B **5** (1984), no. 2, 163–168.

564. G. Zacher, *A normality relation on the lattice of subgroups of a group*, Ann. Mat. Pura Appl. (4) **131** (1982), 57–73.

565. _____, *A lattice characterization of the finiteness of the index of a subgroup in a group*, Atti Accad. Naz. Lincei Rend. Cl. Sci. Fis. Mat. Natur. (8) **69** (1980), no. 6, 317–323. (Italian. English summary)

566. L. Zádori, *Generation of finite partition lattices*, Lectures in Universal Algebra (Szeged, 1983), Colloq. Math. Soc. János Bolyai, vol. 43, North-Holland, Amsterdam and New York, 1986, pp. 573–586.

567. M. I. Zahid and R. W. Heath, *One-point extensions of locally para-H-closed spaces*, J. Austral. Math. Soc. Ser. A **38** (1985), no. 1, 138–142.

568. B. Zelinka, *Tolerances on monounary algebras*, Czechoslovak Math. J. **34(109)** (1984), no. 2, 298–304.

569. P. Zlatoš, *Unitary congruence adjunctions*, Lectures in Universal Algebra (Szeged, 1983), Colloq. Math. Soc. János Bolyai, vol. 43, North-Holland, Amsterdam and New York, 1986, pp. 587–647.

570. _____, *On the ring of the variety of algebras over a ring*, Comment. Math. Univ. Carolin. **24** (1983), no. 2, 325–334.

Translated by BORIS M. SCHEIN

Chapter VI
Partially Ordered Sets. Semilattices. Generalizations of Lattices

V. N. SALIĬ

This survey is a continuation of publications [37] and [OSL,V] which analyzed the topics mentioned in the title and reflected in reviewing journals during July, 1973–1977 and 1978–June, 1982, respectively. See §4.1 of Chapter I-4 for partially ordered sets with orthogonality and, in particular, for orthomodular partially ordered sets.

§VI.1. Partially ordered sets

General topics of the theory of partially ordered sets (posets) were considered in books on lattice theory published in the period covered (see Chapter III). Erné's textbook [156] is an elementary introduction to the theory of partial orderings. In his paper [153] the benefits of studying posets as objects of a different mathematical nature was discussed: lattices as algebras, complete lattices as closure spaces, and quasi-orders from a topological point of view.

Chaudhuri and Mohammad [116] represented an n-element poset by a family of $n \times n$ matrices of zeros and ones, which permitted them to establish new estimates for the number of nonisomorphic partial orderings and the number of distinct partial orderings on an n-element set. Koubek and Rödl [250] showed that if the diagram of a poset, considered as a digraph, does not contain the digraph $K_{m,2}$ (with m vertices, each of them the source of an arc leading in either of the two fixed vertices), then it has no more than $(1 + O(1))\frac{2}{3}\sqrt{m-1}\,n^{3/2}$ arcs. Maryland [268] called a poset P universally connected if P^Q is connected for every poset Q. He found a criterion for universal connectedness of a poset. Jónsson wrote a survey [229] of the theory of ordinal and cardinal operations over posets. He considered arithmetic properties of these operations, decomposability of posets in a cardinal sum or product, in particular, from the point of view of existence of common refinements and cancellation law, etc. Bauer and Wille [77] found simple proofs of Hashimoto's results on the existence of a common refinement for two decompositions of a poset. Miller [275] expressed the property of conditional completeness of a lexicographic product of posets in terms of tree properties of factors. Ginsburg [178] gave an upper estimate for the cardinality of a poset with 0 and 1 using the concept of cardinality of a base (a base is a subset B of an infinite poset P such that each element $x \in P$ is contained between two suitable elements from B). Trombetta [373] considered an axiomatic description of the betweenness relation in posets. Tsirulis [Cirulis] [52] investigated interrelations between a partial ordering and the betweenness relation or the collinearity relation in linearly ordered sets. Krishna Murty and Rao [252] extended the notion of a neutral element of a lattice to arbitrary posets. Vrancken-Mawet [382] studied the congruence lattice of a poset. Milner [277] looked for the possibilities of representing a poset as a union of a sufficiently small number of simpler posets, for example, chains or directed posets. In his joint paper with Pouzet [278] they investigated the property of a poset to contain a cofinal subset isomorphic to a given poset, and also other relations between properties of a poset and its cofinal subsets, while in this joint paper with Prikry [279] they studied the structure of cofinal subsets of posets not containing antichains of a certain given cardinality. Sturm and Rutkowski [363] found conditions under which the join operation is defined on certain subsets of a poset.

Nurtazin [32] produced an elementary classification of complete orders and showed that their theory is not finitely axiomatizable. Schmerl [346] proved that the theory of posets of breadth n has an interpretation in the theory of distributive lattices of $\wedge$-breadth n. Parigot [308] showed that the theory of trees does not possess the independence property. In another paper [309] he constructed a finite axiomatization for the model companion of this theory and established its countable categoricity. Gurevich [188] found a simple proof of undecidability of elementary theories of partial order and lattices. Manaster and Remmel [265] proved the undecidability of the theory of dense n-dimensional (that is, embeddable in the power Q^n of the rational line) partial orderings. For theories of finite-dimensional (for $n > 1$) partial orders model companions were described; they are countably categorical and finitely axiomatizable. Seese, Tuschik, and Weese [353] as well as Baudisch, Seese, and Tuschik [76] studied decidability of certain classes of posets in stationary logic. Steiner and Abian [362] found existence conditions for generic subsets in posets. Seese and Weese [354] showed that every ordinal number is representable in the least class of ordinals, is closed with respect to addition and multiplication, and contains 1, ω, ω_1, ω^ω, and ω_1^ω. Velleman [380] considered a two-person game on an arbitrary poset, in terms of which he stated Jensen's principle. Charreton and Pouzet [115] studied the ordinal type of the chains connected with the Ehrenfeucht-Mostowski theorem.

McNulty [264] wrote a survey of results on recursive posets. It treated topics connected with recursive linear orders and recursive dimension, decidability and finite axiomatizability of elementary theories, computational complexity, and elementary equivalence. A separate section was dedicated to problems. Roy [339] proved existence of recursively enumerably representable linear orders $\leq$ for which $<$ is not recursively enumerable and $=$ is not recursive. Lerman and Rosenstein [260] investigated the complexity of a nontrivial automorphism in various recursive linear orderings. Heck [201] studied properties of orderings on the class of isols (recursive analogs of finite cardinalities). We mention a result from recursive combinatorics obtained by Kierstead and Trotter [238]: a recursive linear ordering of finite breadth n can be covered by $2n-2$ recursive chains, while $3n-3$ recursive chains may not suffice. Grayson [183] considered constructive well-orderings.

Breazu and Stănăşilă [98] showed that the category Ord of posets is naturally isomorphic to the category of topological spaces in which intersection of any family of open sets is open, and investigated connections of such spaces with complete lattices. Sekanian [355] studied families of functors from Ord to Ord that lift the endofunctor Exp. Meseguer [273] considered the subcategory of Ord generated by a system of subsets of a fixed poset. Kvanchilashvili [18] considered a category whose objects are posets and whose morphisms are isotone mappings into lattices of the so-called absolute estimates. Kula, Marshall, and Sladek [255] obtained results on direct limits of finite spaces of orderings. Fuchs [170] found necessary and sufficient conditions under which in the category of all finite chains with partially ordered morphisms an order isomorphism $\mathrm{Hom}(A, B) \cong \mathrm{Hom}(C, D)$ holds for given A, B, C, and D.

Let R be a monoid and K a category. A pair (A, f), where A is an object from K and f a homomorphism of R into the monoid $\mathrm{Hom}_K(A, A)$, is called a K-representation of R. The category of K-representations of R is denoted by K-Act-R. The category K is defined to be fundamental if, for all monoids R and S, the equivalence of the categories K-Act-R and K-Act-S implies equivalence of the categories Set-Act-R and Set-Act-S, where Set is the category of nonempty sets. Skornyakov [39], who introduced these notions, proved, in particular, that Ord is fundamental. Höppner's note [215] was dedicated to injective objects of the category of functors from a poset considered as a category to the category of left modules over an associative ring with identity.

Let $Z(P)$ be a system of subsets associated with every poset P. Objects of the category $\mathrm{Ord}(Z)$ are Z-complete posets (that is, those in which every Z-subset possesses the least upper bound), and whose morphisms are Z-continuous mappings (that is, isotone mappings preserving least upper bounds of Z-subsets). Pasztor, in a series of articles [310]–[312], investigated epimorphisms of the category $\mathrm{Ord}(Z)$ and related categories of partial algebras. Banaschewski and Nelson [70] chose a system $Z(P)$ of subsets of a poset P such that, for every isotone mapping $f: P \to Q$ and every Z-subset $A \subseteq P$, the image $f(A)$ is a Z-subset of Q. They showed that for every such system Z and every isotone mapping $f: P \to Q$ there exists a universal isotone mapping $u_f: P \to P_f$, where P_f is Z-complete and $f \circ u_f$ is Z-continuous. Novak [299] specialized Z as R (the set of all subsets), F (all finite subsets), or $\mathscr{D}$ (all directed subsets) and studied Z-algebraic posets. He established duality of the concepts "finite" and "directed" in the theory of lattices in the language of categories. Bandelt and Erné [72] required that $Z(P)$ contain all one-element subsets of a poset P. They defined Z-continuous posets, Z-morphisms between them, and studied the category so obtained. Erné [157] included in $Z(P)$ all principal ideals of P. He introduced Z-continuous and Z-closed mappings of posets. Universal constructions in the theory of posets were obtained in this way as special cases of a categorial lemma on adjoint functors. In other papers ([155], [158]) of this author (the latter one joint with Wilke) various types of completions were described for quasi-ordered sets: completely distributive, Boolean, etc.

Pouzet and Rival [320] investigated conditions for chain completeness of quotient sets of posets without infinite antichains. Abian and Lihová [59] introduced a notion of a compact poset and proved that every poset without zero possesses a compactification.

Rival's survey [330] was devoted to retracts of posets. He described various types of retractions and certain properties of posets preserved under retractions (completeness, chain completeness, fixed-point property, etc.). Using retractions, he obtained various structural properties of posets. For example, every countable lattice is a retract of a direct product of chains of rationals. Crapo [126] introduced a notion of connection between posets that generalizes that of Galois connection. Here retractions play the role of closure operators. Various results of the theory of Galois connections are fully preserved. Retractions of complete lattices and finite posets were specially considered in this context. Nevermann [291] showed that an order variety generated by a class of posets not coinciding either with the class of all singletons or all trivially ordered posets is not finitely axiomatizable. Pouzet and Rival [321] established that the least order variety containing a class K of posets consists precisely of the retracts of direct products of posets from K. Also, they proved that the order variety generated by the three chains: of natural numbers, negative integers, and the ordinal sum of these two chains—contains all finite and countable lattices. Rival and Wille [332] described the order variety generated by the class of all linearly ordered sets. All its elements are lattices, and it contains every lattice of finite width. Nevermann [292] characterized order varieties generated by classes of $\vee$-semilattices of fixed finite width. In a joint paper with Wille [293] they considered properties of posets preserved under direct products and retractions—the so-called invariants of order varieties. Using new invariants, they were able to clarify the structure of order varieties generated by some classes of posets.

Kalmbach wrote a survey article [232] on homological properties of posets. In particular, she considered homologies of lattices, topological homologies of posets, cohomologies, and right adjoint functors. Brini [100] simplified proofs of earlier results of Quillen and Björner. Bouc [97] computed homology groups for certain posets. Cheng and Mitchell [117] found a necessary and sufficient condition for the cohomological dimension of a poset with finitely many tails not to exceed 1. A poset with 0, which is not a singleton, is called a CW-set if, for each nonzero element x in it, a geometric realization of the simplicial complex corresponding to the interval $(0, x)$ is homeomorphic to a sphere. Björner [84] showed that every CW-set is isomorphic to a poset of cells of a certain regular cell complex and that a Coxeter group with Bruhat's partial order is a CW-set. He established a connection with the shellable property for posets. In [85] he gave an order-theoretic characterization for the set of cells of a regular cell complex with Bruhat's partial order. Walker [383] considered homotopy types and Euler characteristics of posets. Brini and Terrusi [101] pointed out a procedure which, for certain posets, produces a homotopically equivalent simpler poset that is useful for computing the Möbius function of posets. Bayer and Billera devoted their survey article [80] to a calculus of chains in posets connected with convex polytopes and triangulated spheres, and also to Euler complexes of posets. Björner and Walker [89] showed how the homotopic type of a simplicial complex of finite chains in a bounded lattice could be expressed in terms of homotopic types of its intervals of the form $(0, x)$ and $(x, 1)$. A survey [86] of Björner, Garsia, and Stanley is an introduction to Cohen-Macaulay posets with an outline of their applications to combinatorics, algebraic geometry, and topology. Baclawski [68] proved that the canonical module of a Stanley-Reisner ring of a double Cohen-Macaulay poset is isomorphic to a certain ideal of the ring. For an arbitrary finite poset this ideal is isomorphic to a submodule of the canonical module. Using a recursive definition of lexicographically shellable posets, Björner and Wachs ([87], [88]) established that the lattice of faces of a convex polytope, the lattice of bilinear forms, and other posets are shellable.

Liber's survey [23] reflected work on topological partially ordered spaces, topological semilattices and lattices, and fuzzy topological spaces reviewed between July 1973 and 1977. Rudin [340] considered convergence of directed subsets of posets. Dobbs [139] characterized posets that admit a unique topology compatible with their order. Dimitriević [135] considered various types of compatibility between topology and the partial ordering on a poset. Lihová [262] showed that the topologies weakly convexly compatible with a partial order $\leq$ on a set P form a complete sublattice $\beta(P, \leq)$ of the complete lattice $T(P)$ of all topologies on P; the convexly compatible topologies form a complete lattice that is not necessarily a complete sublattice of $T(P)$. The lattice $\beta(P, \leq)$ is distributive. Erné [154] continued studies of topologies on products of posets. In particular, he proved that a product of chains is locally bounded in the order topology if and only if all factors, except finitely many of them, are bounded. For every poset, Hardy and Thornton [197] constructed a T_0-topology and described Green's relations in the semigroup of all closed relations. Novak [298] developed a general theory of continuity in posets. Skucha [358] introduced and studied a concept of a continuous poset that generalizes that of an S-continuous (that is, Scott continuous) lattice. Bosisio [96] considered an analog of S-continuity for posets, where each directed set possesses the least upper bound, and extended concepts and results of classical recursion theory to such posets subject to an additional conditon. Gierz, Lawson, and Stralka [176] suggested the most natural generalization of S-continuity for posets. Isbell [217] produced an example of a complete lattice that admits a balanced topology, but is not balanced in its S-topology. Messano [274] investigated the so-called k-periodic points of the order topology of continuous transformations of linearly ordered sets, both complete and dense in themselves. In terms of antichains, Hanazawa [195] characterized countable paracompactness of the interval topology on ω_1-trees. Earlier, Hart [199] showed that the interval topology on ω_1-trees is locally compact and locally countable and expressed certain properties of this topology in terms of a partial order on the trees. Katz [235] constructed a duality theory for so-called deductive algebras and valuation spaces; it contains Priestley's duality theory for distributive lattices and P-spaces as a special case. Hofmann and Lawson [205] investigated order-theoretical foundations of a theory of quasicompactly generated topological spaces without the separation axiom. Here the principal role is played by a duality theory for complete ∨-semilattices. Pawar and Thakare [313] showed that the space of minimal prime ideals of a 0-distributive semilattice is Hausdorff and obtained a criterion for compactness of this space as well as an analog of Stone's theorem. Let $I(S)$ be the space of all closed ideals of a semilattice S with the Vietoris topology and $Z(S)$ the closed subsemilattice in $I(S)$ generated by the principal ideals of S. Brown and Stepp [102] proved that S and $Z(S)$ are topologically isomorphic. Gierz, Lawson, and Stralka [177] introduced a so-called Z-topology on a semilattice S of finite ∧-breadth and studied its connections with other inner topologies on S. Lystad and Stralka [263] established bialgebraicity of the lattice of closed congruences of a compact topological semilattice with 1 satisfying certain additional conditions. Ruppert [341] produced an example of a separately but not jointly continuous action of a compact semilattice.

Green [184] gave an exposition and complete proofs of principal results on the Möbius function in posets. West's survey [388] was dedicated to extremal problems in posets. Central among them is the problem of determining the cardinality of the largest antichain in a poset. Let $N(a)$ denote the number of antichains that contain an element a in a poset P and N the total number of antichains in P. Here is one of the results obtained by Sands [343]: for every natural number $l > 1$ there exists $r = r(l) > 1$ such that every poset of length $\leq l-1$ possesses an element a for which $N(a)/N \geq 1/r$, where r can be any number satisfying the condition $r \geq 2l$ and $(r-l)^l \leq (2 - 2\frac{1}{r})^r$. West and Tovey [389] discussed conjectures on the maximal size of semiantichains in a direct product of posets. Schmidt [348] investigated a relation between the height of a poset and ordinal types of its chains and antichains (on this subject see Chapter 1 of Zaguia's monograph [394]). Burkill and Smallwood [105] generalized Hanani's theorem (if a poset has at most $n(n+3)/2$ elements, it is representable as a union of n chains or antichains) to the case of several partial orders with a single carrier. A transformation $f \colon P \to P$ of a poset P is called a regression if $f(x) \leq x$ for all $x \in P$. A monotone k-chain in P is a chain $x_1 < x_2 < \cdots < x_k$ such that $f(x_1) \leq f(x_2) \leq \cdots \leq f(x_k)$. West, Trotter, Peck, and Shor [390] proved that if P contains at most $(n+1)^k$ elements and the cardinality of antichains in P does not exceed n, then, for every regression f in P, there exists a monotone $(k+1)$-chain (this estimate cannot be improved). Sauer and Woodrow [344] related lengths of maximal antichains in a poset with properties of its cutsets, that is, subsets that have a nonempty intersection with each maximal antichain. Berenguer, Diaz, and Harper [81] solved a problem of Sperner and Erdös by finding a pointwise determined function on Sperner subsets of a poset. Griggs [185] strengthened the following earlier result of Lee: the family $C(n, k)$ of all subsets of the set $\{1, 2, \ldots, n\}$ that have a nonempty intersection with $\{1, 2, \ldots, k\}$ possesses Sperner's property. Engel's paper [152] was devoted to the so-called strong properties of posets.

Let Π be a property of posets. A survey [95] by Bonnet and Pouzet contains results of various authors which answered the following questions for concrete Π: (1) does a poset with Π have a linear extension with this property? (2) does every linear extension of a poset with Π have this property? They considered topics related to dimension of posets. A poset is called locally chain complete if each of its closed intervals is chain complete (that is, all maximal chains in it are complete). Pouzet and Rival [319] showed that a locally chain complete poset with finite antichains and every countable chain complete poset possess a locally chain complete linear extension. They produced an example of a complete distributive lattice without a complete linear extension. Pinus [35] described countable posets of finite width with a unique linear extension. Rival [331] considered a problem of how, for a finite poset P, one could construct a linear extension $\mathscr{L} = C_1 \oplus C_2 \oplus \cdots \oplus C_n$ that minimized the number n of disjoint chains C_i in P. In his and El-Zahar's joint paper [151] the so-called greedy and optimal linear extensions were specified among all linear extensions of a poset, and various relations between them established. Let $s_{\mathscr{L}}(P)$ denote the number of pairs (x, y) such that y covers x in the linear extension $\mathscr{L}$ of a poset P but y does not cover x in P, while $s(P)$ is the minimum (for all $\mathscr{L}$) of all possible numbers $s_{\mathscr{L}}(P)$. Duffus, Rival, and Winkler [145] proved that $s(P) + 1$ coincides with the width $w(P)$ for a wide class of posets P. Kelly and Trotter [236] surveyed (with proofs) results connected with dimension of posets obtained in the seventies. They described methods for computing the dimension for certain classes of posets and pointed out a connection with graph theory. Trotter [375] proved that the dimension of a direct product of two posets could not exceed the sum of the dimensions of the factors and, for every $n \geq 3$, the dimension of the direct square of the crown S_n^0 is equal to $2n - 2$. In the monograph [394] mentioned earlier Zaguia studied chains of initial intervals of a poset by means of the Krull dimension $\dim_K$ and its generalizations. In his joint paper with Pouzet [322] it was proved that the Krull dimension of a poset is equal to the maximum of the Krull dimensions of its linear extensions. Albu [64] introduced and investigated Gabriel dimension $\dim_G$ for posets. If $\dim_K P$ exists for a poset P, then $\dim_G P$ exists as well and $\dim_K P \leq \dim_G P + 1$. For dual well-ordered sets and certain lattices, $\dim_G$ was computed. The depth of a poset P is the least ordinal number that can be embedded in P without reversing its order. Milner and Prikry [280] studied whether the depth of a given poset could be reduced by means of partitioning it into a few parts. There are linearly ordered sets of countable depth for which this is impossible.

The collection *Linear Algebra and the Theory of Representations* [24] consists of papers that develop the general theory of matrix problems and its applications for solving concrete problems in the representation theory of posets. Nazarova and Roĭter [29] wrote a survey on the development of these methods. They [30] studied representations of completed posets and considered related quadratic forms. Zavadskiĭ and Nazarova [15] found necessary and sufficient conditions under which a poset has finite growth or is tame. Zavadskiĭ [14] described the structure of representations of a poset of finite growth over an arbitrary field. Let a poset P of finite type have width $n = w(P)$ and let the number $c(P)$ of indecomposable representations be finite. For the number $c(n) = \max\{c(P): w(P) = n\}$ Kerner [237] obtained the inequality $c(n) \leq n^3/27 + n^2 + 50n$. Kleiner [239] produced a simpler proof of the fact that the endomorphism ring of an arbitrary indecomposable representation of a poset of finite growth over an arbitrary field is isomorphic to the field ("Schur's Lemma"). Bondarenko [3] described exact posets of infinite growth. Neggers [289] studied linear representations of finite posets as linear operators. Nikulin [31] reduced representations of linear algebras to representations of posets. Müller [282] considered module-theoretic applications of linear representations of posets. Bünermann [103] described Auslander-Reiten quivers of one-parameter posets (they were listed by Otrashevskaya [OSL, V-32]). Roggenkamp [333] characterized Auslander-Reiten quivers of Bäckström partial orders. Vinogradova and Yakovlev [5] classified all pairs of matrices that form a representation of a finite set with two partial orderings each of which is of width ≤ 2. A lattice is called algebraic over a field k if each representation of it (that is, a homomorphism into the lattice of all submodules of a free module of finite rank) over any extension of k is induced by a representation over an algebraic extension of the field. Yakovlev [57] proved that for a transcendental (that is, nonalgebraic) lattice over an algebraically closed field k there exist infinitely many dimensions in each of which there are infinitely many nonisomorphic indecomposable representations of this lattice. If the lattice is of finite length and algebraic over k, there exist only finitely many pairwise nonisomorphic indecomposable representations. English translations [6]–[8] of Gel′fand-Ponomarev's articles on representations of free modular lattices were published. Tsyl′ke [Zielke] [54] confirmed one of the conjectures mentioned in these articles: for each perfect element of a free modular lattice there exists a linearly equivalent element that belongs to the subset $B^+ \cup B^-$ constructed by Gel′fand and Ponomarev. Stekol′shchik [44] constructed invariant elements in modular lattices.

In his survey [78] Baumgartner analyzed results of Lawvere, Galvin, Shelah, his own, and others on ordinal types of real numbers and other uncountable linear orderings obtained in the seventies. Collot [119] continued developing a construction for well-ordering the set of all subsets of the natural series completed by the first transfinite number. Watnick [387] generalized Tennenbaum's theorem on effectively finite recursive linear orderings. Pinus [34] showed that there exists a continuum of linearly ordered sets of cardinality $\aleph_1$ that are equivalent with respect to embedding but pairwise incomparable with respect to epimorphisms. Hagendorf [189] obtained a necessary and sufficient condition for nondissectability of a chain and gave a constructive proof of the fact that every chain possesses an immediate extension. In another paper [190] this author proved that if a chain C is not a singleton, then $C \oplus C$ is not an immediate successor of C and pointed out some unsolved problems of the theory of chains. Todorčević [370] studied the tree $W(C)$ consisting of all well-ordered subsets of a chain C with the partial order of being an initial segment. By means of this and related notions some problems of general topology were solved. In another paper [371] he found necessary and sufficient conditions for the existence of a strictly increasing function on $W(C)$. Kurepa [257] found combinatorial applications of the $W(C)$-construction. Devlin [134] considered properties of reduced degrees of Aronszajn, Kurepa, and Suslin trees. Abraham [61] constructed a model of set theory that possesses no Aronszajn trees of height $\aleph_2$ and $\aleph_3$. Hanazawa [193] discovered an Aronszajn tree that is not countably metacompact. He [194] used the constructivity axiom $V = \mathscr{L}$ to construct an example of a Suslin tree whose direct square contains no stationary antichains, and proved [192] that the existence of a non-Suslin tree with a non-Suslin base of cardinality less than 2^{ω_1} is not deducible from ZFC. Hart [198] gave topological and inner characterizations of R-embeddable ω_1-trees and found certain properties of these trees equivalent to their developability. Margush [266] showed that symmetric difference between corresponding partial orders is essentially the only distance between finite trees that possesses "good" properties.

A series of Fishburn's papers was devoted to interval orders. Let m and n be relatively prime numbers such that $2 \leq m \leq n$. Let $P[m, n]$ denote the class of finite posets $(P, \leq)$ to whose elements there correspond real line intervals with lengths from $[m, n]$ and such that $a \leq b$ holds if and only if the interval representing a is to the left of the interval for b. It was proved in [162] that every class $P[m, n]$ could be axiomatized by a universal elementary sentence. A negative answer to Trotter's problem on two-valued interval orders was given in [164]. Problems of minimization for representations of interval orders on the real line were investigated in [161]. Trotter's article [374] is connected with interval representations of partial orders. Let $S(n)$ be the greatest integer such that each interval order on an n-element set contains a semi-order on a set with $\geq S(n)$ elements. Fishburn [160] showed that $S(15) = S(16) = 9$, thus filling the last lacuna in the list of the first 17 values of $S(n)$. Roubens and Vincke [337] studied numerical and matrix representations of interval orders and, in [338], properties of linear orders and semi-orders at a minimal distance (in the sense of symmetric difference) from a given interval order. Drechsler and Zahn [140] defined completely excisable (abfaserbar) posets and studied their properties. Every completely excisable poset is a lattice, and every planar lattice is completely excisable. For finite distributive lattices complete excisability is equivalent to planarity. Zahn [395] defined and investigated elementary partial orderings, Berghammer and Schmidt [82] considered discrete orderings, Scott [351] uniform partial orderings, and Albert [63] iteratively algebraic partial orderings (the latter always satisfy the ascending chain condition).

Skornyakov [40] defined a class of submonoids of the monoid End P of all isotone transformations of a poset P; he called them admissible monoids. It turned out that, in the case when P possesses 0 or 1, an admissible monoid determines P up to isomorphism or anti-isomorphism. It follows from the elementary equivalence of two posets that they possess elementarily equivalent admissible monoids; the converse holds too. He gave an abstract characterization of admissible monoids and observed that the class of monoids of the form End P is not axiomatizable. These facts generalize many known results (cf. [OSL, V-82]). Babaev ([1], [2]) suggested a generalization of Skornyakov's results, by not assuming the existence of 0 and 1. Anderson and Edwards [66] proved that a lattice-ordered monoid could be embedded in a naturally ordered monoid of isotone transformations of a linearly ordered set if and only if it is distributive. Ulitina [47] found when semigroups of all directed transformations of two posets generate the same semigroup variety. Tsutsura [53] proved that a cardinally decomposable poset is determined up to isomorphism or anti-isomorphism by the partial groupoid of its isotone transformations with a two-element range. He suggested a determinability criterion for cardinally indecomposable posets. On the set G of all partial transformations of a poset P, Edwards and Anderson [150] introduced a partial order $\triangleleft$ defining $\varphi \triangleleft \psi$ if $\text{Dom}\,\varphi \subseteq \text{Dom}\,\psi$ and $\varphi(x) \leq \psi(x)$ for all $x \in P$; they proved that G is a regular semigroup that is a lattice with respect to $\triangleleft$, such that $\varphi(\psi \wedge \chi) = \varphi\psi \wedge \varphi\chi$ and $\varphi(\psi \vee \chi) = \varphi\psi \vee \varphi\chi$. They established also other properties of this semigroup. An isotone transformation of a poset is called its local automorphism if it is one-to-one on every closed interval. Chvalina and Chvalinová [118] found when the monoid of local automorphisms of a locally finite tree acts transitively. Blažková and Chvalina [93] described locally finite trees and forests whose monoid of local automorphisms is regular, inverse, or completely regular.

Morozov [28] proved that the group of all order-preserving recursive permutations of the set of rationals is not constructive. Zervos [17] investigated orbits of the automorphism group of a poset. In particular, he proved that every such orbit either consists of incomparable elements or contains neither minimal nor maximal elements. Jónsson [230] considered the possibility of the representation $\text{Aut}\,P^Q \cong \text{Aut}\,P \times \text{Aut}\,Q$ for the automorphism group of the set P^Q of all isotone mappings of a poset Q into a poset P. For a connection between $\text{Aut}\,P^Q$ and the groups $\text{Aut}\,P$ and $\text{Aut}\,Q$ see the survey [131] of Davey and Duffus. Jónsson and McKenzie [231] found sufficient conditions imposed on P and Q, under which $\text{Aut}\,P^Q$ is an elementary abelian 2-group. A tree is called hereditarily bireflective if each element of the automorphism group of any of its partially ordered subsets is a product of at most two elements of the second order. Moran [281] showed that a tree is hereditarily bireflective if and only if it does not contain a partially ordered subset of a special form. Let T be a tree with the vertices $1, 2, \ldots, n$ and S_n the symmetric group. A permutation $\sigma \in S_n$ is said to be compatible with T if, for every i, $1 \leq i \leq n$, the vertices i and $\sigma(i)$ are adjacent in T. Let $l(\sigma, T)$ denote the minimal number of factors compatible with T, into a product of which σ can be decomposed. Tchuente [366] showed that $\max\{l(\sigma, T) : \sigma \in S_n\} \leq 2n$ for every tree T of the form mentioned. Stanley [361] considered action of a symmetric group on a Boolean algebra, on the partition lattice of a set, etc., and suggested two constructions of representations of a finite set on a finite graded poset. Haile, Larson, and Sweedler [191] classified idempotents of cohomology monoids in the case of Galois cohomologies in terms of Galois group actions on posets. If a group G is homomorphically mapped into the automorphism group of a poset P, then P can be viewed as a partially ordered G-polygon which was studied by Tararin [45] from various points of view. Partially ordered polygons over an arbitrary monoid R were considered by Skornyakov ([41], [42]) who proved, in particular, that all complete (as posets) ordered left R-polygons are injective if and only if R is a group. Jambu-Giraudet ([219], [220]) considered a linearly ordered polygon over a group as a two-carrier model.

Nambooripad and Pastijn [286] considered a certain transformation semigroup associated with a poset and proved that a poset P with 0 and 1 is a complemented modular lattice if and only if this semigroup is regular and coordinatizes P. Thron [369] described the set of maximal elements of a subset X of a poset P in terms of the set of elements of X isolated with respect to a closure operator.

Jónsson and McKenzie [231] investigated problems related to cancellation and refinement properties for posets of the form P^Q. When does $P^Q \cong P^R$ imply $Q \cong R$ and when does $P^R \cong Q^R$ imply $P \cong Q$? Under which conditions does an isomorphism $P^Q = R^S$ imply representations $P \cong T^X$, $Q \cong T^Y$, $R \cong Y \times Z$, $S \cong X \times Z$ for certain T, X, Y, and Z? These topics are considered in the survey [131] by Davey and Duffus that we mentioned earlier. Mikhaĭlov [27] considered an operation $\circ$ in the set P^Q, where $(\varphi \circ \psi)(x) = (\varphi \pi \psi)(x)$ for a fixed isotone mapping $\pi: P \to Q$. Under certain restrictions on π abstract properties of the semigroup constructed in this way determine partial orders on P and $\pi(P)$ up to isomorphism or anti-isomorphism. Speed [360] obtained a formula for the Möbius function of the poset P^Q and pointed out certain applications. Fleischer [166] observed that every Galois connection is the polarity of an antitone relation. Burkill [104] showed that if P is a poset in which any two two-element chains have a common upper bound and a common lower bound and Q is a linearly ordered set, then a function $f: P \to Q$ that increases (decreases) on every three-element chain is increasing (decreasing) on the entire P. Höft and Höft [208] represented the sequence of all k-gonal numbers as a sequence of cardinalities of (modular) lattices of the form $\mathbf{n}^{M_k}$, where M_k is the modular lattice of height 2 with k atoms and $\mathbf{n}$ is an initial segment of the natural series. Winkler [391] used isotone surjections of an n-element poset P onto $\mathbf{n}$ to define the average height $h(x)$ of an element of P and investigated properties of this function. Neggers [290] produced a table of reduced B-decompositions for posets with ≤ 6 elements (these decompositions are determined by isotone surjections of a given poset onto finite chains). Golomb [180] observed that various objects considered in algebra and combinatorics might be interpreted as posets with a fixed mapping ("norm") into the set of natural numbers partly ordered by divisibility. Narushima [287] considered a property of norms on the power set. Turinici [377] used a quasimetric d on a quasi-ordered set to introduce a notion of d-maximal elements and found conditions under which each element is contained in a d-maximal one. Griggs [186] introduced a special measure on subsets of a finite set. Kurepa [256] studied relations between cardinal constructions determined by an isotone mapping of the set of all subsets of a given set into an arbitrary poset.

Gähler [172] used mappings of sets into posets to establish the equivalence of certain properties in topology and optimization theory. Tuschik [378] called a mapping of the direct square of a poset into a finite set coloring of this poset. He investigated so-called additive colorings of linear orders.

Rival [329] surveyed characterizations of posets with the fixed point property([1]) (each isotone transformation possesses a fixed point) and related topics. He established that the fixed point property holds in the following: (1) a downwards directed poset that is compact in its interval topology (for ∨-semilattices, with an additional condition, compactness of the interval topology is also a necessary condition) (Kolibiar [247]); (2) every chain-complete poset that contains a complete coinitial (or a complete cofinal) subset (Höft and Höft [207]); (3) a poset whose comparability graph is connected and satisfies the so-called Helly property (Quilliot [325]). For a connected finite poset of width 2 the fixed point property is equivalent to the tower T_n not being a retract of the poset (Fofanova [50]). Constantin and Fournier [120] generalized the notion of excisable poset and proved the fixed point property for such posets. Wang and Chen [386] found a new example of a chain-complete poset in which every isotone transformation possesses a so-called optimal fixed point. Cousot and Cousot [124] gave a constructive proof of Tarski's theorem on the existence of a fixed point for an isotone transformation of a complete lattice. Fofanova [49] and Rival [329] proved independently that the fixed point property is preserved for upper or lower cones of every nonempty chain. It is not known whether the posets P^Q and $P \times Q$ preserve the fixed point property if this property holds for P and Q. Duffus [144] proved that if P and Q are connected posets, each of which is a product of posets indecomposable into a product, then the product $P \times Q$ possesses an automorphism without fixed points if and only if at least one of the factors has an automorphism without fixed points. For a poset P with 0, in which every chain possesses the least upper bound, Kolibiar [247] studied the structure of the set of fixed points of an isotone transformation. Let $Fp(L)$ denote the set of elements of a lattice L fixed by all automorphisms. Adams and Sichler [62] showed that if L_0 is a lattice, G_0 a group, $|L_0| > 1$ and $|G_0| > 1$, then for every infinite cardinal number $\alpha \geq \max\{|L_0|, |G_0|\}$ there exist 2^α pairwise nonisomorphic lattices L satisfying the conditions: $|L| = \alpha$, L_0 is a sublattice of L, $\operatorname{Aut} L \cong G_0$ and $Fp(L) = L_0$. An analogous result holds for posets. Hanlon [196] proved that the set of fixed points of an automorphism of the partition lattice of the set $\{1, 2, \ldots, n\}$ induced by a fixed permutation is a lattice, and studied its structure. Ueĭskaya [46] characterized posets in which sets of fixed points of directed transformations form a lattice under inclusion, or a chain, or have the same cardinality, etc. A Kleene chain in a poset P with 0 is defined as any increasing chain of the form $\{f^n(0)\}$, where f is an ω-continuous

---

([1])The author is grateful to T. S. Fofanova who made available materials used in preparation of the section on fixed points.

transformation of P (that is, a transformation that preserves the least upper bounds of finite and countable chains). Kamimura and Tang [233] investigated connections between the existence of fixed points for isotone transformations of P and the existence of the least upper bounds of its Kleene chains. Mashburn [269] showed that if a poset of finite width possesses the least fixed point property for ω-continuous transformations, then it is ω-chain complete. Let $\mathscr{L}$ be a continuous lattice, $x \in \mathscr{L}$, and Φ the set of all continuous transformations of $\mathscr{L}$ for which x is the least fixed point. Nguyen and Lassez [295] established that Φ possesses the least element if and only if the minorants of x form an ascending chain in $\mathscr{L}$. Let φ and ψ be transformations of a complete lattice $\mathscr{L}$ such that both superpositions $f = \varphi\psi$ and $g = \psi\varphi$ are isotone. Blair and Roth ([91] and [92]) established that there exist elements $x, y \in \mathscr{L}$ with the properties $x = f(x) \leq \varphi(y)$ and $y = g(y) \leq \psi(x)$. Klimeš ([241] and [242]) considered transformations of posets close to isotone and, in terms of them, characterized completeness of a $\wedge$-semilattice and a $\vee$-semilattice. A transformation f of a poset P is called relatively isotone if $x \leq y$, $f(x) \leq y$, and $x \leq f(y)$ in P imply $f(x) \leq f(y)$. Klimeš [243] showed that (1) a lattice is complete if and only if each relatively isotone transformation f, for which x and $f(x)$ are comparable, has a fixed point, and (2) every relatively isotone transformation of a chain complete poset has a fixed point. Kogalovskiĭ and Soldatova [19] found a characterization for the set of fixed points of sets of isotone transformations of a complete chain. They also obtained the following result: if the set Φ of isotone transformations of a complete lattice is such that the sets of fixed points $F(\varphi)$, $\varphi \in \Phi$, form a downwards directed set with respect to inclusion, then $\bigcap\{F(\varphi)\colon \varphi \in \Phi\} = F(\psi)$ holds for a certain isotone transformation ψ of this lattice. Diviccaro [137] found conditions under which a family of isotone transformations of a poset has a minimal common fixed point in each principal filter of this poset.

Dacić [129] found conditions for the existence and uniqueness of fixed points of antitone transformations of posets of finite length. He obtained various properties of the set of fixed points of monotone transformations of such posets. In [128] he investigated fixed points of antimorphisms of complete lattices, where a transformation $f\colon \mathscr{L} \to \mathscr{L}$ is called a $\wedge$-antimorphism if $f(\inf M) = \sup f(M)$ for all nonempty $B \subseteq \mathscr{L}$, and $\vee$-antimorphisms are defined dually. Let every upper bounded subset of a poset P possess the least upper bound. Abian [58] showed that if an antitone transformation $f\colon P \to P$ is such that for every $x \in P$ with $x \neq f(x)$ we have $f(H) \cap H \neq \varnothing$, where $H = (x, f(x)]$ or $H = [f(x), x)$, then f has a fixed point. Let $\mathscr{L}$ be a bounded distributive lattice and f its dual automorphism determining the structure of a De Morgan algebra on $\mathscr{L}$. Blyth and Varlet [94] established that the set $F(f)$ of fixed points is an antichain. Varlet [379] used $F(f)$ to count the number of fixed points for polarities of finite distributive lattices. An ordered pair (x, y) of elements of a poset P is called a fixed edge of an antitone transformation $f\colon P \to P$ if $x \le y$, $f(x) = y$, and $f(y) = x$. Klimeš [240] showed that every antitone transformation of a complete lattice has a fixed edge and every family of pairwise commuting antitone transformations has a common fixed edge. Tasković [365] obtained necessary and sufficient conditions for conditional completeness of a poset; these are connected with existence of fixed points for isotone transformations and fixed edges for antitone transformations of certain special types.

A relation p in a poset P is called isotone if $a < b$ implies that every $x \in p(a)$ possesses a majorant in $p(b)$ and every $y \in p(b)$ possesses a minorant in $p(a)$. Walker [384] obtained a characterization of finite posets with the fixed point property for isotone relations (x is a fixed point for p if $x \in p(x)$), and established that this property is preserved under retractions of posets having no infinite chains and under direct products of finite posets. Rutkowski [342] showed that if the fixed point property holds for isotone transformations in one of the posets P or Q and for isotone relations in the other, then the direct product $P \times Q$ has the fixed point property for isotone transformations. Dacić [130] proved that an antitone relation p in a complete lattice $\mathscr{L}$ has a fixed edge if it satisfies the condition $\mathrm{supp}(x) \in p(x)$ for all $x \in \mathscr{L}$.

Wang and Li [385] found necessary and sufficient conditions for the existence of a cofinal chain in a directed quasi-ordered set. They studied properties of the minimal order type of such a chain. Premchand [324] suggested two new independent systems of axioms for biordered sets. The starting point for the concept of a biordered set is the set $E(S)$ of idempotents of an arbitrary semigroup S with two quasi-orders $\{(e, f)\colon ef = e\}$ and $\{(e, f)\colon fe = e\}$. Easdown [147] showed that every biordered set is embeddable in a biordered set of the form $E(S)$. In [146] he simplified the proof of Nambooripad's result: every regular biordered set is isomorhic to $E(S)$ for a regular semigroup S. In [148] he produced an abstract characterization of biordered sets of the form $E(S)$ for eventually regular semigroups S. Höft [206] described so-called complete relations which are a natural generalization of order relations on complete lattices. Sokhadze [43] generalized Dilworth's theorem on decomposition of a poset into disjoint chains to an arbitrary set with a transitive binary relation. Novák [300] called a ternary relation of a special form a cyclic order, established an interconnection between cyclic and ordinary partial orders and described representations of cyclic orders. In his joint paper with Novotný [304] notions of width and strong width were introduced for cyclic orders; they investigated complements of cyclic orders (cocyclic orders). They [307] constructed a cyclically ordered set A that is m-universal in the sense that, for every cyclically ordered set B of cardinality m, there exists a cyclically ordered subset of A for which B is a strongly homomorphic image. They [306] studied a binary operation, called power, over cyclically ordered sets. For cyclically ordered sets Novák [303] introduced the notions of cardinal sum and product and cardinal power that satisfy ordinary properties. He [301] constructed an example of a cyclically ordered set with seven elements whose cyclic order cannot be extended to a linear cyclic order. Together with Chajda [112] he found conditions under which such an extension is possible. Novák and Novotný [305] developed a dimension theory for cyclic and cocyclic orders. Width and strong width turned out to be particular cases of dimension. A cut of a cyclic order is defined as a partial order $\leq$ on the same set such that its ternary relation $x < y < z$ coincides with the cyclic order. Novák [302] proved that for every linear order there exists precisely one cyclic order whose cut is the strict part of this linear order. One can get acquainted with the theory of fuzzy orders by Kaufmann's book [21].

Policki [318] observed that in Salamucha's proof of God's existence one cannot drop the requirement of linearity for the order relation used there. Krynicki and Väänänen [254] proved that every partially ordered class whose initial segments are sets is isomorphically embeddable in the class of nontrivial logics closed under the existential quantifier and partially ordered by the relation of being a sublogic. Drugush [12] showed that the lattice join of two superintuitionistic logics could be modelled by a finite forest if each of the logics can be modelled by a finite forest. By definition, two types, probably with different numbers of variables over a model, are in ⊲ relation if the second type can be realized in every elementary extension of this model saturated over countable subsets in which the first type can be realized. Lascar [259] considered this partial order and discovered that in the case of a superstable theory it determines the structure of a distributive lattice freely generated by atoms. Charreton and Pouzet [114] investigated the isomorphic embedding relation on an arbitrary class of structures and carried certain properties of the class of chains over to this case.

Dëgtev [11] proved that elementary theories of ∨-semilattices of disjunctive, positive, linear, and tabular degrees are pairwise disjoint with at most one possible exception of the positive and tabular degrees. Farrington [159] investigated the problem of the form of the lattice of degrees of constructivity in various models of set theory. Groszek and Slaman [187] proved that two statements on the order structure of ∨-semilattices of Turing degrees are compatible with the ZFC axioms. Mal'tsev [25] described the semilattice of degrees of tt-reducibility on tables of length 1. It is continual and isomorphic to the semilattice of all m-degrees. Schwarz [349] decomposed the semilattice R of recursively enumerable degrees into a union of the class P of promptly simple and the class M of cappable recursively enumerable degrees and investigated properties of the factor semilattice R/M. Omanadze [33] showed that a Q-degree of a maximal set is not the least upper bound for any incomparable Q-degrees. Ambos-Spies [65] proved that every countable distributive lattice is $(0, 1)$-embeddable in every interval of p-degrees of Turing (T) and multivalued (m) reducibility, and the ∨-semilattice of p-m-degrees is distributive, while this was not so for p-T-degrees. Selivanov

[38] investigated quasi-orders connected with numerated sets.

Butkovičová [108] constructed special chains of ultrafilters in the Rudin-Frolík partial ordering. She found in the ZFC system a nonminimal ultrafilter on the set of natural numbers that does not have immediate predecessors [106] and studied the existence problem and type of various gaps in the poset of ultrafilters on a countable set [107]. Baumgartner, Taylor, and Wagon [79] gave a structural analysis of α-complete ideals on an uncountable cardinal number α. It turned out that numerous properties of ultrafilters could be transferred to sufficiently saturated ideals, while there are many unexpected deviations. Jech and Prikry [226] considered a certain partial ordering on the set of mappings of ω_1 into ω and showed that in the case of the real-valued measurability of the continuum the least possible cardinality of a cofinal subset in this poset is equal to $2^{\aleph_1}$.

For $n \neq 6$ and $n \neq 7$, Simion [357] proved the absence of nontrivial automorphisms in the lattice $\mathscr{L}_n$ of partitions of a natural number n into a sum of natural numbers. For $\mathscr{L}_6$ and $\mathscr{L}_7$ the automorphism group is isomorphic to the group $Z_2 \times Z_2$. Gansner [174] established that the lattice of order ideals of certain special partial orderings on finite sets of numbers has the Sperner property. Li and Schein [345] produced an example of a poset that is the carrier of a Clifford inverse semigroup but not of a commutative inverse semigroup. Trueman [376] observed that the lattice of fundamental quasi-orders on a completely simple semigroup is not modular. In Freese's survey [169] the problem of describing posets embeddable in free lattices was treated. For example, the existence of an uncountable chain prevents that. There is an analysis of results on finite posets that freely generate either finite lattices or lattices with diagrams that can be easily visualized. Nation [288] considered the problem of embedding posets, that are unions of two infinite antichains, in free lattices. An interval space is defined as an inclusion-ordered poset of closed intervals of a poset, while an interval structure is defined as a relatively atomic conditionally complete poset without 0. Ney [294] pointed out necessary and sufficient conditions for representability of interval structures as interval spaces. A tree algebra is a field of sets generated by the principal filters of a tree, and the interval algebra of a linear order is the Boolean algebra generated by intervals open on the right. Brenner and Monk [99] showed that every tree algebra is embeddable in an interval algebra, but the converse is not true. Bandelt [71] considered the independence problem for a system of axioms for abstract tree algebras. Fujishige [171] suggested a method which, given a lattice of subsets of a finite set, produces two smaller distributive lattices that permit one to reconstruct the original lattice in a unique way. Koh [245] proved that every finite distributive lattice is isomorphic to the lattice of maximal antichains of a suitable finite poset (by definition, $A \leq B \Leftrightarrow (\forall a \in A)(\exists b \in B)(a \leq b)$ for antichains). Gansner [175] pointed out two methods of partitioning the ideal lattice of a finite poset into disjoint chains and found necessary and sufficient conditions for coincidence of these partitions. Pezzoli [316] produced a recursive method for enumeration of ideals in a finite poset. Romanovich [36] found necessary and sufficient conditions for the poset of orthogonality relations on a lattice to be a lattice or a distributive lattice. Studying representations of distributive lattices, Crapo [127] worked out a conceptual system, the most important part of which is the concept of the identity of a binary relation. Hartwig and Drazin [200] introduced a partial ordering on the set of complex $n \times n$ matrices that defines the structure of a $\wedge$-semilattice on it. Baklasary and Kala [69] considered a partial order on the set of Hermitian $n \times n$ matrices.

Birkhoff's survey [83] discussed the role of posets in geometry. Edelman [149] computed the Möbius function for the poset whose elements are regions of the space $\mathbb{R}^n$, obtained as a result of partitioning by finitely many hyperplanes. Vonkomerová [381] introduced a notion of a C-lattice which unites various types of lattices with additional structures met in functional analysis. Because of their applications in functional analysis, Gurvits, Maslyuchenko, and Popov [10] defined and investigated certain characteristics of posets reflecting their extent. Blair [90] proved that every finite distributive lattice could be realized as a lattice of stable matchings. Ito [218] studied classes of automata partially ordered by the relation "is a homomorphic image of". He described certain classes which are lattices in this partial order and produced algorithms for computing exact bounds for two automata. Scott's survey [352] is devoted to applications of posets in computer science. Janowitz [221] considered, among other topics, a model for cluster analysis which leads to interesting problems in the theory of posets. Roubens and Vincke [337] investigated partial orderings connected with decision making theory and utility theory. Using economic models, Kannai and Peleg [234] discussed the possibility of a natural extension of an order defined on a finite set to the set of its subsets. Their results stimulated a lively discussion ([74], [75], [163], [202], [210]).

§VI.2. Semilattices

Gornostaev [9] showed that semilattices, considered as models with a single ternary relation, generate a quasivariety without a finite basis for quasi-identities. Zimmermann and Köhler [396] proved that Mal'tsev's product of two finitely based varieties of Brouwerian semilattices is finitely based. Penner [314] proved that the variety of all semilattices does not possess a finite basis for hyperidentities. For every natural n the set of all semilattice hyperidentities with a single n-ary operation symbol has a finite basis [315]. In this latter article it is observed that, in the lattice of closed sets of hyperidentities, the semilattices can be decomposed into a join of distributive lattices and abelian groups, while diagonal semigroups cover semilattices. Länger [258] characterized the function $f(x_1, x_2, \ldots, x_n) = x_1 \wedge x_2 \wedge \cdots \wedge x_n$ in a $\wedge$-semilattice in terms of identities.

Mamedov [26] showed that a ∨-semilattice is atomic-compact as a poset if and only if it is complete and each of its downwards directed subsets possesses the greatest lower bound.

Slatinsky [359] investigated properties of cardinal sum and product, ordinal and lexicographic sums for ∧-semilattices. Pin [317] proved that a finite monoid divides a semidirect product of two finite semigroups if and only if it satisfies the identities $xyztxz = xyztzx$ and $xyx = xyx^2$.

Fraser and Albert [167] established that the tensor product of ∨-semilattices is associative, and Fraser and Bell [168] solved the word problem in a tensor product of distributive ∨-semilattices.

Grandis [182] used general results on embedding of categories to construct embeddings of certain classes of semilattices in classes of lattices.

A globalization of a semilattice $(S, \cdot)$ is defined as the set $P^*(S)$ of all of its nonempty subsets with the operation $AB = \{xy \colon x \in A, y \in B\}$. Gould, Iskra, and Tsinakis [181] showed that an isomorphism of globalizations of two ∧-semilattices with 1 implies an isomorphism of the semilattices. An analogous result holds for chains (as ∧-semilattices) and lattices. Kobayashi [244] proved that for any semilattices S and T every isomorphism $P^*(S) \cong P^*(T)$ maps S onto T. We note Tamura's results [364]: (1) a ∧-semilattice is complete if and only if its globalization is a lattice with 1; (2) a globalization of a complete semilattice S is a modular lattice if and only if S is dually well ordered.

Ramana Murty and Raman [327] defined a construction of a triple for semilattices with 1 admitting neutral p-closure operators.

Ditor [136] evaluated cardinalities of semilattices of finite ∧-breadth and also obtained results for lattices. For example, a semimodular lattice of finite ∧-breadth, in which principal ideals are finite, is at most countable. A lattice is called a k-lattice if all of its principal ideals are finite and each element covers at most k elements. An example of a countable 2-lattice was given; it is not known whether a continual 3-lattice exists. Shiryaev [55] characterized semilattices of finite breadth with pairwise comparable decomposable elements.

Cornish and Noor [123] investigated standard elements of near-lattices. Krishna Murty [251] showed that an element a of a bounded ∧-semilattice S is central if and only if it is complemented and the principal filter $[a)$ is neutral in the lattice $F(S)$ of filters. Modular pairs were generalized for semilattices. Continuing this research Thakare, Wasadikar, and Maeda [368] carried over to atomic generated ∨-semilattices some of Maeda's results on atomic generated lattices.

Shiryaev [56] described semilattices whose lattices of subsemilattices satisfy the ∨-semidistributivity quasi-identity. Köhler [246] showed that the lattice of all total subalgebras of a Brouwerian semilattice is a maximal distributive sublattice in the subalgebra lattice of this semilattice. Every finite lattice of this class is Boolean.

Fleischer [165] used the classic construction of MacNeille to obtain Hickman's results on a unique extension of a ∧-congruence on a ∧-semilattice S to a lattice congruence on the distributive lattice freely generated by S. Zhitomirskiĭ [13] suggested a new characterization of the congruence lattices of semilattices. The Boolean part of this class, which is known to consist of all lattices isomorphic to the congruence lattices of locally finite trees, is elementarily axiomatizable in the class of all complete lattices. Kurinnoĭ [22] investigated semilattices with a fixed carrier and some congruences. Ramana Murty and Raman [328] showed that congruences of a downwards directed ∨-semilattice corresponding to its distributive ideals are pairwise permutable. A congruence generated by a standard ideal commutes with a congruence generated by an arbitrary ideal. Idziak [216] proved that if the signature of a BCK-algebra A contains at least one of the lattice operations ∨, ∧, then Con A is isomorphic to the lattice $F(A)$ of all filters. Wolniewicz [392] studied, on the algebra of subsets of ∨-semilattices with 1, congruences whose quotients are bounded distributive or Boolean lattices.

Chajda and Nieminen [110] proved direct decomposability of tolerances on a direct product of directed semilattices, and Chajda and Zelinka [113] found a tolerance on a direct product of an infinite family of semilattices which is not decomposable into a direct product of tolerances on the factors. All three of these authors showed [111] that the lattice of tolerances of a chain is isomorphic to the lattice of stable reflexive relations contained in the order relation.

Kolibiar [248] found conditions under which a stable partial order on a semilattice is connected with its subdirect decomposition. For lattices this is equivalent to stability with respect to both lattice operations. Goberstein [179] showed that the set of all stable orders on a semilattice with 0, all of whose nonzero elements are pairwise incomparable, is a modular lattice and that modularity of the lattice of stable orders characterizes this type of semilattice in the class of all semilattices. Bredikhin [4] proved that every semilattice S admits an embedding f in the semilattice of partial orders on a suitable set so that $f(a \vee b) = f(a) \circ f(b)$ if $a \vee b$ exists in S.

Jayaram [223] introduced semiatoms in ∧-semilattices and characterized Boolean algebras as semiatomic uniquely complemented ∧-semilattices. In another paper [224] he axiomatized a finite Boolean algebra as a complemented ∧-semilattice in which each element is representable as a finite meet of prime elements. He observed [225] that a ∧-semilattice with 1 is distributive if and only if each of its filters is a congruence filter. Dobbertin [138] studied distributive semilattices from the viewpoint of special monoids. Hickman and Monro [204] attempted to transfer concepts and results connected with distributivity of ∧-semilattices to arbitrary posets. Thakare and Pawar [367] considered properties of minimal prime ideals in 0-distributive semilattices. Jayaram [222] studied congruences and neutral filters in 0-distributive quasi-complemented semilattices. A semilattice S is called

P-uniform if, for each of its proper prime ideals J, there exist elements $e \in J$ and $f \in S \setminus J$ such that $e \wedge S = f \wedge S$. Hoo and Shum [214] characterized P-uniform semilattices as 0-distributive semilattices in which every proper prime ideal is a minimal prime ideal. They corrected an old result of Balbes. Hickman [203] defined mildly distributive semilattices, gave criteria of mild distributivity of semilattices, and investigated $\vee$-partial congruences on them.

Cirulis [Tsirulis] [51] studied properties of transformations of the form $\varphi_a(x) = (x \to a) \to x$ in an implicative semilattice. Hoo [212] transferred certain results on atoms and $\wedge$-irreducible elements of implicative semilattices to arbitrary $\wedge$-semilattices. In another paper [211] and his joint paper [213] with Ramana Murty that corrects it, the notion of an a-admissible a-implicative semilattice was introduced and analogs of theorems on admissible implicative lattices proved. Ramana Murty and Krishna Murty [326] extended the concept of pseudocomplements in $\wedge$-semilattices with 0 and studied properties of the so-called weakly distributive $*$-semilattices.

Wolniewicz [393] formalized Wittgenstein's ideas from his *Tractatus* in the language of $\vee$-semilattices with 1. Zaretskiĭ [16] investigated the injective hull of the $\cup$-semilattice of all subsets of a nonempty finite set. Bandelt and Schmid [73] showed that the semilattice of quotients of a semilattice S is isomorphic to the set of all of its Q-sets under the operation $\cap$. Schmid [347] described the maximal semilattice of quotients $Q(S)$ of a finite semilattice S and found conditions equivalent to its rational completeness (that is, to the equality $Q(S) = S$). Ježek and Kepka [228] defined an idempotent semimodule over a commutative semiring and found all finite simple idempotent semimodules. Cao [109] considered semirings arising from semilattices with a commutative multiplication distributive with respect to addition. Shoji [356] showed that every selfinjective nonsingular semilattice is a complete Boolean algebra and vice versa. Fofanova [48] described the main results and problems related to polygons over distributive lattices. Kornienko [20] showed that every ideal of an arbitrary distributive lattice L with 0 and 1 is a flat L-polygon. Cornish ([121], [122]) studied connections between near-lattices and BCK-algebras. Abid [60] introduced and studied a metric on $\vee$-semilattices. Ježek [227] defined the real line in the language of simple semilattices with two fixed permutable automorphisms. Cowen [125] connected the notion of compactness of a subset, introduced by him for special semilattices, with certain problems of graph theory. Markowsky [267] proved that the problem of extension of semilattices suggested by Arbib and Manese is NP-complex.

§VI.3. Generalizations of lattices

A poset is called a hypolattice if each of its closed intervals is a lattice and the lattice operations coincide on overlapping intervals. Miller [276] constructed a representation for Boolean hypolattices that generalized Stone's representation for Boolean algebras (a hypolattice is Boolean, by definition, if it is relatively complemented and all its sublattices are distributive). Powell [323] applied properties of Boolean hypolattices to l-groups.

Romanowska [334] showed that a free bisemilattice with one distributive law and n generators is finite, and she described a 60-element free o-distributive bisemilattice with three generators. In another paper [336] she suggested a method for constructing bisemilattices from lattices and semilattices that generalizes the construction of a rigid semilattice of algebras. Dudek ([141], [142]) showed that a bisemilattice is a lattice if and only if in its polynomials $(x \vee y) \wedge y$ and $(x \wedge y) \vee y$ not both variables are essential, and gave a lower estimate of the number of those polynomials on a bisemilattice that depend essentially on n variables ($> 2 + n!$ for $n \geq 3$). In his joint paper [143] with Romanowska they proved that bisemilattices with four essentially binary polynomials could be divided into five natural classes. Romanowska [335] considered in the set $\mathscr{L}^P$ of all mappings of a finite poset P into a finite distributive lattice $\mathscr{L}$ pointwise lattice operations as well as a binary operation of convolution which is associative if and only if P is a forest. In this case the algebra obtained is a distributive bisemilattice. Gałuszka [173] observed that if one of the distributive laws and an anomal identity hold in a bisemilattice, then the latter is a lattice. Various classes of generalized lattices were introduced by weak laws of absorption in bisemilattices and an ascending chain of varieties of such generalized lattices was constructed.

Schweigert [350] studied generalizations of lattices that appeared as a result of weakening associative and commutative laws. In such a case a connection with posets turns out to be many-faceted.

Anderson [67] introduced seven generalizations of the concept of an exact bound in a finite poset, and studied properties of 49 thus-defined algebras with two binary operations.

Leutöla and Nieminen [261] considered posets P in which every two-element subset has minimal upper bounds and maximal lower bounds. Using a choice function χ they chose one minimal upper bound and one maximal lower bound. Thus they obtained an algebra $(P, \vee, \wedge)$ which is called a χ-lattice. They investigated identities in χ-lattices and their relation to the original partial order. Nieminen [296] defined notions of ideal, congruence, modularity, and distributivity for χ-lattices and showed that the congruence lattice of a χ-lattice is always distributive. In another paper [297] he showed that the inclusion-ordered set CSub of convex sublattices of a lattice is a χ-lattice. A lattice $\mathscr{L}$ is modular (distributive) if and only if the χ-lattice CSub$\mathscr{L}$ is modular (distributive). Observe that χ-lattices are connected with multilattices in the sense of Klaucová. Tomková [372] found certain sufficient conditions under which graphs of two downwards directed multilattices are isomorphic. If the multilattices are lower semimodular, then these conditions become necessary too.

Davey and McCarthy [132] constructed a topological representation theory for the variety of weakly associative lattices generated by the three-element set $\{0, 1, 2\}$ with the relation $0 < 1$, $1 < 2$, $2 < 0$ on which, by definition, $x \wedge y = x \Leftrightarrow x \leq y \Leftrightarrow x \vee y = y$. They described finite algebras and congruence lattices of algebras from this variety.

A pseudosemilattice is an idempotent groupoid that satisfies the identities $(xy)(xz) = (xy)z$, $(xy)((xz)(xt)) = ((xy)(xz))(xt)$ and the dual ones(^{2}). Meakin ([270], [271]) described the structure of a free pseudosemilattice with an arbitrary number of generators and, jointly with Pastijn [272], showed how pseudosemilattices could be obtained from semilattices, posets, and semilattice homomorphisms. Nambooripad [283] characterized pseudosemilattices that are bi-ordered sets—they form a subvariety in the variety of all pseudosemilattices. He [285] represented pseudosemilattices of the general form as certain subsets of the product of two posets.

Kröger [253] used two right normal bands to construct an algebra, called a generalized lattice. Höft [209] defined a sum of a double system of posets and studied its properties.

Konstantinidou and Mittas [249] found properties of hyperlattices, which are algebras with two hyperoperations that generalize the lattice operations join and meet.

Bibliography

1. E. A. Babaev, *On semigroups of isotone mappings*, Preprint No. 202-85, deposited at VINITI, 1985. (Russian)

2. _____, *An abstract characterization of endomorphism semigroups of an ordered set*, Preprint No. 5527-85, deposited at VINITI, 1985. (Russian)

3. V. M. Bondarenko, *Exact partially ordered sets of infinite growth*, Linear Algebra and the Theory of Representations, Akad. Nauk Ukrain. SSR Inst. Mat., Kiev, 1983, pp. 68–85. (Russian)

4. D. A. Bredikhin, *A representation theorem for semilattices*, Proc. Amer. Math. Soc. **90** (1984), no. 2, 219–220.

5. L. P. Vinogradova and A. V. Yakovlev, *Representation of a set with two partial order relations*, Zap. Nauchn. Sem. Leningrad. Otdel. Mat. Inst. Steklov. (LOMI) **132** (1983), 69–75. (Russian)

6. I. M. Gel'fand and V. A. Ponomarev, *Free modular lattices and their representations*, Uspekhi Mat. Nauk **29** (1974), no. 6 (180), 3–58; English transl., London Math. Soc. Lecture Note Ser. vol. 69, Cambridge Univ. Press, Cambridge, 1982, pp. 173–228.

7. _____, *Lattices, representations, and algebras connected with them.* I, Uspekhi Mat. Nauk **31** (1976), no. 5 (191), 71–88; English transl., London Math. Soc. Lecture Note Ser., vol. 69, Cambridge Univ. Press, Cambridge, 1982, pp. 229–247.

8. _____, *Lattices, representations, and algebras connected with them.* II, Uspekhi Mat. Nauk vol. 32 (1977) no. 1 (193), 85–106; English transl., London Math. Soc. Lecture Note Ser. vol. 69, Cambridge Univ. Press, Cambridge, 1982, pp. 249–272.

(^{2})*Translator's note.* For the definition and properties of semilattices see Boris M. Schein, *Pseudosemilattices and pseudolattices*, Izvestiya Vysshikh Uchebnykh Zavedeniĭ, Matematika 1972, no. 2, 81–97 [Russian]; MR **46** #5203 [English translation in: Transl. of the Amer. Math. Society (II) **119** (1983), 1–16].

9. O. M. Gornostaev, *Quasivarieties generated by classes of models*, Preprint No. 2206-85, deposited at VINITI by the editors of Sibirsk. Mat. Zh., 1985. (Russian)

10. L. N. Gurvits, V. K. Maslyuchenko, and M. M. Popov, *Ordinal characteristics of partially ordered sets*, Preprint No. 311-83, deposited at VINITI, 1983. (Russian)

11. A. N. Dëgtev, *Semilattices of disjunctive and linear degrees*, Mat. Zametki **38** (1985), no. 2, 310–316, 350; English transl. in Math. Notes **38** (1985), no. 1–2.

12. Ya. M. Drugush, *Union of logics modeled by finite trees*, Algebra i Logika **21** (1982), no. 2, 149–161; English transl. in Algebra and Logic **21** (1982), no. 2.

13. G. I. Zhitomirskiĭ, *A characterization of the congruence lattices on semilattices*, Preprint No. 8142-84, deposited at VINITI by the editors of Izv. Vyssh. Uchebn. Zaved. Mat., 1984. (Russian)

14. A. G. Zavadskiĭ, *The structure of representations of partially ordered sets of finite growth*, Linear Algebra and the Theory of Representations, Akad. Nauk Ukrain. SSR Inst. Mat., Kiev, 1983, pp. 55–67. (Russian)

15. A. G. Zavadskiĭ and L. A. Nazarova, *Partially ordered sets of finite growth*, Funktsional. Anal. i Prilozhen. **16** (1982), no. 2, 72–73; English transl. in Functional Anal. Appl. **16** (1982), no. 2.

16. K. A. Zaretskiĭ, *Injective hull of a semilattice of sets*, Ordered Sets and Lattices, No. 8, Izdat. Saratov Univ., Saratov, 1982, pp. 32–39. (Russian)

17. S. P. Zervos, *Cardinals as orbits of groups of automorphisms of ordered sets*, Trudy Mat. Inst. Steklov. **154** (1983), 118–123; English transl. in Proc. Steklov Inst. Math. **154** (1984).

18. M. T. Kvanchilashvili, *Absolute estimates over ordered sets*, Soobshch. Akad. Nauk Gruzin. SSR **114** (1984), no. 3, 493–496. (Russian)

19. S. R. Kogalovskiĭ and V. V. Soldatova, *Remarks on fixed points of sets of isotone transformations*, Preprint No. 850-84, deposited at VINITI, 1984. (Russian)

20. V. S. Kornienko, *Distributive structures with plane ideals*, Ordered Sets and Lattices, No. 8, Izdat. Saratov Univ., Saratov, 1982, pp. 63–70. (Russian)

21. A. Kaufmann, *Introduction à la théorie des sous-ensembles flous à l'usage des ingénieurs*, Tome I: *Éléments théoriques de base*, Masson, Paris, New York, and Barcelona, 1977. (French)

22. G. Ch. Kurinnoĭ, *Semilattices with shared congruences*, Preprint No. 28Uk-85, deposited at the Ukrainian NIINTI, 1985. (Russian)

23. S. A. Liber, *Topology and order*, Ordered Sets and Lattices, No. 7, Izdat. Saratov Univ., Saratov, 1983, pp. 106–121. (Russian)

24. Yu. A. Mitropol'skiĭ, (ed.), *Linear Algebra and the Theory of Representations*, Akad. Nauk Ukrain. SSR Inst. Mat., Kiev, 1983. (Russian)

25. An. A. Mal'tsev, *Structure of the semilattice of $tt1$-degrees*, Sibirsk. Mat. Zh. **26** (1985), no. 2, 132–139, 223; English transl. in Siberian Math. J. **26** (1985), no. 2.

26. O. M. Mamedov, *Atomic-compact semilattices*, Akad. Nauk Azerbaĭdzhan. SSR Dokl. **39** (1983), no. 7, 7–9. (Russian)

27. A. B. Mikhaĭlov, *On semigroups of mappings*, Preprint No. 850-85, deposited at VINITI, 1984. (Russian)

28. A. S. Morozov, *The group $Aut_r(\mathbf{Q}, \leq)$ is not constructivizable*, Mat. Zametki **36** (1984), no. 4, 473–478; English transl. in Math. Notes **36** (1984), no. 3–4.

29. L. A. Nazarova and A. V. Roĭter, *The linear-algebraic method in the theory of representations*, Linear Algebra and the Theory of Representations, Akad. Nauk Ukrain. SSR Inst. Mat., Kiev, 1983, pp. 3–18. (Russian)

30. _____, *Representations and forms of weakly completed partially ordered sets*, Linear Algebra and the Theory of Representations, Akad. Nauk Ukrain. SSR Inst. Mat., Kiev, 1983, pp. 19–54. (Russian)

31. A. V. Nikulin, *Weakly primary algebras of finite type*, Preprint No. 1729Uk-84, deposited at the Ukrainian NIINTI, 1984. (Russian)

32. A. T. Nurtazin, *An elementary classification and some properties of complete orders*, Boundary Value Problems for Differential Equations and their Applications in Mechanics and Technology, "Nauka" Kazakh. SSR, Alma-Ata, 1983, pp. 122–125. (Russian)

33. R. Sh. Omanadze, *The upper semilattice of recursively enumerable Q-degrees*, Algebra i Logika **23** (1984), no. 2, 175–184, 240–241; English transl. in Algebra and Logic **23** (1984), no. 2.

34. A. G. Pinus, *"Embeddability" and "endomorphy" relations on linear orders*, Ordered Sets and Lattices, no. 8, Izdat. Saratov Univ., Saratov, 1982, pp. 81–91. (Russian)

35. _____, *Uniqueness of linear extensions of partial orders*, Sibirsk. Mat. Zh. **24** (1983), no. 4, 131–137; English transl. in Siberian Math. J. **24** (1983), no. 4.

36. V. A. Romanovich, *On structural properties of the semi-orthogonality relation on lattices*, Ordered Sets and Lattices, no. 8, Izdat. Saratov Univ., Saratov, 1982, pp. 96–108. (Russian)

37. V. N. Salii, *Ordered sets. Semilattices. Generalizations of lattices*, Ordered Sets and Lattices, no. 7, Izdat. Saratov Univ., Saratov, 1983, pp. 121–142. (Russian)

38. V. L. Selivanov, *The structure of degrees of generalized index sets*, Algebra i Logika **21** (1982), no. 4, 472–491; English transl. in Algebra and Logic **21** (1982), no. 4.

39. L. A. Skornyakov, *A general view of representations of monoids*, Algebra i Logika **20** (1981), no. 5, 571–574; English transl. in Algebra and Logic **20** (1981), no. 5.

40. _____, *Monoids of isotone mappings*, Mat. Sb. **123(165)** (1984), no. 1, 50–68; English transl. in Math. USSR-Sb. **51** (1985), no. 1.

41. _____, *Injective objects of categories of representations of monoids*, Mat. Zametki **36** (1984), no. 2, 159–170; English transl. in Math. Notes **36** (1984), no. 1–2.

42. _____, *Injectivity of all ordered left polygons over a monoid*, Vestnik Moskov. Univ. Ser. I Mat. Mekh. **1986**, no. 3, 17–19. (Russian)

43. G. A. Sokhadze, *A generalization of Dilworth's theorem*, Preprint No. 3774-85, deposited at VINITI by the editors of Izv. Vyssh. Uchebn. Zaved. Mat., 1985. (Russian)

44. R. B. Stekol'shchik, *Invariant elements in a modular lattice*, Funktsional. Anal. i Prilozhen. **18** (1984), no. 1, 82–83; English transl. in Functional Anal. Appl. **18** (1984), no. 1.

45. V. M. Tararin, *On automorphism groups of linearly ordered sets*, Preprint No. 677-84, deposited at VINITI, 1984. (Russian)

46. N. B. Ueiskaya, *Some related fixed point properties of semigroups of transformations*, Theory of Semigroups and Its Applications, no. 7, Izdat. Saratov Univ., Saratov, 1984, pp. 82–92. (Russian)

47. N. S. Ulitina, *Ordered sets with identical identities in semigroups of directed transformations*, Associative Actions, Leningrad. Gos. Ped. Inst., Leningrad, 1983, pp. 106–115. (Russian)

48. T. S. Fofanova, *Polygons over distributive lattices*, Universal Algebra (Esztergom, 1977), Colloq. Math. Soc. János Bolyai, vol. 29, North-Holland, Amsterdam and New York, 1982, pp. 289–292.

49. _____, *On the fixed point property of partially ordered sets*, Contributions to Lattice Theory (Szeged, 1980), Colloq. Math. Soc. János Bolyai, vol. 33, North-Holland, Amsterdam and New York, 1983, pp. 401–406.

50. _____, *The fixed points property in ordered sets of width two*, Order **4** (1987), no. 2, 101–106.

51. Ya. P. Tsirulis [J. P. Cirulis], *Relative pseudocomplements in semilattices*, Izv. Vyssh. Uchebn. Zaved. Mat. (1982), no. 2, 78–80; English transl. in Soviet Math. (Iz. VUZ) **26** (1982), no. 2.

52. _____, *Mutual definability of linear order, betweenness, and codirectedness*, Latv. Mat. Ezhegodnik **1983**, no. 27, 252–259. (Russian)

53. A. I. Tsutsura, *Definability of ordered sets by a partial groupoid of endomorphisms of rank 2*, Associative Actions, Leningrad. Gos. Ped. Inst., Leningrad, 1983, pp. 124–137. (Russian)

54. A. Tsyl'ke, *On perfect elements of free modular lattices*, Funktsional. Anal. i Prilozhen. **16** (1982), no. 1, 87–88; English transl. in Functional Anal. Appl. **16** (1982), no. 1.

55. V. M. Shiryaev, *Semilattices of finite width with mutually comparable decomposable elements*, Ordered Sets and Lattices, no. 8, Izdat. Saratov Univ., Saratov, 1982, pp. 108–120. (Russian)

56. _____, *Semilattices with semidistributive lattices of subsemilattices*, Vestnik Beloruss. Gos. Univ. Ser. I (1985), no. 1, 61–64, 80. (Russian)

57. A. V. Yakovlev, *Representations of lattices over division rings*, Trudy Mat. Inst. Steklov. **165** (1984), 220–228; English transl. in Proc. Steklov Inst. Math. **165** (1985).

58. A. Abian, *A fixed point theorem for image-intersecting mappings*, Proc. Amer. Math. Soc. **73** (1979), no. 3, 300–302.

59. A. Abian and J. Lihová, *Compact partially ordered sets and compactification of partially ordered sets*, Math. Slovaca **32** (1982), no. 4, 321–325.

60. Z. Abid, *Semi-valuation et métrique associée*, Math. Sci. Humaines **22** (1984), no. 87, 67–82.

61. U. Abraham, *Aronszajn trees on $\aleph_2$ and $\aleph_3$*, Ann. Pure Appl. Logic **24** (1983), no. 3, 213–230.

62. M. E. Adams and J. Sichler, *Automorphism groups of posets and lattices with a given subset of fixed points*, Monatsh. Math. **93** (1982), no. 3, 173–190.

63. M. H. Albert, *Iteratively algebraic posets have the ACC*, Semigroup Forum **30** (1984), no. 3, 371–373.

64. T. Albu, *Gabriel dimension of partially ordered sets. II*, Bull. Math. Soc. Sci. Math. R. S. Roumanie (N. S.) **28(76)** (1984), no. 3, 199–205.

65. K. Ambos-Spies, *On the structure of polynomial time degrees*, Lecture Notes in Comput. Sci., vol. 166, Springer-Verlag, Berlin and New York, 1984, pp. 198–208.

66. M. Anderson and C. C. Edwards, *A representation theorem for distributive l-monoids*, Canad. Math. Bull. **27** (1984), no. 2, 238–240.

67. Ph. H. Anderson, *Latticoids defined with generalized suprema and infima*, Algebra Universalis **16** (1983), no. 3, 304–311.

68. K. Baclawski, *Canonical modules of partially ordered sets*, J. Algebra **83** (1983), no. 1, 1–5.

69. J. K. Baklasary and R. Kala, *Partial orderings between matrices one of which is of rank one*, Bull. Polish Acad. Sci. Math. **31** (1983), no. 1–2, 5–7.

70. B. Banaschewski and E. Nelson, *Completions of partially ordered sets*, SIAM J. Comput. **11** (1982), no. 3, 521–528.

71. H.-J. Bandelt, *Ein Axiomensystem für Baum-Algebren*, Časopis Pěst. Mat. **108** (1983), no. 4, 353–355. (Russian summary)

72. H.-J. Bandelt and Marcel Erné, *The category of $\mathbb{Z}$-continuous posets*, J. Pure Appl. Algebra **30** (1983), no. 3, 219–226.

73. H.-J. Bandelt and J. Schmid, *Multipliers on semilattices and semigroups of quotients*, Houston J. Math. **9** (1983), no. 3, 333–343.

74. S. Barberá, C. R. Barrett, and P. K. Pattanaik, *On some axioms for ranking sets of alternatives*, J. Econom. Theory **33** (1984), no. 2, 301–308.

75. S. Barberá and P. K. Pattanaik, *Extending an order on a set to the power set: Some remarks on Kannai and Peleg's approach*, J. Econom. Theory **32** (1984), no. 1, 185–191.

76. A. Baudisch, D. G. Seese, and H.-P. Tuschik, *ω-trees in stationary logic*, Fund. Math. **119** (1983), no. 3, 205–215.

77. H. Bauer and R. Wille, *A proof of Hashimoto's refinement theorem*, Contributions to General Algebra, 2 (Klagenfurt, 1982), Hölder-Pichler-Tempsky, Vienna, 1983, pp. 35–41.

78. J. E. Baumgartner, *Order types of real numbers and other uncountable orderings*, Ordered Sets (Banff, Alta., 1981), NATO Adv. Study Inst. Ser. C: Math. Phys. Sci., vol. 83, Reidel, Dordrecht, Boston, 1982, pp. 239–277.

79. J. E. Baumgartner, A. D. Taylor, and S. Wagon, *Structural properties of ideals*, Dissertationes Math. (Rozprawy Mat.) **197** (1982).

80. M. M. Bayer and L. J. Billera, *Counting faces and chains in polytopes and posets*, Contemp. Math. **34** (1984), 207–252.

81. X. Berenguer, J. Diaz, and L. Harper, *A solution of the Sperner-Erdös problem*, Theoret. Comput. Sci. **21** (1982), no. 1, 99–103.

82. R. Berghammer and G. Schmidt, *Discrete ordering relations*, Discrete Math. **43** (1983), no. 1, 1–7.

83. G. Birkhoff, *Ordered sets in geometry*, Ordered Sets (Banff, Alta., 1981), NATO Adv. Study Inst. Ser. C: Math. Phys. Sci., vol. 83, Reidel, Dordrecht, Boston, 1982, pp. 407–443.

84. A. Björner, *Posets and regular CW complexes*, Rep. Dept. Math. Univ. Stockholm (1982), no. 33.

85. _____, *Posets, regular CW complexes and Bruhat order*, European J. Combin. **5** (1984), no. 1, 7-16.

86. A. Björner, A. M. Garsia, and R. P. Stanley, *An introduction to Cohen-Macaulay partially ordered sets*, Ordered Sets (Banff, Alta., 1981), NATO Adv. Study Inst. Ser. C: Math. Phys. Sci., vol. 83, Reidel, Dordrecht, Boston, 1982, pp. 583-615.

87. A. Björner and M. Wachs, *On lexicographically shellable posets*, Rep. Dept. Math. Univ. Stockholm (1982), no. 9.

88. _____, *On lexicographically shellable posets*, Trans. Amer. Math. Soc. **277** (1983), no. 1, 323-341.

89. A. Björner and J. W. Walker, *A homotopy complementation formula for partially ordered sets*, European J. Combin. **4** (1983), no. 1, 11-19.

90. Ch. Blair, *Every finite distributive lattice is a set of stable matching*, J. Combin. Theory **37** (1984), no. 3, 353-356.

91. Ch. Blair and A. E. Roth, *An extension and simple proof of a constrained lattice fixed point theorem*, Algebra Universalis **9** (1979), no. 1, 131-132.

92. _____, *Erratum: "An extension and simple proof of a constrained lattice fixed point theorem"*, Algebra Universalis **12** (1981), no. 1, 134.

93. R. Blažková and J. Chvalina, *Regularity and transitivity of local-automorphism semigroups of locally finite forests*, Arch. Math. (Brno) **20** (1984), no. 4, 183-194.

94. T. S. Blyth and J. C. Varlet, *Fixed points in MS-algebras*, Bull. Soc. Roy. Sci. Liège **53** (1984), no. 1, 3-8.

95. R. Bonnet and M. Pouzet, *Linear extensions of ordered sets*, Ordered Sets (Banff, Alta., 1981), NATO Adv. Study Inst. Ser. C: Math. Phys. Sci., vol. 83, Reidel, Dordrecht, Boston, 1982, pp. 125-170.

96. A. Bosisio, *Una nota su proprietà topologiche e d'ordine e la ricorsività generalizzata*, Boll. Un. Mat. Ital. D (6) **2** (1983), no. 3, 835-855.

97. S. Bouc, *Homologie de certaines ensembles ordonnés*, C. R. Acad. Sci. Paris Sér. I Math. **299** (1984), no. 2, 49-52.

98. V. Breazu and C. Stănăşilă, *Overtopologies and their algebraic significance*, Bul. Inst. Politehn. Bucureşti Ser. Electrotehn. **43** (1981), no. 3, 3-10.

99. G. Brenner and D. Monk, *Tree algebras and chains*, Lecture Notes in Math., vol. 1004, Springer-Verlag, New York and Berlin, 1983.

100. A. Brini, *Some homological properties of partially ordered sets*, Adv. in Math. **43** (1982), no. 2, 197-201.

101. A. Brini and A. Terrusi, *Homotopically invariant reductions of partially ordered sets*, Ann. Scuola Norm. Sup. Pisa Cl. Sci. (4) **11** (1984), no. 3, 381-393.

102. D. R. Brown and J. W. Stepp, *The structure semilattice of a compact UDC semigroup*, Semigroup Forum **31** (1985), no. 2, 235-250.

103. D. Bünermann, *Auslander-Reiten quivers of exact one-parameter partially ordered sets*, Lecture Notes in Math., vol. 903, Springer-Verlag, New York and Berlin, 1981, pp. 55-61.

104. H. Burkill, *Monotonic functions on partially ordered sets*, J. Combin. Theory **37** (1984), no. 3, 248-256.

105. H. Burkill and C. V. Smallwood, *A generalization of Hanani's theorem on partial order*, Monatsh. Math. **95** (1983), no. 3, 177-179.

106. E. Butkovičová, *Ultrafilters without immediate predecessors in Rudin-Frolík order*, Comment. Math. Univ. Carolin. **23** (1982), no. 4, 757-766.

107. _____, *Gaps in Rudin-Frolík order*, General Topology and its Relations to Modern Analysis and Algebra, V (Prague, 1981), Sigma Ser. Pure Math., vol. 3, Heldermann, Berlin, 1983, pp. 56-58.

108. _____, *Long chains in Rudin-Frolík order*, Comment. Math. Univ. Carolin. **24** (1983), no. 3, 563-570.

109. Zhi Qiang Cao, *Comparison between two kinds of semilattice-semigroups*, Acta Math. Sci. (English Ed.) **4** (1984), no. 3, 311-317.

110. I. Chajda and J. Nieminen, *Direct decomposability of tolerances on lattices, semilattices and quasilattices*, Czechoslovak Math. J. **32** (1982), no. 1, 110-115.

111. I. Chajda, J. Nieminen, and Bohdan Zelinka, *Tolerances and orderings on semilattices*, Arch. Math. (Basel) **19** (1983), no. 3, 125–131.

112. I. Chajda and V. Novák, *On extensions of cyclic orders*, Časopis Pěst. Mat. **110** (1985), no. 2, 116–121.

113. I. Chajda and B. Zelinka, *Directly decomposable tolerances on direct products of lattices and semilattices*, Czechoslovak Math. J. **33** (1983), no. 4, 519–521.

114. Ch. Charretton and M. Pouzet, *Comparaison des structures engendrées par des chaînes*, Seminarberichte **49**, Humboldt Univ., Berlin, 1983, pp. 17–27.

115. _____, *Chains in Ehrenfeucht-Mostowski models*, Fund. Math. **118** (1983), no. 2, 109–122.

116. Niranjan-Prasad Chaudhuri and Abdul-Aali Jasim Mohammad, *Representation of a poset of n elements by a matrix of the family $M_p^{(n)}$ and the computation of the number of nonisomorphic posets of n elements*, Libyan J. Sci. **11** (1981), 25–36.

117. Charles Ching-an Cheng and B. Mitchell, *Posets of cohomological dimension one with finitely many tails*, J. Algebra **77** (1982), no. 2, 382–391.

118. J. Chvalina and L. Chvalinová, *Transitively acting monoids of local automorphisms of locally finite trees*, Arch. Math. (Basel) **19** (1983), no. 2, 71–82.

119. F. Collot, *Complement à la construction d'un "bon ordre"*, Rev. Bio-math., no. 83 (1983), 1–52.

120. J. Constantin and G. Fournier, *Ordonnés escamotables et points fixes*, Discrete Math. **53** (1985), 21–33.

121. W. H. Cornish, *The free implicative BCK-extension of a distributive nearlattice*, Math. Japon. **27** (1982), no. 3, 279–286.

122. _____, *Conversion of nearlattices into implicative BCK-algebras*, Math. Sem. Notes Kobe Univ. **10** (1982), no. 1, 1–8.

123. W. H. Cornish and A. S. A. Noor, *Standard elements in a near-lattice*, Bull. Austral. Math. Soc. **26** (1982), no. 2, 185–213.

124. P. Cousot and R. Cousot, *Constructive versions of Tarski's fixed point theorems*, Pacific J. Math. **82** (1979), no. 1, 43–57.

125. R. H. Cowen, *Compactness via prime semilattices*, Notre Dame J. Formal Logic **24** (1983), no. 2, 199–204.

126. H. Crapo, *Ordered sets: Retracts and connections*, J. Pure Appl. Algebra **23** (1982), no. 1, 13–28.

127. _____, *Unities and negation: On the representation of finite lattices*, J. Pure Appl. Algebra **23** (1982), no. 2, 109–135.

128. R. Dacić, *Fixed points of antimorphisms*, Publ. Inst. Math. (Beograd) (N. S.) **26(40)** (1979), 91–96.

129. _____, *Properties of monotone mappings in partially ordered sets*, Publ. Inst. Math. (Beograd) (N. S.) **30(44)** (1981), 33–39.

130. _____, *On fixed edges of antitone self-mappings of complete lattices*, Publ. Inst. Math. (Beograd) (N. S.) **34(48)** (1983), 49–53.

131. B. A. Davey and D. Duffus, *Exponentiation and duality*, Ordered Sets (Banff, Alta., 1981), NATO Adv. Study Inst. Ser. C: Math. Phys. Sci., vol. 83, Reidel, Dordrecht, Boston, 1982, pp. 43–95.

132. B. A. Davey and M. J. McCarthy, *A representation theory for the variety generated by the triangle*, Acta Math. Acad. Sci. Hungar. **38** (1981), no. 1–4, 241–255.

133. K. J. Devlin, *A new construction of a Kurepa tree with no Aronszajn subtree*, Fund. Math. **118** (1983), no. 2, 123–127.

134. _____, *Reduced powers of $\aleph_2$-trees*, Fund. Math. **118** (1983), no. 2, 129–134.

135. R. Dimitriević, *On compatibility of some topologies with the ordering*, Math. Balkanica (1977), no. 7, 51–57.

136. S. Z. Ditor, *Cardinality questions concerning semilattices of finite breadth*, Discrete Math. **48** (1984), no. 1, 47–59.

137. M. L. Diviccaro, *Common fixed-point theorems for families of functions in ordered sets*, Riv. Mat. Univ. Parma (4) **9** (1983), 87–93.

138. H. Dobbertin, *Measurable refinement monoids and applications to distributive semilattices, Heyting algebras and Stone spaces*, Math. Z. **187** (1984), no. 1, 13–21.

139. D. Dobbs, *Posets admitting a unique order-compatible topology*, Discrete Math. **41** (1982), no. 3, 235–240.

140. K. Drechsler and W. Zahn, *Abfaserbare teilweise geordnete Mengen*, Beiträge Algebra Geom. **12** (1982), 73–76.

141. J. Dudek, *On bisemilattices* I, Colloq. Math. **47** (1982), no. 1, 1–5.

142. _____, *On bisemilattices* II, Demonstratio Math. **15** (1982), no. 2, 465–475.

143. J. Dudek and A. Romanowska, *Bisemilattices with four essentially binary polynomials*, Contributions to Lattice Theory (Szeged, 1980), Colloq. Math. Soc. János Bolyai, vol. 33, North-Holland, Amsterdam and New York, 1983, pp. 337–360.

144. D. Duffus, *Automorphisms and products of ordered sets*, Algebra Universalis **19** (1984), no. 3, 366–369.

145. D. Duffus, I. Rival, and P. Winkler, *Minimizing setups for cycle-free ordered sets*, Proc. Amer. Math. Soc. **85** (1982), no. 4, 509–513.

146. D. Easdown, *A new proof that regular biordered sets come from regular semigroups*, Proc. Roy. Soc. Edinburgh Sect. A **96** (1984), no. 1–2, 109–116.

147. _____, *Biordered sets are biordered subsets of idempotents of semigroups*, J. Austral. Math. Soc. Ser. A **37** (1984), no. 2, 258–268.

148. _____, *Biordered sets of eventually regular semigroups*, Proc. London Math. Soc. **49** (1984), no. 3, 483–503.

149. P. H. Edelman, *A partial order on the regions of $\mathbb{R}^n$ dissected by hyperplanes*, Trans. Amer. Math. Soc. **283** (1984), no. 2, 617–631.

150. C. C. Edwards and M. Anderson, *Lattice properties of the symmetric weakly inverse semigroup on a totally ordered set*, J. Austral. Math. Soc. Ser. A **31** (1981), no. 4, 395–404.

151. M. H. El-Zahar and I. Rival, *Greedy linear extensions to minimize jumps*, Discrete Appl. Math. **11** (1985), no. 2, 143–156.

152. K. Engel, *Strong properties in partially ordered sets* I, Discrete Math. **47** (1983), no. 2–3, 229–234.

153. M. Erné, *Isomorphismen und Identifikationen in der Ordnungstheorie*, Math. Semesterber. **28** (1981), no. 1, 74–91.

154. _____, *Topologies on products of partially ordered sets* III: *Order convergence and order topology*, Algebra Universalis **13** (1981), no. 1, 1–23.

155. _____, *Distributivegesetze und Dedekindsche Schnitte*, Abh. Braunschweig. Wiss. Ges. **33** (1982), no. 117–145.

156. _____, *Einführung in die Ordnungstheorie*, Bibliographisches Institut, Mannheim, 1982.

157. _____, *Adjunctions and standard constructions for partially ordered sets*, Contributions to General Algebra, 2 (Klagenfurt, 1982), Hölder-Pichler-Tempsky, Vienna, 1983, pp. 77–106.

158. M. Erné and G. Wilke, *Standard completions for quasiordered sets*, Semigroup Forum **27** (1983), no. 1–4, 351–356.

159. C. P. Farrington, *Constructible lattices of c-degrees*, J. Symbolic Logic **47** (1982), no. 4, 739–754.

160. P. C. Fishburn, *Aspects of semiorders within interval orders*, Discrete Math. **40** (1982), no. 2–3, 181–191.

161. _____, *Interval lengths for interval orders: A minimization problem*, Discrete Math. **47** (1983), no. 1, 63–82.

162. _____, *Threshold-bounded interval orders and a theory of picycles*, SIAM J. Algebraic Discrete Methods **4** (1983), no. 3, 290–305.

163. _____, *Comment on the Kannai-Peleg impossiblity theorem for extending orders*, J. Econom. Theory **32** (1984), no. 1, 176–179.

164. _____, *Paradoxes of two-length interval orders*, Discrete Math. **52** (1984), no. 2–3, 165–175.

165. I. Fleischer, *Congruence extension from a semilattice to the freely generated distributive lattice*, Czechoslovak Math. J. **32(107)** (1982), no. 4, 623–626.

166. _____, *Every Galois connection is the polarity of an antitone relation*, J. Pure Appl. Algebra **32** (1984), no. 1, 49–50.

167. G. A. Fraser and J. P. Albert, *Associativity of the tensor product of semilattices*, Proc. Edinburgh Math. Soc. **27** (1984), no. 3, 337–340.

168. G. A. Fraser and A. M. Bell, *The word problem in the tensor product of distributive semilattices*, Semigroup Forum **30** (1984), no. 1, 117–120.

169. R. Freese, *Some order-theoretic questions about free modular lattices*, Ordered Sets (Banff, Alta., 1981), NATO Adv. Study Inst. Ser. C: Math. Phys. Sci., vol. 83, Reidel, Dordrecht, Boston, 1982, pp. 355–377.

170. E. Fuchs, *On the 0-category of finite chains*, Z. Math. Logik Grundlag. Math. **28** (1982), no. 1, 63–65.

171. S. Fujishige, *A decomposition of distributive lattices*, Discrete Math. **55** (1985), no. 1, 35–55.

172. S. Gähler, *On equivalent existence properties in topology and in optimization theory*, Lecture Notes in Math., vol. 1060, Springer-Verlag, Berlin and New York, 1984, pp. 267–270.

173. J. Gałuszka, *Generalized absorption laws in bisemilattices*, Algebra Universalis **19** (1984), no. 3, 304–318.

174. E. R. Gansner, *On the lattice of order ideals of an up-down poset*, Discrete Math. **39** (1982), no. 2, 113–122.

175. _____, *Parenthesizations of finite distributive lattices*, Algebra Universalis **16** (1983), no. 3, 287–303.

176. G. Gierz, J. D. Lawson, and A. D. Stralka, *Quasicontinuous posets*, Houston J. Math. **9** (1983), no. 2, 191–208.

177. _____, *Intrinsic topologies on semilattices of finite breadth*, Semigroup Forum **31** (1985), no. 1, 1–17.

178. J. Ginsburg, *A note on the cardinality of infinite partially ordered sets*, Pacific J. Math. **106** (1983), no. 2, 265–270.

179. S. M. Goberstein, *On the modularity of the lattice of fundamental orders on a semilattice*, Semigroup Forum **24** (1982), no. 1, 83–86.

180. S. W. Golomb, *Normed division domains*, Amer. Math. Monthly **88** (1981), no. 9, 680–686.

181. M. Gould, J. A. Iskra, and C. Tsinakis, *Globally determined lattices and semilattices*, Algebra Universalis **19** (1984), no. 2, 137–141.

182. M. Grandis, *Concrete representations for inverse and distributive exact categories*, Rend. Accad. Naz. Sci. XL Mem. Mat. (5) **8** (1984), no. 1, 99–120.

183. R. J. Grayson, *Constructive well-orderings*, Z. Math. Logik Grundlag. Math. **28** (1982), no. 6, 495–504.

184. C. Green, *The Möbius function of a partially ordered set*, Ordered Sets (Banff, Alta., 1981), NATO Adv. Study Inst. Ser. C: Math. Phys. Sci., vol. 83, Reidel, Dordrecht, Boston, 1982, pp. 555–581.

185. J. R. Griggs, *Collections of subsets with the Sperner property*, Trans. Amer. Math. Soc. **269** (1982), 575–591.

186. _____, *Poset measure and saturated partitions*, Stud. Appl. Math. **66** (1982), no. 1, 91–93.

187. M. J. Groszek and Th. A. Slaman, *Independence results on the global structure of the Turing degrees*, Trans. Amer. Math. Soc. **277** (1983), no. 2, 579–588.

188. Yuri Gurevich, *Existential interpretation* II, Arch. Math. Logik Grundlag. **22** (1982), no. 3–4, 103–120.

189. J. G. Hagendorf, *Extensions immédiates de chaînes*, Z. Math. Logik Grundlag. Math. **28** (1982), no. 1, 15–44.

190. _____, *Quelques résultats et conjectures de la théorie des chaînes*, C. R. Acad. Sci. Paris Sér. I Math. **299** (1984), no. 17, 839–841.

191. D. E. Haile, R. G. Larson, and M. E. Sweedler, *A new invariant for $\mathbb{C}$ over $\mathbb{R}$: Almost invertible cohomology theory and the classification of idempotent cohomology classes and algebras by partially ordered sets with a Galois group action*, Amer. J. Math. **105** (1983), no. 3, 689–814.

192. M. Hanazawa, *On Aronszajn trees with a non-Souslin base*, Tsukuba J. Math. **6** (1982), no. 2, 177–185.

193. _____, *Countable metacompactness and tree topologies*, J. Math. Soc. Japan **35** (1983), no. 1, 59–70.

194. _____, *On the product of Souslin trees*, Saitama Math. J. **1** (1983), 1–7.

195. _____, *Note on countable paracompactness of collectionwise Hausdorff tree topologies*, Saitama Math. J. **2** (1984), 7–20.

196. Ph. Hanlon, *The fixed-point partition lattices*, Pacific J. Math. **96** (1981), no. 2, 319–341.

197. D. W. Hardy and M. C. Thornton, *Partial orders and their semigroups of closed relations*, Semigroup Forum **25** (1982), no. 1–2, 171–184.

198. K. P. Hart, *Characterizations of **R**-embeddable and developable ω_1-trees*, Nederl. Akad. Wetensch. Indag. Math. **44** (1982), no. 3, 277–283.

199. _____, *More remarks on Souslin properties and tree topologies*, Topology Appl. **15** (1983), no. 2, 151–158.

200. R. E. Hartwig and M. P. Drazin, *Lattice properties of the ∗-order for complex matrices*, J. Math. Anal. Appl. **86** (1982), no. 2, 359–378.

201. W. S. Heck, *Large families of incomparable A-isols*, J. Symbolic Logic **48** (1983), no. 2, 250–252.

202. R. A. Heiner and D. J. Packard, *A uniqueness result for extending orders, with application to collective choice as inconsistency resolution*, J. Econom. Theory **32** (1984), no. 1, 180–184.

203. Robert Hickman, *Mildly distributive semilattices*, J. Austral. Math. Soc. Ser. A **36** (1984), no. 3, 287–315.

204. R. C. Hickman and G. P. Monro, *Distributive partially ordered sets*, Fund. Math. **120** (1984), no. 2, 151–166.

205. K. H. Hoffman and J. D. Lawson, *On the order-theoretical foundation of a theory of quasicompactly generated spaces without separation axiom*, J. Austral. Math. Soc. Ser. A **36** (1984), no. 2, 194–212.

206. Hartmut F. W. Höft, *Crossed and complete binary relations*, Rev. Roumaine Math. Pures Appl. **28** (1983), no. 8, 703–708.

207. Hartmut F. W. Höft and Margret H. Höft, *Coinitial sets and fixed points in partially ordered sets*, Algebra Universalis **16** (1983), no. 2, 258–260.

208. _____, *An order-theoretic representation of the polygonal numbers*, Fibonacci Quart. **22** (1984), no. 4, 318–323.

209. Margret H. Höft, *Sums of double systems of partially ordered sets*, Demonstratio Math. **16** (1983), no. 1, 229–238.

210. R. Holzman, *An extension of Fishburn's theorem on extending orders*, J. Econom. Theory **32** (1984), no. 1, 192–196.

211. C. S. Hoo, *Pseudocomplemented and implicative semilattices*, Canad. J. Math. **34** (1982), no. 2, 423–437.

212. _____, *Atoms, primes and implicative lattices*, Canad. Math. Bull. **27** (1984), no. 3, 279–285.

213. C. S. Hoo and P. V. Ramana Murty, *Modular and admissible semilattices*, Canad. J. Math. **36** (1984), no. 5, 795–799.

214. C. S. Hoo and K. P. Shum, *0-distributive and P-uniform semilattices*, Canad. Math. Bull. **25** (1982), no. 3, 317–324.

215. M. Höppner, *A note on the structure of injective diagrams*, Manuscripta Math. **44** (1983), no. 1–3, 45–50.

216. P. M. Idziak, *Filters and congruence relations in BCK-semilattices*, Math. Japon. **29** (1984), no. 6, 975–980.

217. J. Isbell, *Completion of a construction of Johnstone*, Proc. Amer. Math. Soc. (1982), no. 3, 333–334.

218. M. Ito, *Some classes of automata as partially ordered sets*, Math. Systems Theory **15** (1982), no. 4, 357–370.

219. M. Jambu-Giraudet, *Bi-interpretable groups and lattices*, Trans. Amer. Math. Soc. **278** (1983), no. 1, 253–269.

220. _____, *Quelques remarques sur l'équivalence élémentaire entre groupes ou treillis d'automorphismes de chaînes 2-homogènes*, Discrete Math. **53** (1985), 117–124.

221. M. J. Janowitz, *Applications of the theory of partially ordered sets to cluster analysis*, Universal Algebra and Applications (Warsaw, 1978), Banach Center Publ., 9, PWN, Warsaw, 1982, pp. 305–319.

222. C. Jayaram, *Quasicomplemented semilattices*, Acta Math. Hungar. **39** (1982), no. 1–3, 39–47.

223. _____, *Semiatoms in semilattices*, Math. Sem. Notes Kobe Univ. **10** (1982), no. 2, 351–366.

224. _____, *Semilattices and finite Boolean algebras*, Algebra Universalis **16** (1983), no. 3, 390–394.

225. _____, *Congruence filters in semilattices*, Pure Appl. Math. Sci. **18** (1983), no. 1–2, 13–17.

226. Th. Jech and K. Prikry, *Cofinality of the partial ordering of functions from ω_1 into ω under eventual domination*, Math. Proc. Cambridge Philos. Soc. **95** (1984), no. 1, 25–32.

227. J. Ježek, *Simple semilattices with two commuting automorphisms*, Algebra Universalis **15** (1982), no. 2, 162–175.

228. J. Ježek and T. Kepka, *Simple semimodules over commutative semirings*, Acta Sci. Math. (Szeged) **46** (1983), no. 1–4, 17–27.

229. B. Jónsson, *Arithmetic of ordered sets*, Ordered Sets (Banff, Alta., 1981), NATO Adv. Study Inst. Ser. C: Math. Phys. Sci., vol. 83, Reidel, Dordrecht, Boston, 1982, pp. 3–41.

230. _____, *Powers of partially ordered sets: The automorphism group*, Math. Scand. **51** (1982), no. 1, 121–141.

231. B. Jónsson and R. McKenzie, *Powers of partially ordered sets: Cancellation and refinement properties*, Math. Scand. **51** (1982), no. 1, 87–120.

232. G. Kalmbach, *Ordered sets and homology*, Contributions to General Algebra, 2 (Klagenfurt, 1982), Hölder-Pichler-Tempsky, Vienna, 1983, pp. 163–178.

233. T. Kamimura and A. Tang, *Kleene chain completeness and fixed-point properties*, Theoret. Comput. Sci. **23** (1983), no. 3, 317–331.

234. Y. Kannai and B. Peleg, *A note on the extension of an order on a set to the power set*, J. Econom. Theory **32** (1984), no. 1, 172–175.

235. M. Katz, *Deduction systems and valuation spaces*, Logique et Anal. (N.S.) **26** (1983), no. 102, 157–175.

236. D. Kelly and W. T. Trotter, *Dimension theory for ordered sets*, Ordered Sets (Banff, Alta., 1981), NATO Adv. Study Inst. Ser. C: Math. Phys. Sci., vol. 83, Reidel, Dordrecht, Boston, 1982, pp. 171–211.

237. O. Kerner, *On the number of indecomposable representations of a partially ordered set*, Comm. Algebra **11** (1983), no. 10, 1123–1143.

238. H. A. Kierstead and W. T. Trotter, *An extremal problem in recursive combinatorics*, Congr. Numer. **33** (1981), 143–153.

239. M. Kleiner, *Schur's lemma for partially ordered sets of finite type*, J. Algebra **88** (1984), no. 2, 435–437.

240. J. Klimeš, *Fixed edge theorems for complete lattices*, Arch. Math. (Basel) **17** (1981), no. 4, 227–234.

241. _____, *Characterizations of completeness for semilattices by using of fixed points*, Scripta Fac. Sci. Natur. Univ. Purk. Brun. **12** (1982), no. 10, 507–513.

242. _____, *Characterizations of a semilattice completeness*, Scripta Fac. Sci. Natur. Univ. Purk. Brun. **14** (1984), no. 8, 399–407.

243. _____, *Fixed point characterization of completeness on lattices for relatively isotone mappings*, Arch. Math. (Basel) **20** (1984), no. 3, 125–132.

244. Y. Kobayashi, *Semilattices are globally determined*, Semigroup Forum **29** (1984), no. 1–2, 217–222.

245. K. M. Koh, *On the lattice of maximum-sized antichains of a finite poset*, Algebra Universalis **17** (1983), no. 1, 73–86.

246. P. Köhler, *Brouwerian semilattices: The lattice of total subalgebras*, Universal Algebra and Applications (Warsaw, 1978), Banach Center Publ., 9, PWN, Warsaw, 1982, pp. 47–56.

247. M. Kolibiar, *Fixed point theorems for ordered sets*, Studia Sci. Math. Hungar. **17** (1982), no. 1-4, 45-50.

248. _____, *Compatible orderings in semilattices*, Contributions to General Algebra, 2 (Klagenfurt, 1982), Hölder-Pichler-Tempsky, Vienna, 1983, pp. 215-220.

249. M. Konstantinidou and J. Mittas, *An introduction to the theory of hyperlattices*, Math. Balkanica (1977), no. 7, 187-193.

250. V. Koubek and V. Rödl, *On number of covering arcs in orderings*, Comment. Math. Univ. Carolin. **22** (1981), no. 4, 721-733.

251. M. Krishna Murty, *Neutrality in semilattices*, Math. Sem. Notes Kobe Univ. **10** (1982), no. 1, 143-155.

252. M. Krishna Murty and G. C. Rao, *Neutrality in posets*, Algebra Universalis **13** (1981), no. 3, 401-404.

253. H. Kröger, *Verallgemeinerte Verbände mit innerer Verbandsstruktur*, Mat. Vesnik **5** (1981), no. 3, 253-263.

254. M. Krynicki and J. Väänänen, *On orderings of the family of all logics*, Arch. Math. Logik Grundlag. **22** (1982), no. 304, 141-158.

255. M. Kula, M. A. Marshall, and A. Sladek, *Direct limits of finite spaces of orderings*, Pacific J. Math. **112** (1984), no. 2, 391-406.

256. D. Kurepa, *Measure and order*, Proceedings of the Conference Topology and Measure, II, Part 2 (Rostock/Warnemünde, 1977), Ernst-Moritz-Arndt Univ., Greifswald, 1980, pp. 107-118.

257. _____, *A link between ordered sets and trees. On the rectangle tree hypothesis*, Publ. Inst. Math. (Beograd) (N.S.) **31(45)** (1982), 121-128.

258. H. Länger, *A characterization of infimum functions*, Algebra Universalis **16** (1983), no. 1, 124-125.

259. D. Lascar, *Ordre de Rudin-Keisler et poids dans les théories stables*, Z. Math. Logik Grundlag. Math. **28** (1982), no. 5, 413-430.

260. M. Lerman and J. G. Rosenstein, *Recursive linear orderings*, Patras Logic Symposium (Patras, 1980), North-Holland, Amsterdam and New York, 1982, pp. 123-136.

261. K. Leutöla and J. Nieminen, *Posets and generalized lattices*, Algebra Universalis **16** (1983), no. 3, 344-354.

262. J. Lihová, *On the lattice of convexly compatible topologies on a partially ordered set*, Contributions to Lattice Theory (Szeged, 1980), Colloq. Math. Soc. János Bolyai, vol. 33, North-Holland, Amsterdam and New York, 1983, pp. 609-625.

263. G. S. Lystad and A. R. Stralka, *Lawson semilattices with bialgebraic congruence lattices*, General Topology and Modern Analysis (Proc. Conf., Univ. California, Riverside, CA, 1980), Academic Press, New York, 1981, pp. 247-254.

264. G. F. McNulty, *Infinite ordered sets, a recursive perspective*, Ordered Sets (Banff, Alta., 1981), NATO Adv. Study Inst. Ser. C: Math. Phys. Sci., vol. 83, Reidel, Dordrecht, Boston, 1982, pp. 299-330.

265. A. B. Manaster and J. B. Remmel, *Partial orderings of fixed finite dimension: Model companions and density*, J. Symbolic Logic **46** (1981), no. 4, 789-802.

266. T. Margush, *Distances between trees*, Discrete Math. **4** (1982), no. 4, 281-290.

267. G. Markowsky, *Extending semilattices is hard*, Algebra Universalis **17** (1983), no. 3, 406-407.

268. W. Maryland, *Result on finite connected partially ordered sets*, J. Combin. Inform. System Sci. **6** (1981), no. 2, 129-136.

269. J. D. Mashburn, *The least fixed point property for ω-chain continuous function*, Houston J. Math. **9** (1983), no. 2, 231-244.

270. J. Meakin, *Local semilattices on two generators*, Semigroup Forum **24** (1982), no. 2-3, 95-116.

271. _____, *The free local semilattice on a set*, J. Pure Appl. Algebra **27** (1983), no. 3, 263-275.

272. J. Meakin and F. Pastijn, *The structure of pseudo-semilattices*, Algebra Universalis **13** (1981), no. 3, 355-372.

273. J. Meseguer, *Order completion monads*, Algebra Universalis **16** (1983), no. 1, 63-82.

274. B. Messano, *Existence of periodic points of a mapping of a totally ordered set into itself*, Matematiche (Catania) **35** (1980), no. 1-2, 287-300. (Italian. English summary)

275. G. G. Miller, *Lexicographic products and completeness*, Indian J. Math. **23** (1981), no. 1-3, 129-137.

276. J. B. Miller, *Representation of Boolean hypolattices*, Bull. Austral. Math. Soc. **24** (1981), no. 3, 389-404.

277. E. C. Milner, *On the decomposition of partially ordered sets into directed sets*, Proc. Internat. Math. Conf., Singapore 1981, North-Holland, Amsterdam and New York, 1982, pp. 85-90.

278. E. C. Milner and M. Pouzet, *On the cofinality of partially ordered sets*, Ordered Sets (Banff, Alta., 1981), NATO Adv. Study Inst. Ser. C: Math. Phys. Sci., vol. 83, Reidel, Dordrecht, Boston, 1982, pp. 279-298.

279. E. C. Milner and K. Prikry, *The cofinality of a partially ordered set*, Proc. London Math. Soc. **46** (1983), no. 3, 454-470.

280. _____, *Some results on the depth and width of partial orders*, C. R. Math. Rep. Acad. Sci. Canada **6** (1984), no. 3, 139-144.

281. G. Moran, *Trees and the bireflection property*, Israel J. Math. **41** (1982), no. 3, 244-260.

282. W. Müller, *Kleinersche Fasersummen unzerlegbarer Moduln*, Bayreuth. Math. Schr. (1983), no. 14, 79-108.

283. K. S. S. Nambooripad, *Pseudosemilattices and biordered sets*, I, Simon Stevin **55** (1981), no. 3, 103-110.

284. _____, *Pseudosemilattices and biordered sets*, II. *Pseudo-inverse semigroups*, Simon Stevin **56** (1982), no. 3, 143-159.

285. _____, *Pseudosemilattices and biordered sets*, III. *Regular locally testable semigroups*, Simon Stevin **56** (1982), no. 4, 239-256.

286. K. S. S. Nambooripad and F. J. Pastijn, *The fundamental representation of a strongly regular Baer semigroup*, J. Algebra **92** (1985), no. 2, 283-302.

287. H. Narushima, *Principle of inclusion-exclusion on partially ordered sets*, Discrete Math. **42** (1982), no. 2-3, 243-250.

288. J. B. Nation, *On partially ordered sets embeddable in a free lattice*, Algebra Universalis **18** (1984), no. 3, 327-333.

289. J. Neggers, *Representation of finite posets as linear operators*, J. Combin. Inform. System Sci. **6** (1981), no. 3, 279-293.

290. _____, *Reduced Brylawski decompositions of finite posets*, J. Combin. Inform. System Sci. **7** (1982), no. 3, 180-196.

291. Peter Nevermann, *A note on axiomatizable order varieties*, Algebra Universalis **17** (1983), no. 1, 129-131.

292. _____, *Order varieties generated by $\vee$-semilattices of finite width*, Discrete Math. **53** (1985), 167-171.

293. P. Nevermann and R. Wille, *The strong selection property and ordered sets of finite length*, Algebra Universalis **18** (1984), no. 1, 18-28.

294. P. Ney, *Interval space representation of interval structures*, Math. Nachr. **117** (1984), 135-140.

295. V. L. Nguyen and J.-L. Lassez, *A dual problem to least fixed points*, Theoret. Comput. Sci. **16** (1981), no. 2, 211-221.

296. J. Nieminen, *On distributive and modular χ-lattices*, Yokohama Math. J. **31** (1983), no. 1-2, 13-20.

297. _____, *On χ_{mub}-lattices and convex substructures of lattices and semilattices*, Acta Math. Hungar. **44** (1984), no. 3-4, 229-236.

298. D. Novak, *Generalization of continuous posets*, Trans. Amer. Math. Soc. **272** (1982), no. 2, 645-667.

299. _____, *On a duality between the concepts "finite" and "directed"*, Houston J. Math. **8** (1982), no. 4, 545-563.

300. V. Novák, *Cyclically ordered sets*, Czechoslovak Math. J. **32** (1982), no. 3, 460-473.

301. _____, *On some minimal problem*, Arch. Math. (Basel) **20** (1984), no. 2, 95-99.

302. _____, *Cuts in cyclically ordered sets*, Czechoslovak Math. J. **34** (1984), no. 2, 322-333.

303. _____, *Operations on cyclically ordered sets*, Arch. Math. (Basel) **20** (1984), no. 3, 133–139.

304. V. Novák and M. Novotný, *On determination of cyclic order*, Czechoslovak Math. J. **33** (1983), no. 4, 555–563.

305. _____, *Dimension theory for cyclically and cocyclically ordered sets*, Czechoslovak Math. J. **33** (1983), no. 4, 647–653.

306. _____, *On a power of cyclically ordered sets*, Časopis Pěst. Mat. **109** (1984), no. 4, 421–424.

307. _____, *Universal cyclically ordered sets*, Czechoslovak Math. J. **35** (1985), no. 1, 158–161.

308. M. Parigot, *Théories d'arbres*, J. Symbolic Logic **47** (1982), no. 4, 841–853.

309. _____, *Le modèle compagnon de la théorie des arbres*, Z. Math. Logik Grundlag. Math. **29** (1983), no. 2, 137–150.

310. A. Pasztor, *The epis of $Pos(Z)$*, Comment. Math. Univ. Carolin. **23** (1982), no. 2, 285–299.

311. _____, *Epis of some categories of Z-continuous partial algebra*, Acta Cybernet. **6** (1983), no. 1, 111–123.

312. _____, *Surjections of complete posets and of continuous algebras*, Algebra Universalis **17** (1983), no. 1, 65–72.

313. Y. S. Pawar and N. K. Thakare, *The space of minimal prime ideals in a 0-distributive semilattice*, Period. Math. Hungar. **13** (1982), no. 4, 309–319.

314. P. Penner, *Hyperidentities of lattices and semilattices*, Algebra Universalis **13** (1981), no. 3, 307–314.

315. _____, *Hyperidentities of semilattices*, Houston J. Math. **10** (1984), no. 1, 81–108.

316. L. Pezzoli, *On D-complementation*, Adv. in Math. **51** (1984), no. 3, 226–239.

317. J.-E. Pin, *On semidirect products of two finite semilattices*, Semigroup Forum **28** (1984), no. 1–3, 73–81.

318. P. Kornelius Policki, *Zum Problem des Beweises ex motu der Existenz Gottes*, Logique et Anal. (N.S.) **26** (1983), no. 101, 37–50.

319. M. Pouzet and I. Rival, *Which ordered sets have a complete linear extension?*, Canad. J. Math. **33** (1981), no. 5, 1245–1254.

320. _____, *Quotients of complete ordered sets*, Algebra Universalis **17** (1983), no. 3, 393–405.

321. _____, *Every countable lattice is a retract of a direct product of chains*, Algebra Universalis **18** (1984), no. 3, 295–307.

322. M. Pouzet and N. Zaguia, *Dimension de Krull des ensembles ordonnés*, Discrete Math. **53** (1985), 173–192.

323. W. B. Powell, *Boolean hypolattices with applications to ℓ-groups*, Arch. Math. (Basel) **39** (1982), no. 6, 535–540.

324. S. Premchand, *Independence of axioms for biordered sets*, Semigroup Forum **28** (1984), no. 1–3, 249–263.

325. A. Quilliot, *An application of the Helly property to the partially ordered sets*, J. Combin. Theory Ser. A **35** (1983), no. 2, 185–198.

326. P. V. Ramana Murty and M. Krishna Murty, *Some remarks on certain classes of semilattices*, Internat. J. Math. Sci. **5** (1982), no. 1, 21–30.

327. P. V. Ramana Murty and V. Raman, *Triple construction of semilattices with 1 admitting neutral p-closure operators*, Math. Slovaca **32** (1982), no. 4, 367–378.

328. _____, *Permutability of distributive congruence relations in a join semilattice directed below*, Math. Slovaca **35** (1985), no. 1, 43–49.

329. I. Rival, *The problem of fixed points in ordered sets*, Discrete Math. **8** (1980), 283–292.

330. _____, *The retract construction*, Ordered Sets (Banff, Alta., 1981), NATO Adv. Study Inst. Ser. C: Math. Phys. Sci., vol. 83, Reidel, Dordrecht, Boston, 1982, pp. 97–122.

331. _____, *Optimal linear extensions by interchanging chains*, Proc. Amer. Math. Soc. **89** (1983), no. 3, 387–394.

332. I. Rival and R. Wille, *The smallest order variety containing all chains*, Discrete Math. **35** (1981), 203–212.

333. K. W. Roggenkamp, *Auslander-Reitan species of Bäckström orders*, J. Algebra **65** (1983), no. 2, 449–476.

334. A. Romanowska, *On distributivity of bisemilattices with one distributive law*, Universal Algebra (Esztergom, 1977), Colloq. Math. Soc. János Bolyai, vol. 29, North-Holland, Amsterdam and New York, 1982, pp. 653–661.

335. _____, *Algebras of functions from partially ordered sets into distributive lattices*, Universal Algebra and Lattice Theory (Puebla, 1982), Lecture Notes in Math., vol. 1004, Springer-Verlag, New York and Berlin, 1983, pp. 245–256.

336. _____, *Building bisemilattices from lattices and semilattices*, Contributions to General Algebra, 2 (Klagenfurt, 1982), Hölder-Pichler-Tempsky, Vienna, 1983, pp. 343–358.

337. M. Roubens and Ph. Vincke, *Représentation et ajustement des ordres d'intervalles*, Cah. Geogr. Besancon. Sémin. et Notes Rech. (1981), no. 21, 55–75.

338. _____, *Linear orders and semiorders close to an interval order*, Discrete Appl. Math. **6** (1983), no. 3, 311–314.

339. Dev Kumar Roy, *R.e. presented linear orders*, J. Symbolic Logic **48** (1983), no. 2, 369–376.

340. M. E. Rudin, *Directed sets which converge*, General Topology and Modern Analysis (Proc. Conf., Univ. California, Riverside, CA, 1980), Academic Press, New York, 1981, pp. 305–307.

341. W. Ruppert, *A separately, but not jointly continuous action of a compact semilattice*, Semigroup Forum **30** (1984), no. 2, 241–242.

342. A. Rutkowski, *Multifunctions and the fixed point property for products of ordered sets*, Order **2** (1985), no. 1, 61–67.

343. B. Sands, *Counting antichains in finite partially ordered sets*, Discrete Math. **35** (1981), 213–228.

344. N. Sauer and R. E. Woodrow, *Finite cutsets and finite antichains*, Order **1** (1984), no. 1, 35–46.

345. B. M. Schein and Lide Li, *Ordered sets which support inverse semigroups: Three examples*, Semigroup Forum **30** (1984), no. 2, 234–236.

346. J. H. Schmerl, $\aleph_0$-*categorical distributive lattices of finite breadth*, Proc. Amer. Math. Soc. **87** (1983), no. 4, 707–713.

347. J. Schmid, *The maximal semigroup of quotients of a finite semilattice*, Semigroup Forum **29** (1984), no. 1–2, 35–49.

348. D. Schmidt, *The relation between the height of a well-founded partial ordering and the order types of its chains and antichains*, J. Combin. Theory Ser. B **31** (1981), no. 2, 183–189.

349. S. Schwarz, *The quotient semilattice of the recursively enumerable degrees modulo the cappable degrees*, Trans. Amer. Math. Soc. **283** (1984), no. 1, 315–328.

350. D. Schweigert, *Near lattices*, Math. Slovaca **32** (1982), no. 3, 313–317.

351. B. M. Scott, *Local bases and product partial order.* II, Math. Centre Tracts (1983), no. 169, 73–80.

352. D. Scott, *Some ordered sets in computer sciences*, Ordered Sets (Banff, Alta., 1981), NATO Adv. Study Inst. Ser. C: Math. Phys. Sci., vol. 83, Reidel, Dordrecht, Boston, 1982, pp. 677–718.

353. D. Seese, H.-P. Tuschik, and M. Weese, *Undecidable theories in stationary logic*, Proc. Amer. Math. Soc. **84** (1982), no. 4, 563–567.

354. D. Seese and M. Weese, $L(aa)$-*elementary types of well-orderings*, Z. Math. Logik Grundlag. Math. **28** (1982), no. 6, 557–564.

355. M. Sekanina, *On system of subobject functors in the category of ordered sets*, Universal Algebra and Applications (Warsaw, 1978), Banach Center Publ., 9, PWN, Warsaw, 1982, pp. 225–232.

356. K. Shoji, *Note on self-injective non-singular semigroups*, Semigroup Forum **24** (1982), no. 2–3, 189–194.

357. R. Simion, *The lattice automorphisms of the dominance ordering*, Discrete Math. **49** (1984), no. 1, 89–93.

358. J. Skucha, *Topological properties of C-nets*, Demonstratio Math. **15** (1982), no. 2, 405–419.

359. E. Slatinsky, *Die arithmetische Operation der Summe*, Arch. Math. (Basel) **20** (1984), no. 1, 9–20.

360. T. P. Speed, *On the Möbius function of* Hom (P, Q), Bull. Austral. Math. Soc. **29** (1984), no. 1, 39–46.

361. R. P. Stanley, *Some aspects of groups acting on finite posets*, J. Combin. Theory Ser. A **32** (1982), no. 2, 132–161.

362. D. D. Steiner and A. Abian, *Nonexistence of nonmolecular generic sets*, Publ. Inst. Math. (Beograd) (N.S.) **36(50)** (1984), 29–34.

363. T. Sturm and A. Rutkowski, *A note on the least upper bound operation in the partially ordered set*, Demonstratio Math. **17** (1984), no. 1, 79–84.

364. T. Tamura, *On chains whose power semigroups are lattices*, Semigroup Forum **30** (1984), no. 1, 35–40.

365. M. R. Tasković, *Monotone mappings of ordered sets*, Proc. 3rd Algebraic Conf. (Beograd, 1982), Univ. Novi Sad, Novi Sad, pp. 153–154.

366. M. Tchuente, *Des permutations de degré n compatibles avec des arbres d'ordre n*, Publ. Centre Rech. Math. Pures (I) (Neuchâtel) (1981), no. 16, 11–16.

367. N. K. Thakare and Y. S. Pawar, *Minimal prime ideals in 0-distributive semilattices*, Period. Math. Hungar. **13** (1982), no. 3, 237–246.

368. N. K. Thakare, M. P. Wasadikar, and S. Maeda, *On modular pairs in semilattices*, Algebra Universalis **19** (1984), no. 2, 255–256.

369. R. Thron, *The maximal elements of a partially ordered set and a modification of the theorem of Krein-Milman for closure operators*, Ann. Univ. Sci. Budapest. Eötvös Sect. Math. **26** (1983), 17–25.

370. S. Todorčević, *Stationary sets, trees and continuums*, Publ. Inst. Math. (Beograd) (N.S.) **29(43)** (1981), 249–262.

371. _____, *Real functions on the family of all well-ordered subsets of a partially ordered set*, J. Symbolic Logic **48** (1983), no. 1, 91–96.

372. M. Tomková, *On multilattices with isomorphic graphs*, Math. Slovaca **1982** (32), no. 1, 63–73.

373. M. Trombetta, *Un'assiomatica per la relazione "fra"*, Rend. Istit. Mat. Univ. Trieste **15** (1983), no. 1-2, 96–107.

374. W. T. Trotter, *Stacks and splits of partially ordered sets*, Discrete Math. **35** (1981), 229–256.

375. _____, *The dimension of the cartesian product of partial orders*, Discrete Math. **53** (1985), 255–263.

376. D. C. Trueman, *Fundamental preorders and partial orders on completely* $[0-]$ *simple semigroups*, Semigroup Forum **26** (1983), no. 1-2, 139–150.

377. M. Turinici, *A generalization of Altman's ordering principle*, Proc. Amer. Math. Soc. **90** (1984), no. 1, 128–132.

378. H.-P. Tuschik, *The Ramsey theorem for additive colourings*, Seminarberichte **49**, Humboldt Univ., Berlin, 1983, pp. 129–136.

379. J. C. Varlet, *Fixed points in finite De Morgan algebras*, Discrete Math. **53** (1985), 265–280.

380. D. Velleman, *On a generalization of Jensen's* $\Box_\varkappa$, *and strategic closure of partial orders*, J. Symbolic Logic **48** (1983), no. 4, 1046–1052.

381. M. Vonkomerová, *A note on C-lattices*, Acta Math. Univ. Comenian. (1982), no. 40-41, 33–44.

382. L. Vrancken-Mawet, *Sur des congruences d'un ensemble ordonné. Application à l'étude du lattis des sous-algèbres d'un demi-lattis de Brouwer fini*, Bull. Soc. Roy. Sci. Liège **51** (1982), no. 5-8, 174–187.

383. J. W. Walker, *Homotopy type and Euler characteristic of partially ordered sets*, European J. Combin. **2** (1981), no. 4, 373–384.

384. _____, *Isotone relations and the fixed point property for posets*, Discrete Math. **48** (1984), no. 2-3, 275–288.

385. Shang Zhi Wang and Bo Yu Li, *On the minimal cofinal subsets of a directed quasi-ordered set*, Discrete Math. **48** (1984), no. 2-3, 289–306.

386. Ti Xiang Wang and Kang Yan Chen, *On the problem of the existence of optimal fixed points*, J. Math. Res. Exposition **3** (1983), no. 3, 47–50.

387. R. Watnick, *A generalization of Tennenbaum's theorem on effectively finite recursive linear orderings*, J. Symbolic Logic **49** (1984), no. 2, 563-569.

388. D. West, *Extremal problems in partially ordered sets*, Ordered Sets (Banff, Alta., 1981), NATO Adv. Study Inst. Ser. C: Math. Phys. Sci., vol. 83, Reidel, Dordrecht, Boston, 1982, pp. 473-521.

389. D. West and C. A. Tovey, *Semiantichains and antichain coverings in direct products of partial orders*, SIAM J. Algebraic Discrete Methods **2** (1981), no. 3, 295-305.

390. D. West, W. T. Trotter, G. W. Peck, and P. Shor, *Regressions and monotone chains: a Ramsey-type extremal problem for partial orders*, Combinatorica **4** (1984), no. 1, 117-119.

391. P. Winkler, *Average height in a partially ordered set*, Discrete Math. **39** (1982), no. 3, 337-341.

392. B. Wolniewicz, *An algebra of subsets for join-semilattice with unit*, Polish Acad. Sci. Inst. Philos. Sociol. Bull. Sect. Logic **13** (1984), no. 1, 21-24.

393. _____, *A topology for logical space*, Polish Acad. Sci. Inst. Philos. Sociol. Bull. Sect. Logic **13** (1984), no. 4, 255-259.

394. N. Zaguia, *Chaînes d'ideaux et de sections initiales d'un ensemble ordonné*, Publ. Dép. Math. Nouvelle Sér. D (1983), no. 7.

395. W. Zahn, *Elementare Halbordnungen*, Beiträge Algebra Geom. **15** (1983), 109-120.

396. U. Zimmermann and P. Köhler, *Products of finitely based varieties of Brouwerian semilattices*, Algebra Universalis **18** (1984), 110-116.

Translated by BORIS M. SCHEIN

Recent Titles in This Series

(Continued from the front of this publication)

113 **A. F. Lavrik,** Twelve Papers in Logic and Algebra
112 **D. A. Gudkov and G. A. Utkin,** Nine Papers on Hilbert's 16th Problem
111 **V. M. Adamjan, et al.,** Nine Papers on Analysis
110 **M. S. Budjanu, et al.,** Nine Papers on Analysis
109 **D. V. Anosov, et al.,** Twenty Lectures Delivered at the International Congress of Mathematicians in Vancouver, 1974
108 **Ja. L. Geronimus and Gábor Szegő,** Two Papers on Special Functions
107 **A. P. Mišina and L. A. Skornjakov,** Abelian Groups and Modules
106 **M. Ja. Antonovskiĭ, V. G. Boltjanskiĭ, and T. A. Sarymsakov,** Topological Semifields and Their Applications to General Topology
105 **R. A. Aleksandrjan, et al.,** Partial Differential Equations, Proceedings of a Symposium Dedicated to Academician S. L. Sobolev
104 **L. V. Ahlfors, et al.,** Some Problems on Mathematics and Mechanics, On the Occasion of the Seventieth Birthday of Academician M. A. Lavrent'ev
103 **M. S. Brodskiĭ, et al.,** Nine Papers in Analysis
102 **M. S. Budjanu, et al.,** Ten Papers in Analysis
101 **B. M. Levitan, V. A. Marčenko, and B. L. Roždestvenskiĭ,** Six Papers in Analysis
100 **G. S. Ceĭtin, et al.,** Fourteen Papers on Logic, Geometry, Topology and Algebra
99 **G. S. Ceĭtin, et al.,** Five Papers on Logic and Foundations
98 **G. S. Ceĭtin, et al.,** Five Papers on Logic and Foundations
97 **B. M. Budak, et al.,** Eleven Papers on Logic, Algebra, Analysis and Topology
96 **N. D. Filippov, et al.,** Ten Papers on Algebra and Functional Analysis
95 **V. M. Adamjan, et al.,** Eleven Papers in Analysis
94 **V. A. Baranskiĭ, et al.,** Sixteen Papers on Logic and Algebra
93 **Ju. M. Berezanskiĭ, et al.,** Nine Papers on Functional Analysis
92 **A. M. Ančikov, et al.,** Seventeen Papers on Topology and Differential Geometry
91 **L. I. Barklon, et al.,** Eighteen Papers on Analysis and Quantum Mechanics
90 **Z. S. Agranovič, et al.,** Thirteen Papers on Functional Analysis
89 **V. M. Alekseev, et al.,** Thirteen Papers on Differential Equations
88 **I. I. Eremin, et al.,** Twelve Papers on Real and Complex Function Theory
87 **M. A. Aĭzerman, et al.,** Sixteen Papers on Differential and Difference Equations, Functional Analysis, Games and Control
86 **N. I. Ahiezer, et al.,** Fifteen Papers on Real and Complex Functions, Series, Differential and Integral Equations
85 **V. T. Fomenko, et al.,** Twelve Papers on Functional Analysis and Geometry
84 **S. N. Černikov, et al.,** Twelve Papers on Algebra, Algebraic Geometry and Topology
83 **I. S. Aršon, et al.,** Eighteen Papers on Logic and Theory of Functions
82 **A. P. Birjukov, et al.,** Sixteen Papers on Number Theory and Algebra
81 **K. K. Golovkin, V. P. Il'in, and V. A. Solonnikov,** Four Papers on Functions of Real Variables
80 **V. S. Azarin, et al.,** Thirteen Papers on Functions of Real and Complex Variables
79 **V. I. Arnol'd, et al.,** Thirteen Papers on Functional Analysis and Differential Equations
78 **A. V. Arhangel'skiĭ, et al.,** Eleven Papers on Topology
77 **L. A. Balašov, et al.,** Fourteen Papers on Series and Approximation
76 **Geng Ji, et al.,** Thirteen Papers on Algebra and Analysis
75 **A. A. Andronov, et al.,** Seven Papers on Equations Related to Mechanics and Heat

(See the AMS catalog for earlier titles)